T0133105

BASIC PRINCIPLES
of
PLASMA PHYSICS

A STATISTICAL APPROACH

FRONTIERS IN PHYSICS

David Pines, Editor

S. L. Adler and R. F. Dashen
Current Algebras and Applications to Particle Physics, 1968

P. W. Anderson *Concepts in Solids: Lectures on the Theory of Solids, 1965 (2nd printing, 1971)*

V. Barger and D. Cline
Phenomenological Theories of High Energy Scattering: An Experimental Evaluation, 1967

N. Bloembergen *Nonlinear Optics: A Lecture Note and Reprint Volume, 1965*

N. Bloembergen *Nuclear Magnetic Relaxation: A Reprint Volume, 1961*

R. Brout *Phase Transitions 1965*

E. R. Caianiello *Combinatorics and Renormalization in Quantum Field Theory, 1973*

G. F. Chew *S. Matrix Theory of Strong Interactions: A Lecture Note and Reprint Volume, 1961*

P. Choquard *The Anharmonic Crystal, 1967*

P. G. de Gennes *Superconductivity of Metals and Alloys, 1966*

R. P. Feynman *Quantum Electrodynamics: A Lecture Note and Reprint Volume, 1961 (2nd printing, with corrections, 1962)*

R. P. Feynman *The Theory of Fundamental Processes: A Lecture Note Volume, 1961*

R. P. Feynman *Statistical Mechanics: A Set of Lectures, 1972*

R. P. Feynman *Photon-Hadron Interactions, 1972*

H. Frauenfelder *The Mössbauer Effect: A Review—with a Collection of Reprints, 1966*

S. C. Frautschi *Regge Poles and S-Matrix Theory, 1963*

H. L. Frisch and J. L. Lebowitz
The Equilibrium Theory of Classical Fluids: A Lecture Note and Reprint Volume, 1964

M. Gell-Mann and Y. Ne'eman
The Eightfold Way: (A Review—with a Collection of Reprints), 1964

W. A. Harrison *Pseudopotentials in the Theory of Metals, 1966 (2nd printing, 1971)*

R. Hofstadter *Electron Scattering and Nuclear and Nucleon Structure: A Collection of Reprints with an Introduction, 1963*

D. Horn and F. Zachariasen
Hadron Physics at Very High Energies, 1973

S. Ichimaru *Basic Principles of Plasma Physics: A Statistical Approach, 1973*

BASIC PRINCIPLES of PLASMA PHYSICS

A STATISTICAL APPROACH

S. ICHIMARU *University of Tokyo*

CRC Press
Taylor & Francis Group
Boca Raton London New York

CRC Press is an imprint of the
Taylor & Francis Group, an **informa** business

First published 1973 by W. A. Benjamin, Inc.

Published 2018 by CRC Press
Taylor & Francis Group
6000 Broken Sound Parkway NW, Suite 300
Boca Raton, FL 33487-2742

CRC Press is an imprint of the Taylor & Francis Group, an informa business

Visit the Taylor & Francis Web site at
http://www.taylorandfrancis.com

and the CRC Press Web site at
http://www.crcpress.com

Library of Congress Cataloging in Publication Data:

Ichimaru, Setsuo.
Basic principles of plasma physics.

(Frontiers in physics)
Includes bibliographical references.
1. Plasma (Ionized gases) 2. Statistical physics.
I. Title. II. Series.
QC718.I22 530.4'4 73-9768
ISBN 0-805-38752-2
ISBN 0-805-38753-0 (pbk.)

ISBN 13: 978-0-8053-8753-7 (pbk)

M. Jacob and G. F. Chew

Strong-Interaction Physics: A Lecture Note Volume, 1964

L. P. Kadanoff and G. Baym

Quantum Statistical Mechanics: Green's Function Methods in Equilibrium and Nonequilibrium Problems, 1962 (2nd printing, 1971)

I. M. Khalatnikov

An Introduction to the Theory of Superfluidity, 1965

J. J. J. Kokkedee

The Quark Model, 1969

A. M. Lane *Nuclear Theory: Pairing Force Correlations to Collective Motion, 1963*

T. Loucks *Augmented Plane Wave Method: A Guide to Performing Electronic Structure Calculations–A Lecture Note and Reprint Volume, 1967*

A. B. Migdal and V. Krainov

Approximation Methods in Quantum Mechanics, 1969

A. B. Migdal *Nuclear Theory: The Quasiparticle Method, 1968*

Y. Ne'eman *Algebraic Theory of Particle Physics: Hadron Dynamics in Terms of Unitary Spin Currents, 1967*

P. Nozières *Theory of Interacting Fermi Systems, 1964*

R. Omnès and M. Froissart

Mandelstam Theory and Regge Poles: An Introduction for Experimentalists, 1963

G. E. Pake and T. L. Estle

The Physical Principles of Electron Paramagnetic Resonance, 2nd edition, completely revised, enlarged, and reset, 1973

D. Pines *The Many-Body Problem: A Lecture Note and Reprint Volume, 1961 (3rd printing, 1973 [with corrections taken over from 2nd printing, 1962])*

R. Z. Sagdeev and A. A. Galeev

Nonlinear Plasma Theory, 1969

J. R. Schrieffer *Theory of Superconductivity, 1964 (2nd printing, 1971)*

J. Schwinger *Quantum Kinematics and Dynamics, 1970*

E. J. Squires *Complex Angular Momenta and Particle Physics: A Lecture Note and Reprint Volume, 1963*

L. Van Hove, N. M. Hugenholtz, and L. P. Howland

Problems in Quantum Theory of Many-Particle Systems: A Lecture Note and Reprint volume, 1961

FRONTIERS IN PHYSICS

David Pines, Editor

IN PREPARATION :

R. C. Davidson *Theory of Nonneutral Plasmas*

S. Doniach and E. H. Sondheimer
Green's Functions for Solid State Physicists

G. B. Field, H. C. Arp, and J. N. Bahcall
The Redshift Controversy

CONTENTS

EDITOR'S FOREWORD

The problem of communicating in a coherent fashion the recent developments in the most exciting and active fields of physics seems particularly pressing today. The enormous growth in the number of physicists has tended to make the familiar channels of communication considerably less effective. It has become increasingly difficult for experts in a given field to keep up with the current literature; the novice can only be confused. What is needed is both a consistent account of a field and the presentation of a definite "point of view" concerning it. Formal monographs cannot meet such a need in a rapidly developing field, and, perhaps more important, the review article seems to have fallen into disfavor. Indeed, it would seem that the people most actively engaged in developing a given field are the people least likely to write at length about it.

FRONTIERS IN PHYSICS has been conceived in an effort to improve the situation in several ways: first, to take advantage of the fact that the leading physicists today frequently give a series of lectures, a graduate seminar, or a graduate course in their special fields of interest. Such lectures serve to summarize the present status of a rapidly developing field and may well constitute the only coherent account available at the time. Often, notes on lectures exist (prepared by the lecturer himself, by graduate students, or by postdoctoral fellows) and have been distributed in mimeographed form on a limited basis. One of the principal purposes of the FRONTIERS IN PHYSICS Series is to make such notes available to a wider audience of physicists.

It should be emphasized that lecture notes are necessarily rough and informal, both in style and content, and those in the series will prove no exception. This is as it should be. The point of the series is to offer new, rapid, more informal, and it is hoped, more effective ways for physicists to teach one another. The point is lost if only elegant notes qualify.

The second way to improve communication in very active fields of physics is by the publication of collections of reprints of recent articles. Such collections are themselves useful to people working in the field. The value of the reprints would, however, seem much enhanced if the collection would be accompanied by an introduction of moderate length which would serve to tie the collection together and, necessarily, constitute a brief survey of the present status of the field. Again, it is appropriate that such an introduction be informal, in keeping with the active character of the field.

A third possibility for the series might be called an informal monograph, to connote the fact that it represents an intermediate step between lecture notes and formal monographs. It would offer the author an opportunity to present his views of a field that has developed to the point at which a summation might prove extraordinarily fruitful, but for which a formal monograph might not be feasible or desirable.

Fourth, there are the contemporary classics—papers or lectures which constitute a particularly valuable approach to the teaching and learning of physics today. Here one thinks of fields that lie at the heart of much of present-day research, but whose essentials are by now well understood, such as quantum electrodynamics or magnetic resonance. In such fields some of the best pedagogical material is not readily available, either because it consists of papers long out of print or lectures that have never been published.

The above words, written in August, 1961, seem equally applicable today. During the past decade, plasma physics has undergone a period of particularly rapid growth; today it is an active, mature field of physics, which has in turn influenced greatly development in other fields, including geophysics, space physics, and astrophysics. Setsuo Ichimaru's introductory text-monograph, "Basic Principles of Plasma Physics," in which considerable stress is placed on the underlying unity of the fundamental theories in plasma physics, should prove of great assistance both to the beginning graduate student and the experineced researcher. Professor Ichimaru has made a number of important contributions to fundamental plasma theory, and especially to our understanding of fluctuations and turbulence in plasma. In the present volume he presents a coherent and self-contained account of plasma theory, an account in which plasma theory is viewed not as an isolated subject, but as an integral part of statistical physics. In addition to providing the reader with an excellent introduction to the fundamental concepts of plasma physics, Professor Ichimaru takes him on a guided tour of a number of the present-day frontiers of the field, including some of the most recent considerations on fluctuations, relaxation, and diffusion in both weakly turbulent and strong turbulent plasma.

DAVID PINES

Urbana, Illinois
September 1973

PREFACE

Plasma physics is concerned with the equilibrium and nonequilibrium properties of a statistical system containing many charged particles. The forces of interaction between the particles are electromagnetic, extending themselves over wide ranges. The system is characterized by an enormous number of microscopic degrees of freedom arising from the motion of individual particles. The statistics thus enters into a theoretical treatment of the macroscopic behavior of such a system. The theory of plasmas, therefore, basically involves the many-body problem. The main aim of this book is to give a coherent and self-contained account of the fundamental theories in plasma physics, not as an isolated subject, but as an integral part of statistical physics and the many-body problem.

The author has taught these subjects to graduate students at both the University of Tokyo and the University of Illinois for the past several years; the material presented here is an outgrowth of those lectures. This book stresses the fundamentals of statistical theory rather than attempting to give a broad survey of the field as a whole. As a result, a number of highly interesting and important areas are not discussed here; an example is magnetohydrodynamics and its application to stability analyses of plasmas.

The book begins with the fundamental concepts of nonequilibrium statistical mechanics and electrodynamics. The kinetic theories of plasmas, such as the Vlasov equation, are considered in Chapter 2; electromagnetic response functions, such as the dielectric response function, are introduced and calculated in Chapter 3. Various aspects of plasma theories are then developed as special limiting cases of those fundamental concepts. The simplest cases, that is, the longitudinal and transverse properties of a plasma in thermodynamic equilibrium, are considered in Chapters 4 and 5, which reveal the varied classes of electromagnetic phenomena involved in plasmas. Transient processes associated with the spatial and temporal propagation of plasma waves, including echo phenomena, are discussed in Chapter 6. These problems are closely related to the experimental aspects of plasma studies. The cases of various instabilities in homogeneous and inhomogeneous plasmas are considered in Chapters 7 and 8. These instabilities appear when the plasma deviates significantly from the state of thermodynamic equilibrium.

From Chapter 4 to Chapter 8, the basic equations involved in the theoretical analyses are the Vlasov and Maxwell equations; fluctuations and correlations in plasmas are not taken into consideration. In the last three chapters, we go beyond a Vlasov description of plasmas, and study the effects associated with fluctuations and correlations: The general formulations of fluctuation phenomena in plasmas are described in Chapter 9, where kinetic-theoretical treatments of fluctuations and their applications to specific problems are considered. Relaxation and transport phenomena in plasmas are directly related to fluctuations and correlations. In Chapter 10, these relationships are studied; the rates of temperature relaxations and of spatial diffusion across the magnetic field are calculated as specific examples. Finally, in Chapter 11, the problems associated with enhanced fluctuations in an unstable plasma are considered; these are the plasma turbulence problems. A survey of various theoretical approaches is presented; physical features that require qualitative distinction between a turbulent plasma and a stable quiescent plasma are stressed; consequences of such theories of plasma turbulence are described. Studies of fluctuations, relaxations, and turbulence constitute major frontiers in the modern development of plasma theory.

The presentation in this book is self-contained and should be read without difficulty by those who have adequate preparation in classical mechanics, electricity and magnetism, and elementary kinetic theory and statistics. The level of treatment is intermediate to advanced. References are cited mainly for the purpose of assisting the reader in finding relevant discussions extending and supplementing the material presented here. The book will be useful to graduate students (at all levels) and research physicists in plasma physics, and in such related fields as solid-state physics, geophysics, and astrophysics.

I should like to express my sincere gratitude to Professor David Pines, who introduced me to this field and has given me much helpful advice and kind encouragement over the years as well as during the writing of this book. I wish to thank Professor John Bardeen and Professor Marshall N. Rosenbluth, whose guidance and help have had a deep influence on my scientific and human evolution. Finally, I wish to thank Professor Ralph O. Simmons, head of the physics department at the University of Illinois, for his hospitality during the course of this work.

SETSUO ICHIMARU

Tokyo, Japan
July 1973

BASIC PRINCIPLES of PLASMA PHYSICS

A STATISTICAL APPROACH

CHAPTER 1

INTRODUCTION

1.1 INTRODUCTORY SURVEY

A *plasma* may be defined as any statistical system containing mobile charged particles. Vague as it may sound, the foregoing statement is sufficient to define what is known as plasma in physics and engineering.

We can trace the history of plasma research back to the time of Michael Faraday's study of the "dark discharges." From the middle of the last century until the earlier part of the present century, many prominent physicists engaged in research on electric discharges in gases; among them were J. J. Thomson, J. S. Townsend, and I. Langmuir. It was Langmuir and Tonks [1] who coined the word "plasma" to mean that part of a gaseous discharge which contained almost equal densities of electrons and positive ions. They did so in connection with the oscillatory behavior observed in it, the so-called plasma oscillation.

Many important contributions to the basic understanding of plasma phenomena have been made by astronomers and geophysicists. Indeed, many problems in astronomy and geophysics, such as the dynamic behavior of the ionized matter and magnetic field near the surfaces of the sun and stars, the origin of cosmic rays, the emission mechanisms of pulsars and radio sources, the dispersion and broadening of signals traveling through interstellar space, the dynamics of the magnetosphere, and the propagation of electromagnetic radiation through the upper atmosphere, are closely related to the fundamental aspects of plasma physics.

Work on plasmas was confined to a comparatively few individuals and laboratories until the early 1950s, when intensive work was begun in the United States, the United Kingdom, and the Soviet Union on the realization of controlled release of nuclear fusion energy. Although it has become clear that many years will pass before this objective is successfully achieved, the study of plasmas, which was once regarded as one of the old-fashioned fields in physics, has under this impetus become one of the central topics in science and engineering. In addition, interest in plasmas has been spurred in recent years by other possible technological applications, such as the direct conversion of heat energy into electricity by a magnetohydrodynamic means, the propulsion of space vehicles, and the development of new electronic devices.

Let us note that, according to the definition presented at the beginning of this section, plasmas may be found not only in gases but also in solids; electronic phenomena in semiconductors, semimetals, and metals can all be viewed as examples of plasma effects in solids. In the cases of the electrons in metals and semimetals, the density is so high that we must take account of the degeneracy brought about by the Pauli principle. The degenerate conduction electrons found in such metals are accordingly called a *quantum plasma*.

In Table 1.1, we list approximate magnitudes of densities and temperatures for a number of typical plasmas.

Table 1.1

Approximate Electron Densities and Temperatures for Typical Plasmas

Plasma	Density (cm^{-3})	Temperature (°K)
Interstellar gas	10^0	10^4
Gaseous nebula	10^2	10^4
Ionosphere (F layer)	10^6	10^3
Solar corona	10^6	10^6
Tenuous laboratory plasma	10^{11}	10^4
Solar atmosphere	10^{14}	10^4
Dense laboratory plasma	10^{15}	10^5
Thermonuclear plasma	10^{16}	10^8
Metal	10^{23}	10^2
Stellar interior	10^{27}	10^7

1.2 SIMPLE CONSEQUENCES OF THE COULOMB INTERACTION

A plasma is a collection of charged particles. The Coulomb force with which the charged particles interact is well known to be a long-range force. As a consequence, the physical properties of a plasma exhibit remarkable differences from those of an ordinary neutral gas.

Most salient features of the plasma can be understood by investigating the behavior of the electrons, by far the most mobile charged component of the plasma. We therefore assume a smeared-out background of positive charges, so that the average field of space charge in the plasma may be canceled. Such a simple model for a plasma is called the *electron gas*. In this section, we shall mostly adopt this electron gas model to study a number of basic consequences arising from the Coulomb interaction; the results will also be extended to cases that include various species of charged components.

We begin this section with a consideration of the Coulomb cross section for momentum transfer. This example is intended to illustrate how a naive substitution of Coulomb potential in the calculation of plasma properties would lead to a false prediction. It will also point up the necessity of taking into account the organized or *collective* behavior of many charged particles brought about by the long-range interactions. The Debye screening and plasma oscillations are typical examples of such collective phenomena; the plasma thus strongly exhibits a mediumlike character.

A. Coulomb Cross Section for Momentum Transfer

Suppose that a charged particle with mass m and electric charge q is injected toward a Coulomb scattering center q_0 with initial speed v and impact parameter b. (See Fig. 1.1.) The angle χ of scattering is related to the impact parameter via

$$\cot\frac{\chi}{2}=\frac{bmv^2}{q_0q}. \tag{1.1}$$

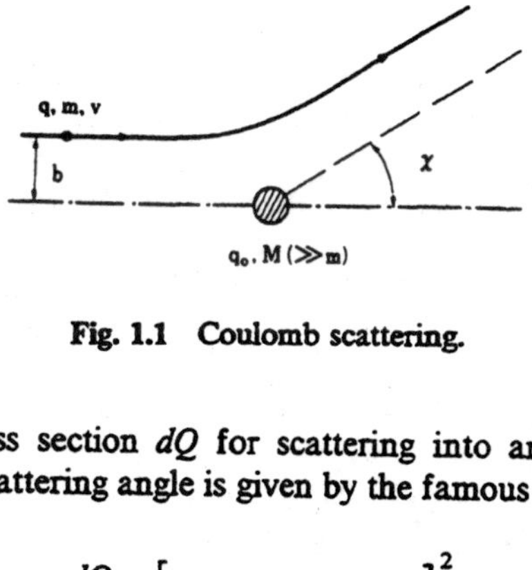

Fig. 1.1 Coulomb scattering.

The differential cross section dQ for scattering into an infinitesimal solid angle do around a scattering angle is given by the famous Rutherford formula

$$\frac{dQ}{do}=\left[\frac{q_0q}{2mv^2\sin^2(\chi/2)}\right]^2. \tag{1.2}$$

The cross section Q_m for momentum transfer can then be calculated by integrating (1.2) over solid angles with a weighting function $(1-\cos\chi)$ which represents the fractional change of momentum on scattering. If for the moment we deliberately choose a finite angle $\chi_{\min}$ for the lower limit of the χ

integration, we calculate

$$Q_m = \int_{\chi_{min}}^{\pi} (1-\cos\chi)\{q_0q/[2mv^2\sin^2(\chi/2)]\}^2 2\pi \sin\chi\, d\chi$$

$$= 4\pi(q_0q/mv^2)^2 \ln\{[\sin(\chi_{min}/2)]^{-1}\}$$

$$\cong 4\pi(q_0q/mv^2)^2 \ln(2/\chi_{min})$$

$$\cong 4\pi(q_0q/mv^2)^2 \ln(b_{max}mv^2/|q_0q|). \qquad (1.3)$$

In the last steps of (1.3), we have assumed that

$$\chi_{min} \ll 1, \qquad (1.4)$$

and made use of (1.1); b_{max} is the impact parameter corresponding to χ_{min}. We thus find that Q_m is proportional to $\ln(2/\chi_{min})$ or $\ln(b_{max}mv^2/|q_0q|)$. The logarithmic factor, whose appearance is typical of the Coulomb interaction, is called the *Coulomb logarithm*. The cross section is seen to diverge logarithmically as χ_{min} approaches zero.

This anomaly points to the inadequacy of the treatment above when it is applied directly to the charged particles in the plasma. The cross section for the momentum transfer is virtually proportional to the electric resistivity of the plasma; experiment tells us that ordinary plasmas are characterized by finite values of resistivity, not infinite ones.

The origin of the foregoing logarithmic divergence can be traced to those scattering acts which take place with large impact parameters and thus with small scattering angles. Those charged particles in the plasma which are located at long distances from a scattering center will undoubtedly also be influenced by many other scattering centers with similar strengths of interactions. We may thus conclude that a simple picture of binary Coulomb scattering cannot correctly describe the behavior of charged particles interacting at large distances in the plasma.

The appearance of such a Coulomb anomaly is not limited to cases of transport processes alone. Another famous example is found in a calculation of the ground state energy of a degenerate electron gas [2]. Regarding Coulomb interactions as a perturbation, we find that a divergence of similar character starts to appear from the second order of the perturbation calculations. The origin of this divergence is again traced to those scattering acts involving small momentum transfers.

B. Debye Screening

Having thus recognized a failure of a simple picture of binary collisions in a plasma, we now take up the problem of determining the effective interaction between charged particles in a plasma. We shall thus calculate an effective potential field around a Coulomb scattering center by taking explicit account of the statistical distribution of other charged particles. This calculation will lead to the notion of Debye screening. The potential field around a charged particle is effectively screened by the cloud of other charged particles; its force range is now confined within a certain characteristic length determined by the density and temperature of the plasma. A number of corollaries which immediately follow from such a calculation will then be discussed; we shall see that the logarithmic divergence of the Coulomb cross section can be cured by cutting off the interaction at that characteristic length and that a plasma tends to stay neutral when viewed on a macroscopic scale.

Consider a point charge q_0 located at the origin ($\mathbf{r}=0$); in vacuum, it produces a potential field

$$\varphi_0(\mathbf{r})=q_0/r. \tag{1.5}$$

In the plasma the spatial distribution of charged particles is affected by the presence of such a potential field and deviates from a uniform distribution. The space-charge field so induced around the point charge in turn produces an extra potential field, which should be added to the original potential $\varphi_0(\mathbf{r})$; a new effective potential $\varphi(\mathbf{r})$ is thus constructed as a summation of the two. The space-charge distribution induced in the plasma is determined not from the bare potential $\varphi_0(\mathbf{r})$, but from the effective potential in a self-consistent fashion. A calculation along these lines was originally carried out by Debye and Hückel [3] in 1923 in connection with the theory of screening in a strong electrolyte.

We consider an electron gas with density n and temperature T in energy units.† Since the point charge is located at the origin, we may write the Poisson equation for the effective field as

$$\nabla^2\varphi(\mathbf{r})=-4\pi q_0\delta(\mathbf{r})+4\pi e\langle\rho(\mathbf{r})\rangle \tag{1.6}$$

where $\delta(\mathbf{r})$ is the three-dimensional delta function; e denotes the unit electric charge, 4.803×10^{-10} esu. The average density variation $\langle\rho(\mathbf{r})\rangle$ of the electrons from the uniform distribution is calculated by applying the usual methods of statistical mechanics; assuming Maxwell–Boltzmann statistics for

† Throughout the book, we express the temperature in energy units unless otherwise specified. Conversion may be achieved with the aid of the relation

$$1\,\text{eV}=1.6021\times10^{-12}\,\text{erg}=1.1605\times10^{4}\,{}^\circ\text{K}.$$

the electrons, we have

$$\langle \rho(\mathbf{r}) \rangle = n \exp(e\varphi(\mathbf{r})/T) - n. \tag{1.7}$$

If the potential energy can be assumed to be much smaller on the average than the kinetic energy, we may expand (1.7) with respect to $e\varphi(\mathbf{r})/T$:

$$\langle \rho(\mathbf{r}) \rangle \cong ne\varphi(\mathbf{r})/T. \tag{1.8}$$

Substituting this in (1.6), we find

$$-\nabla^2\varphi(\mathbf{r}) + (4\pi ne^2/T)\varphi(\mathbf{r}) = 4\pi q_0\delta(\mathbf{r}). \tag{1.9}$$

The solution of this equation with the boundary condition that $\varphi(\mathbf{r})$ vanishes at infinity is

$$\varphi(\mathbf{r}) = (q_0/r)\exp(-r/\lambda_D). \tag{1.10}$$

The parameter

$$\lambda_D = (T/4\pi ne^2)^{1/2} \tag{1.11}$$

introduced here has the dimension of length and is called the *Debye length*; it is an important parameter for the study of plasmas. Putting numbers for the known physical parameters in (1.11), we find the following convenient numerical formula for the Debye length:

$$\lambda_D = 6.9(T/n)^{1/2} \quad [\text{cm}] \tag{1.12}$$

where T and n are to be measured in units of °K and cm^{-3}. For an electron gas with $n = 10^{10}$ and $T = 10^4$, λ_D is computed to be approximately 7×10^{-3} cm; generally, the Debye length takes on a value smaller than any macroscopic scale of distance.

The physical meaning of (1.10) is clear. For distances much smaller than λ_D the effective potential is essentially equivalent to the bare Coulomb potential, while for distances larger than λ_D the potential field decreases exponentially; the potential field around a point charge is effectively screened out by the induced space-charge field in the electron gas for distances greater than the Debye length.

The calculation of the Debye screening can be easily extended to cases that involve a variety of mobile charged components. If we use suffix σ to distinguish between various constituents of the plasma and let q_σ and n_σ denote the charge and density of a given kind of particle, the Debye length of such a plasma is now expressed as

$$\lambda_D = \left(\sum_\sigma \frac{4\pi n_\sigma q_\sigma^2}{T} \right)^{-1/2} \tag{1.13}$$

Based on such an idea of Debye screening, let us return to the problem of Coulomb scattering in the plasma. As we have just observed, the Coulomb field of a point charge is effectively screened out at distances larger than the Debye length. In the calculation of the Coulomb cross section in (1.3), rather than letting the lower limit of the χ integration approach zero, we may hold it at a finite angle $\chi_{\min}$ corresponding to a maximum impact parameter $b_{\max}$ equal to the Debye length; we thus avoid the logarithmic divergence. In most cases, the Coulomb logarithm determined in this way assumes a large number (typically, 15); in actual computation, whether the impact parameter should be chosen equal to the Debye length or some multiple of it is immaterial because such ambiguity affects only the argument in the logarithm.

Hence, within such a logarithmic accuracy, we can choose $|q_0 q| \cong e^2$ and $mv^2 \cong 3T$, a statistical average, to estimate the logarithm in the final expression of (1.3). The Coulomb logarithm, denoted by $\ln \Lambda$, is thereby obtained as

$$\ln \Lambda = \ln(12\pi n \lambda_D^3). \tag{1.14}$$

The calculations carried out thus far on Debye screening have relied upon the assumption that the average potential energy per electron is much smaller than the average kinetic energy, so that the expansion (1.8) may be used. When the potential and density distribution are given by (1.10) and (1.8) [$q_0 = -e$ in (1.10)], we calculate the average potential energy to be $-e^2/2\lambda_D$ (Problem 1.3). We thus compare

$$\frac{|\text{average potential energy}|}{|\text{average kinetic energy}|} = \frac{e^2/2\lambda_D}{\frac{3}{2}T} = \frac{1}{12\pi n \lambda_D^3}. \tag{1.15}$$

It is clear that $n\lambda_D^3$ measures the average number of electrons contained in the cube of volume λ_D^3. Generally this number takes on a value much greater than unity; for instance, when $T = 10^4\,^\circ\text{K}$ and $n = 10^{10}$ cm^{-3}, we compute $n\lambda_D^3 \cong 3 \times 10^3$. We may therefore conclude that the expansion (1.8) is consistent in ordinary circumstances.

The analysis just presented concerns the screening of the potential field around a point charge in a plasma. Similar arguments are also applicable to the space-charge distributions produced by spontaneous fluctuations. Suppose that the space-charge neutrality of the plasma is destroyed locally by some mechanism (e.g., by thermal agitation) and that a local distribution of space charge appears; other charged particles in the plasma then act to neutralize the spontaneous fluctuation of space charge. In the light of the foregoing consideration we may argue that it is energetically quite unfavorable for a space-charge distribution to appear over a distance larger than the Debye length. Since any plasma contains a collection of randomly moving charged particles, we see that the macroscopic neutrality is maintained by the foregoing statistical mechanism.

C. Plasma Oscillation

In the preceding section we arrived at the notion of Debye screening by considering the static equilibrium between a charged particle and an electron gas. Let us now extend our considerations to the dynamic behavior of an electron gas when an external electric charge is suddenly introduced in it.

When a nonequilibrium space-charge field is produced in the electron gas, a group of electrons starts to move in such a way as to screen the static electric field. Since the electron possesses a finite mass, those electrons which begin to move cannot stop at the exact state of equilibrium; they overshoot the target and produce another nonequilibrium distribution in the opposite direction. The electrons begin to move in the reverse direction, overshoot the equilibrium again, and so on; the electrons perform a sort of oscillatory motion, with the Coulomb interaction acting as the restoring force and the mass of the electron as the inertia. This is called a *plasma oscillation*; as is clear from the foregoing argument, it is a phenomenon closely related to the Debye screening.

The frequency of the plasma oscillation may be calculated by considering the displacement of a portion of the electron gas of thickness ξ, as shown in Fig. 1.2. The strength of the electric field thereby produced is

$$E = 4\pi n e \xi.$$

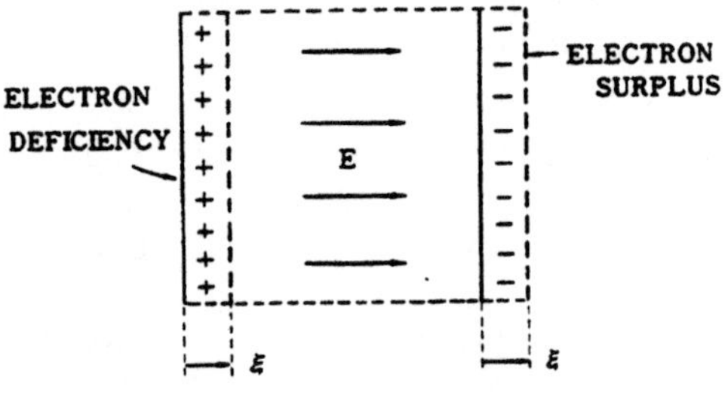

Fig. 1.2 Plasma oscillation.

Substituting this strength in the equation of motion $m\ddot{\xi}=-eE$, we find an oscillatory equation for ξ to be

$$\ddot{\xi}+(4\pi ne^2/m)\xi=0, \tag{1.16}$$

whence the angular frequency of the oscillation is determined to be

$$\omega_p=(4\pi ne^2/m)^{1/2}. \tag{1.17}$$

This is called the *plasma frequency*. Putting numbers for the known physical parameters in (1.17), we find

$$\omega_p=5.6\times 10^4 n^{1/2} \quad [\text{sec}^{-1}] \tag{1.18}$$

where n is to be measured in units of cm^{-3}. For $n=10^{12}$, we compute $\omega_p=5.6\times 10^{10}$; generally, the plasma oscillations take place at fairly high frequencies.

For a multicomponent plasma, the frequency of plasma oscillation is similarly given by

$$\omega_p=\left(\sum_\sigma \frac{4\pi n_\sigma q_\sigma^2}{m_\sigma}\right)^{1/2} \tag{1.19}$$

where m_σ denotes the mass of a particle of the σ species. In this case it is also possible to consider other modes of oscillations because there are additional degrees of freedom corresponding to relative motion between different kinds of particles. Thus, for a two-component plasma consisting of equal densities of electrons and ions, the oscillation with frequency (1.19) corresponds to an out-of-phase oscillation between the two constituents, which may be viewed as an *optical mode* in such a system; the other branch represents an in-phase oscillation and may be called an *acoustic mode*. We shall study the nature of those various oscillations in Chapter 4.

1.3 PARAMETERS DESCRIBING THE PLASMA

Through a consideration of the static and dynamic consequences of Coulomb interaction, we have obtained two important parameters, λ_D and ω_p, which describe mediumlike properties of the plasma. On the other hand, the plasma consists of a large number of discrete particles. In fact, the interplay between these two distinct features, namely, mediumlike character and individual particlelike behavior, is one of the most intriguing aspects of plasma physics. In this section we wish to study the way in which various parameters describing the plasma may be associated with these distinct features. We shall thereby be led to introduce an important dimensionless parameter, the plasma parameter.

A. Discreteness Parameters and Fluid Limit

The Debye length (1.11) and the plasma frequency (1.17) can be computed when the values of the four parameters m, e, $1/n$, and T are known for a given system of charged particles. These may be regarded as the basic quantities characterizing a classical plasma. Let us note also that each of these parameters represents a quantity related to the discrete nature of the individual particles constituting the plasma; they represent, respectively, the mass, electric charge, average volume occupied, and average kinetic energy *per particle*. We may therefore call them the *discreteness parameters* of the plasma.

Having thus recognized the nature of these parameters, we can now introduce a theoretical limiting procedure by which the individuality of the particles is suppressed and the mediumlike properties of the plasma emphasized. Suppose a process in which we cut each particle into finer and finer pieces; in the limit, m, e, $1/n$, and T all approach zero. If in this limit we keep the fluidlike parameters, such as the mass density nm, charge density ne, and kinetic energy density nT, finite, we must regard the discreteness parameters m, e, $1/n$, and T as infinitesimal quantities of the same order. Such a limiting procedure is called the *fluid limit*; it was introduced by Rostoker and Rosenbluth [4]. In this limit the plasma becomes a continuous fluid.

It is important to note that the Debye length and the plasma frequency are kept finite in this fluid limit. They therefore provide appropriate scales by which length and time may be measured in the plasma.

B. The Plasma Parameter

The notion of the discreteness parameters, as we shall see in the following chapter, is especially useful in a theoretical treatment of plasma kinetic equations. In practical applications, however, it is often more convenient and mathematically clear to deal with dimensionless parameters rather than those quantities which have finite physical dimensions.

We wish, therefore, to investigate what kind of dimensionless parameters we can construct out of the four discreteness parameters. We thus write an equation

$$[m^x(1/n)^y T^z e] = 1, \tag{1.20}$$

where the square brackets mean that we are concerned only with the dimensions of the quantity inside them. Since the electric charge has the dimensions of $[\text{mass}]^{1/2}[\text{length}]^{3/2}[\text{time}]^{-1}$, Eq. (1.20) may be decomposed into three equations

$$x+z+\tfrac{1}{2}=0, \qquad 3y+2z+\tfrac{3}{2}=0, \qquad -2z-1=0,$$

whence we find

$$x=0, \qquad y=-\tfrac{1}{6}, \qquad z=-\tfrac{1}{2}. \tag{1.21}$$

This is the only solution obtainable from (1.20).

We would like to find a dimensionless parameter that is of the same order as the discreteness parameters. Hence, we take the cubic power of (1.20)

$$(en^{1/6}T^{-1/2})^3=(8\pi^{3/2}n\lambda_D^3)^{-1}.$$

This calculation indicates that

$$g\equiv 1/n\lambda_D^3 \tag{1.22}$$

is the required dimensionless parameter. We call it the *plasma parameter*; in the literature its definition may sometimes differ by a constant factor.

As we saw in Section 1.2B, the plasma parameter measures the ratio between the average potential and kinetic energies for a plasma in thermal equilibrium. Being ordinarily a small number, it provides a convenient expansion parameter in the theoretical treatments of plasmas.

C. Coulomb Collisions

The usefulness of the concept of the discreteness parameters or the plasma parameter may be illustrated through a consideration of the frequency of Coulomb collisions. Adopting the Coulomb logarithm (1.14) in (1.3), the collision frequency $\nu_m(v)$ of an electron for momentum transfer may be written as

$$\nu_m(v)=nQ_m v=\frac{4\pi ne^4}{m^2v^3}\ln\Lambda. \tag{1.23}$$

For the electrons with a Maxwellian distribution,

$$f(\mathbf{v})=(m/2\pi T)^{3/2}\exp(-mv^2/2T), \tag{1.24}$$

we find that the average value of v^3 is

$$\langle v^3\rangle=8(2/\pi)^{1/2}(T/m)^{3/2}.$$

Consequently, the average collision frequency ν_m may be obtained as

$$\nu_m = \left(\frac{\pi}{2}\right)^{3/2} \frac{ne^4}{m^{1/2}T^{3/2}} \ln \Lambda. \tag{1.25}$$

We here observe that, apart from the dependence on the Coulomb logarithm, the collision frequency is a quantity of the first order in the discreteness parameters. Coulomb collisions essentially involve discreteness of the particles; in the fluid limit, the collision frequency vanishes. These arguments provide a foundation for a collisionless description of the plasma in the fluid limit.

In passing, we compare the collision frequency with the plasma frequency. From (1.17) and (1.25), we find

$$\frac{\nu_m}{\omega_p} = \left[32(2\pi)^{1/2} n\lambda_D^3\right]^{-1} \ln \Lambda. \tag{1.26}$$

Hence, as long as $n\lambda_D^3 \gg 1$, we have $\omega_p \gg \nu_m$. Similar comparison may be made between the Debye length and the Coulomb mean-free path (Problem 1.5).

1.4 COLLECTIVE VERSUS INDIVIDUAL-PARTICLE ASPECTS OF DENSITY FLUCTUATIONS

As we remarked in the previous discussions, the plasma is characterized by a fascinating interplay between its mediumlike character, which arises in turn from the long-range nature of the Coulomb interaction, and the individual-particlelike behavior; like an ordinary gas, it consists of an extremely large number of distinct particles.

In order to appreciate the essential features associated with the simultaneous appearance of such collective and individual-particlelike behaviors of a plasma, we find it instructive to investigate the equation of motion for the density fluctuations of an electron gas [5]. We assume that we are dealing with point particles, so that the density field of the electrons is given by a superposition of the three-dimensional δ functions

$$\rho(\mathbf{r}) = \sum_{i=1}^{n} \delta(\mathbf{r} - \mathbf{r}_i) \tag{1.27}$$

where $\mathbf{r}_i$ refers to the position of the ith electron. We are working with the periodic boundary conditions for a cube of unit volume, so that the summation in (1.27) runs from $i=1$ to $i=n$, the average number of electrons in the unit volume. The spatial Fourier components of the density fluctuations are then calculated to be

$$\rho_{\mathbf{k}}=\int \rho(\mathbf{r})\exp(-i\mathbf{k}\cdot\mathbf{r})\,d\mathbf{r}=\sum_i \exp(-i\mathbf{k}\cdot\mathbf{r}_i). \tag{1.28}$$

We now differentiate (1.28) twice with respect to time:

$$\ddot{\rho}_{\mathbf{k}}=-\sum_i\left[(\mathbf{k}\cdot\mathbf{v}_i)^2+i\mathbf{k}\cdot\dot{\mathbf{v}}_i\right]\exp(-i\mathbf{k}\cdot\mathbf{r}_i). \tag{1.29}$$

Here, $\mathbf{v}_i$ designates the velocity of the ith electron. The acceleration $\dot{\mathbf{v}}_i$ of the ith electron is calculated from the force acting on that electron from all the other electrons plus the positive charge background

$$\begin{aligned}\dot{\mathbf{v}}_i&=-\frac{1}{m}\frac{\partial}{\partial\mathbf{r}_i}\sum_{j(\neq i)}\frac{e^2}{|\mathbf{r}_i-\mathbf{r}_j|}\\ &=-i\frac{4\pi e^2}{m}\sum_{j(\neq i)}\sum_{\mathbf{k}}{}'\frac{\mathbf{k}}{k^2}\exp[i\mathbf{k}\cdot(\mathbf{r}_i-\mathbf{r}_j)]\\ &=-i\frac{4\pi e^2}{m}\sum_{\mathbf{k}}{}'\frac{\mathbf{k}}{k^2}\rho_{\mathbf{k}}\exp(i\mathbf{k}\cdot\mathbf{r}_i)\end{aligned} \tag{1.30}$$

where, as indicated by the prime, the term with $\mathbf{k}=0$ is to be omitted in the $\mathbf{k}$ summation because of the neutralizing background of positive charges. Substituting this formula in (1.29), we find

$$\ddot{\rho}_{\mathbf{k}}=-\sum_i(\mathbf{k}\cdot\mathbf{v}_i)^2\exp(-i\mathbf{k}\cdot\mathbf{r}_i)-\frac{4\pi e^2}{m}\sum_{\mathbf{q}}{}'\frac{\mathbf{k}\cdot\mathbf{q}}{q^2}\rho_{\mathbf{k}-\mathbf{q}}\rho_{\mathbf{q}}. \tag{1.31}$$

The first term on the right-hand side represents the influence of the translational motion of the individual electrons; the second term arises from their mutual interaction. We may rewrite (1.31) by separating out the term with $\mathbf{q}=\mathbf{k}$ in the second term on the right-hand side; we then have

$$\ddot{\rho}_{\mathbf{k}}+\omega_{\mathrm{p}}^2\rho_{\mathbf{k}}=-\sum_i(\mathbf{k}\cdot\mathbf{v}_i)^2\exp(-i\mathbf{k}\cdot\mathbf{r}_i)-\frac{4\pi e^2}{m}\sum_{\mathbf{q}(\neq\mathbf{k})}{}'\frac{\mathbf{k}\cdot\mathbf{q}}{q^2}\rho_{\mathbf{k}-\mathbf{q}}\rho_{\mathbf{q}} \tag{1.32}$$

where ω_{p} is the plasma frequency given by (1.17).

We see that the fluctuations in electron density oscillate at a frequency ω_{p} provided we can neglect the two terms on the right-hand side of (1.32). Let us see in what circumstances we may do so. In order to estimate the first term, let us assume that the velocity distribution is a Maxwellian (1.24). We may then calculate the average of the first term as

$$\sum_i (\mathbf{k}\cdot\mathbf{v}_i)^2 \exp(-i\mathbf{k}\cdot\mathbf{r}_i) = \sum_i \exp(-i\mathbf{k}\cdot\mathbf{r}_i) \int (\mathbf{k}\cdot\mathbf{v})^2 f(\mathbf{v})\, d\mathbf{v}$$

$$= k^2 (T/m)\rho_{\mathbf{k}}. \tag{1.33}$$

The second term involves a product of two density fluctuations. Since the density fluctuation

$$\rho_{\mathbf{q}} = \sum_i \exp(-i\mathbf{q}\cdot\mathbf{r}_i)$$

for $\mathbf{q}\neq 0$ is a sum of exponential terms with randomly varying phases (and since the expectation value $\langle \rho_{\mathbf{q}} \rangle$ vanishes for a translationally invariant system), we might hope that this term represents a small correction to the equation of motion of $\rho_{\mathbf{k}}$ and may be neglected in first approximation. Bohm and Pines [5] called such an approximation the *random phase approximation*, in view of the randomly varying phases of terms which contribute to a given $\rho_{\mathbf{k}}$, as distinguished from the coherent addition of terms for $\rho_0 = n$.

Within the random phase approximation, we see that the condition for collective oscillatory behavior of the $\rho_{\mathbf{k}}$, and hence the entire electron gas, is that

$$k^2 \ll k_{\mathrm{D}}^2 \tag{1.34}$$

where

$$k_{\mathrm{D}} = (4\pi n e^2 / T)^{1/2} \tag{1.35}$$

is the inverse of the Debye length (1.11) and will henceforth be called the *Debye wave number*. Thus, in general, we may expect that the electron gas behaves collectively for $k \ll k_{\mathrm{D}}$, in that it will exhibit oscillations at a frequency near ω_{p}. On the other hand, for short-wavelength phenomena, for which $k \gg k_{\mathrm{D}}$, the plasma behaves like a system of free individual particles, since then the first term on the right-hand side of (1.32) governs the system behavior. In the region of k near k_{D} the behavior will be more complicated, since we deal with a transition from single-particle to collective behavior.

We have thus seen that whether a plasma behaves collectively or like an assembly of individual particles depends on the wavelengths of the phenomena involved. A plasma is, in general, capable of exhibiting both kinds of behavior.

References

1. L. Tonks and I. Langmuir, *Phys. Rev.* 33, 195 (1929); I. Langmuir, *Phys. Rev.* 33, 954 (1929); L. Tonks and I. Langmuir, *Phys. Rev.* 34, 876 (1929).
2. D. Pines, *Elementary Excitations in Solids* (W. A. Benjamin, New York, 1963).
3. P. Debye and E. Hückel, *Physik.Z.* 24, 185 (1923).
4. N. Rostoker and M. N. Rosenbluth, *Phys. Fluids* 3, 1 (1960).
5. D. Pines and D. Bohm, *Phys. Rev.* 85, 338 (1952).

Problems

1.1. Derive (1.1) and (1.2).

1.2. Compare the temperature and the Fermi energy (at $T=0$) for each electron plasma listed in Table 1.1. Then, compute the Coulomb logarithm (1.14) for those cases in which the classical statistics may be applicable.

1.3. Calculate the average potential energy per electron from (1.8) and (1.10).

1.4. Construct a dimensionless parameter out of m, e, $1/n$, and $\hbar$, the quantities characterizing the degenerate electron gas. Show that the parameter thus obtained is equivalent to $r_s \equiv r_0/a_0$, where $r_0=(3/4\pi n)^{1/3}$ is the average interparticle distance, and $a_0=\hbar^2/me^2$ is the Bohr radius.

1.5. Calculate the average value l_m of the Coulomb free path $(nQ_m)^{-1}$ and show that λ_D/l_m is of the same order in the plasma parameter as ν_m/ω_p.

CHAPTER 2

KINETIC EQUATIONS

Plasma physics involves a rich class of phenomena related to the dynamical processes in statistical mechanics. It is therefore essential first to study the structure and properties of the basic kinetic equations which govern the dynamical behavior of the plasma.

We shall begin this chapter with a discussion of the Vlasov equation: It will be derived from the Bogoliubov–Born–Green–Kirkwood–Yvon hierarchy; in the process, its validities and limitations will also be studied. This is particularly important in view of the fact that the Vlasov equation is used most frequently in the theoretical investigation of the plasma phenomena. We shall next turn to a consideration of the collision term in the kinetic equation; this involves a study of the particle correlations, which are not included in a treatment based on the Vlasov equation. In this connection, such important concepts as Bogoliubov's hierarchy of relaxation times and the dielectric response function will be introduced. Finally, an explicit expression for the collision term will be obtained and its properties investigated. Some of the formal aspects of the plasma kinetic equations are discussed also in Appendix A; these are intended for those readers who are more interested in the formal structure of the plasma kinetic equations.

2.1 THE VLASOV EQUATION

We consider a classical system containing N identical particles of charge q and mass m in a box of volume V; $n=N/V$ is the average number density. We assume a smeared-out background of opposite charge such that the average space-charge field of the system vanishes.

The dynamical behavior of such a system of N interacting particles may generally be investigated through the Liouville equation [1, 2]. Equivalently, we could adopt a microscopic distribution function, such as Eq. (1.27), to describe the system behavior; an extension of (1.27) to a six-dimensional phase space is called the Klimontovich distribution function [3]. This distribution function obeys a continuity equation in the phase space, which is the Klimontovich equation. (For more detail, see Appendix A.)

The microscopic distribution functions, consisting of a summation of N six-dimensional delta functions, would not by themselves correspond to the coarse-grained quantities which would be observed in the macroscopic world. To establish the necessary correspondence, we carry out a statistical average of a quantity involving microscopic distributions. A joint distribution function of s particles may thus be defined through such a statistical average of a product of s Klimontovich functions.

The system of charged particles can then be described by the Bogoliubov–Born–Green–Kirkwood–Yvon (BBGKY) hierarchy equations [1, 2].

$$\left[\frac{\partial}{\partial t}+\sum_{i=1}^{s}\mathsf{L}(i)-\frac{1}{n}\sum_{i\neq j}^{s}\mathsf{V}(i,j)\right]f_s(1,\ldots,s)$$

$$=\sum_{i=1}^{s}\int \mathsf{V}(i,s+1)f_{s+1}(1,\ldots,s+1)\,d(s+1). \qquad (2.1)$$

Here, $i\equiv(\mathbf{r}_i,\mathbf{v}_i)$ represents the position and velocity of the ith particle; $f_s(1,\ldots,s)$ is the s-particle distribution function; $\mathsf{L}(i)$ is a single-particle operator

$$\mathsf{L}(i)=\mathbf{v}_i\cdot\frac{\partial}{\partial \mathbf{r}_i}+\frac{q}{m}\left[\mathbf{E}_{\text{ext}}(\mathbf{r}_i)+\frac{\mathbf{v}_i}{c}\times\mathbf{B}_{\text{ext}}(\mathbf{r}_i)\right]\cdot\frac{\partial}{\partial \mathbf{v}_i} \qquad (2.2)$$

acting upon the ith particle under the influence of externally applied fields $\mathbf{E}_{\text{ext}}(\mathbf{r})$ and $\mathbf{B}_{\text{ext}}(\mathbf{r})$†; and $\mathsf{V}(i,j)$ is a two-particle operator

$$\mathsf{V}(i,j)=\frac{q^2n}{m}\left(\frac{\partial}{\partial \mathbf{r}_i}\frac{1}{|\mathbf{r}_i-\mathbf{r}_j|}\right)\cdot\frac{\partial}{\partial \mathbf{v}_i} \qquad (2.3)$$

which represents the effects of Coulomb interaction. A derivation of the BBGKY hierarchy from the Klimontovich equation [3, 4] is described in Appendix A.

The distribution functions are normalized in such a way that

$$V^{-1}\int f_1(1)d1=1, \qquad V^{-1}\int f_{s+1}(1,\ldots,s+1)\,d(s+1)=f_s(1,\ldots,s). \qquad (2.4)$$

We assume that these functions are symmetric with respect to interchange between the coordinates of any two particles.

†For simplicity in notation, the time variables are not explicitly written in this section.

As Eq. (2.1) clearly demonstrates, the structure of the BBGKY hierarchy is such that it does not close in itself: The equation for the single-particle distribution depends on the two-particle distribution; the equation for the two-particle distribution in turn requires knowledge of the three-particle distribution; and so forth. We must, therefore, find a method of truncating such an infinite series of equations.

Such a truncation may be achieved if we consider the fluid limit of the plasma as described in Section 1.3A. Since the plasma parameter g defined by (1.22) takes on a small value for an ordinary plasma, we may try to solve the BBGKY hierarchy through a power-series expansion with respect to the plasma parameter. Thus, we expand

$$f_s = f_s^{(0)} + f_s^{(1)} + f_s^{(2)} + \cdots. \tag{2.5}$$

In view of the definitions (2.2) and (2.3), the operators $\mathsf{L}(i)$ and $\mathsf{V}(i,j)$ may be regarded as zeroth order in the discreteness parameters. Hence, in the fluid limit, the BBGKY hierarchy reduces to

$$\left[\frac{\partial}{\partial t} + \sum_{i=1}^{s} \mathsf{L}(i)\right] f_s^{(0)}(1,\ldots,s) = \sum_{i=1}^{s} \int \mathsf{V}(i,s+1) f_{s+1}^{(0)}(1,\ldots,s+1)\, d(s+1). \tag{2.6}$$

We now make an *ansatz* for the solution in the form

$$f_s^{(0)}(1,\ldots,s) = \prod_{i=1}^{s} F(i), \tag{2.7}$$

a simple product of s single-particle distribution functions, and substitute it in (2.6). It is not difficult then to find that (2.7) indeed satisfies (2.6) if the single-particle distribution function is determined as a solution of the equation

$$\left[\frac{\partial}{\partial t} + \mathsf{L}(1) - \int \mathsf{V}(1,2) F(2;t)\, d2\right] F(1;t) = 0. \tag{2.8}$$

This equation is called the *Vlasov equation* [5]. As may be clear from the foregoing derivation, the Vlasov equation offers a precise description of the plasma in the fluid limit as long as the validity of the factorization in the form of (2.7) is assured in the same limit.

It is interesting to note a formal similarity between the Vlasov equation (2.8) and the Klimontovich equation (A.8) discussed in Appendix A. Yet it is quite important also to note the substantial difference in physical contents involved between the two equations: While the Klimontovich equation deals with the microscopic distribution function (A.2), containing all the fine structures arising from the individuality of the particles, the Vlasov equation concerns itself with a coarse-grained distribution function which is obtained from a statistical average of the microscopic distribution function. The fluctuations due to discreteness of the particles, for example, are not included in the Vlasov equation.

The Vlasov equation may be rewritten in a little more conventional form if we substitute (2.2) and (2.3) into (2.8). Furthermore, we may wish to extend the equation so that those general cases which involve various kinds of charged particles (multicomponent plasmas) can also be treated; such an extension can be achieved in quite a formal manner. Denoting by $f_\sigma(\mathbf{r},\mathbf{v};t)$ the single-particle distribution function for the particles of σ species normalized according to (2.4), we have

$$\left\{\frac{\partial}{\partial t}+\mathbf{v}\cdot\frac{\partial}{\partial \mathbf{r}}+\frac{q_\sigma}{m_\sigma}\left[\mathbf{E}(\mathbf{r},t)+\frac{\mathbf{v}}{c}\times\mathbf{B}(\mathbf{r},t)\right]\cdot\frac{\partial}{\partial \mathbf{v}}\right\}f_\sigma(\mathbf{r},\mathbf{v};t)=0 \tag{2.9}$$

where

$$\mathbf{E}(\mathbf{r},t)=\mathbf{E}_{\text{ext}}(\mathbf{r},t)-\sum_\sigma q_\sigma n_\sigma\frac{\partial}{\partial \mathbf{r}}\int d\mathbf{r}'\int d\mathbf{v}'\frac{f_\sigma(\mathbf{r}',\mathbf{v}';t)}{|\mathbf{r}-\mathbf{r}'|}, \tag{2.10a}$$

$$\mathbf{B}(\mathbf{r},t)=\mathbf{B}_{\text{ext}}(\mathbf{r},t). \tag{2.10b}$$

Equation (2.9) has been called by a number of different names in addition to the Vlasov equation. It is called the *time-dependent Hartree equation*, because the last term of (2.10a) in fact represents the electrostatic Hartree field produced by the average distribution of charged particles; in (2.9), particle interactions are taken into account only through such an average self-consistent field. We may also note in this connection that (2.7) corresponds to the Hartree factorization.

The electromagnetic fields, as Eqs. (2.10) stand, enable us to take account of only the electrostatic interactions between the particles. We can easily extend the equations, however, so that the electromagnetic interactions may also be included in the Vlasov formalism. The modification required is to add terms on the right-hand sides of (2.10) which describe the effects of the average magnetic field produced by the current distribution of charged particles. [Equations (3.41) in Chapter 3 determine those interaction fields.]

Equation (2.9) may also be called the *collisionless Boltzmann equation*, since the terms describing the particle collisions are apparently missing on its right-hand side. Collisional effects arise as a consequence of the field fluctuations away from the average Hartree field; in the fluid limit, such fluctuations, being of the order of g, are to be neglected. We may recall that a similar conclusion has already been reached in Section 1.3C based on a comparison between the frequency of Coulomb collisions and the plasma frequency.

Because of this failure to include the collisional effects, the Vlasov equation is unable to describe an approach of the system toward thermodynamic equilibrium. Indeed, in the absence of external fields ($\mathbf{E}_{ext}=0$, $\mathbf{B}_{ext}=0$), we can easily prove that any velocity distribution function, uniform in space and stationary in time, can satisfy (2.9); the average Hartree field always vanishes for such a homogeneous situation because of the overall neutrality of the system. To obtain a kinetic theory of the plasma capable of describing an approach to equilibrium, we must depart from the fluid limit and include the effects associated with the discreteness of the particles. We shall study these in the following sections.

2.2 PAIR CORRELATION FUNCTION AND COLLISION TERM

In connection with the s-particle distribution function $f_s(1,\ldots,s)$, it is also convenient to introduce the s-particle correlation function $C_s(1,\ldots,s)$ along lines quite analogous to the Mayer cluster expansion [6]. Physically, C_s represents the totally correlated part of f_s. It can be successively defined as

$$C_1(1)=f_1(1),$$

$$C_2(1,2)=f_2(1,2)-C_1(1)C_1(2),$$

$$C_3(1,2,3)=f_3(1,2,3)-C_1(1)C_1(2)C_1(3)$$
$$-C_1(1)C_2(2,3)-C_1(2)C_2(3,1)-C_1(3)C_2(1,2),$$

and so on. The correlation functions are also symmetric with respect to interchange of two particles. To avoid repetitious subscripts, we write the single-particle distribution function as $F(1)=C_1(1)$, the pair correlation function as $G(1,2)=C_2(1,2)$, and the ternary correlation function as $H(1,2,3)=C_3(1,2,3)$. Thus, we have expansions

$$f_1(1)=F(1), \tag{2.11a}$$

$$f_2(1,2)=F(1)F(2)+G(1,2), \tag{2.11b}$$

$$f_3(1,2,3)=F(1)F(2)F(3)+F(1)G(2,3)$$
$$+F(2)G(3,1)+F(3)G(1,2)+H(1,2,3). \qquad (2.11c)$$

We have assumed the factorization (2.7) to be true in the fluid limit. It then follows that $G(1,2)/F(1)F(2)$ should vanish in the same limit; let us then assume that this ratio is of the order of g; that is,

$$G(1,2)/F(1)F(2)\sim g^1. \qquad (2.12)$$

We may also assume a systematic trend, so that

$$H(1,2,3)/F(1)F(2)F(3)\sim g^2. \qquad (2.13)$$

Although these are the assumptions commonly adopted for a kinetic theory of the plasma near thermodynamic equilibrium, we must be aware that there are some instances in which these conditions are simply violated. A most typical example is the short-range correlations; as $|\mathbf{r}_1-\mathbf{r}_2|$ approaches zero, the pair correlation function $G(1,2)$ usually tends to become of the same order as the product $F(1)F(2)$ of the single-particle distribution functions. Short-range correlations cannot be analyzed through a theory based on assumptions (2.12) and (2.13). Similar violation occurs in the theory of turbulent plasmas, in which plasma-wave instabilities act to enhance the fluctuations and correlations tremendously. In either of these cases we are faced with problems involving strong correlations; some aspects of such problems will be treated in Chapter 11.

With these reservations in mind, let us accept (2.12) and (2.13) and study the consequences arising from them. We therefore substitute (2.11) into the first two equations of the BBGKY hierarchy (2.1) and retain the terms up to the first order in the plasma parameter. The first equation thus reads

$$\left[\frac{\partial}{\partial t}+\mathrm{L}(1)-\int \mathrm{V}(1,2)F(2;t)\,d2\right]F(1;t)=\int \mathrm{V}(1,2)G(1,2;t)\,d2. \qquad (2.14)$$

The left-hand side of this equation is identical to that of the Vlasov equation, (2.8). Equation (2.14) contains, on its right-hand side, an additional term depending on the pair correlation between the particles. This new term is of the first order in the plasma parameter; it describes the collisional effects brought about by the discreteness of the particles. We may call it a *collision term*.

The second equation of the BBGKY hierarchy contains F and G; the terms involving H as well as some others drop out because they are of the second order in the plasma parameter. The equation is then simplified with the aid of (2.14); we thus obtain

$$\left\{\frac{\partial}{\partial t}+L(1)+L(2)-\int[V(1,3)+V(2,3)]F(3;t)\,d3\right\}G(1,2;t)$$

$$-\int V(1,3)F(1;t)G(2,3;t)\,d3-\int V(2,3)F(2;t)G(3,1;t)\,d3$$

$$=\frac{1}{n}[V(1,2)+V(2,1)]F(1;t)F(2;t). \tag{2.15}$$

The left-hand side of this equation is homogeneous with respect to G; the right-hand side may be regarded as a driving term which acts to create the particle correlations through Coulomb interactions. Equations (2.14) and (2.15) thereby form a set of coupled equations between the single-particle and pair correlation functions.

A formal solution to (2.15) is not very difficult to obtain: we first write

$$G(1,2;t)=\int d1'\int d2'\int_{-\infty}^{t} dt'\,G_0(1',2';t')\,U(1,1';t-t')\,U(2,2';t-t'). \tag{2.16}$$

By so doing, we have expressed the correlation between 1 and 2 (in the phase space) at time t as a superposition of the correlation effects created between $1'$ and $2'$ at t' which propagate to 1 and 2 during the time interval $t-t'$. Hence, we may regard the function U as a particle propagator for the pair correlation function. The particular choice of integration domain $[-\infty,t]$ for t' reflects the notion of causality; a detailed discussion on this subject will be given in the following chapter in connection with the response functions.

We now substitute (2.16) in (2.15); we then find that (2.16) is a solution to (2.15) if G_0 is given by the right-hand side of (2.15), that is,

$$G_0(1,2;t)=(1/n)[V(1,2)+V(2,1)]F(1;t)F(2;t), \tag{2.17}$$

and if the propagator U satisfies the equation

$$\left[\frac{\partial}{\partial t}+L(1)-\int V(1,2)F(2;t)\,d2\right]U(1,1';t-t')$$

$$-\int V(1,2)F(1;t)\,U(2,1';t-t')\,d2=0 \tag{2.18}$$

with the initial condition

$$U(1,1';0)=\delta(1-1'). \tag{2.19}$$

We may therefore have a solution in the form

$$G(1,2;t)=(1/n)\int d1'\int d2'\int_{-\infty}^{t} dt'\{[V(1',2')+V(2',1')]F(1';t')F(2';t')$$
$$\times U(1,1';t-t')U(2,2';t-t')\}. \qquad (2.20)$$

Let us note that (2.18) is a linearized version of the Vlasov equation. In fact, if we make a replacement $F \rightarrow F+\delta f$ in (2.8) and linearize it with respect to δf, then we obtain (2.18), in which U occupies the places of δf. Consequently, the propagator U obeys the linearized Vlasov equation.

Although we have succeeded in finding a formal solution for the pair correlation function, this is perhaps almost as far as we can proceed at the moment. We are still left with a coupled problem in that to solve the linearized Vlasov equation (2.18) for U we must know the time-dependent behavior of F and that the time evolution of F in turn depends on U through G. To resolve this coupling, we need to introduce additional physical arguments, due to Bogoliubov, on the various time scales involved in the dynamical processes.

2.3 BOGOLIUBOV'S HIERARCHY OF CHARACTERISTIC TIME SCALES

When treating dynamical problems in statistical physics, we often find that there exists a well-defined hierarchy among the magnitudes of the various time scales involved. This was first noted by Bogoliubov [1].

We begin with the relaxation time τ_0 associated with a hydrodynamic quantity; it may be estimated as a macroscopic distance L divided by the sound velocity c_s which is a characteristic velocity in a hydrodynamic problem:

$$\tau_0 \sim L/c_s. \qquad (2.21)$$

The characteristic time τ_1 for the single-particle distribution function to relax to its local equilibrium values may be calculated as the mean free path l divided by the mean velocity of the particles, $\langle v\rangle \sim (T/m)^{1/2}$:

$$\tau_1 \sim l/\langle v\rangle. \qquad (2.22)$$

Since $c_s \sim \langle v\rangle$, we have, for slowly varying disturbances ($L \gg l$),

$$\tau_0 \gg \tau_1. \qquad (2.23)$$

We may similarly consider a characteristic time scale τ_2 associated with the pair correlation function; it may be determined as an average time for a particle to travel over a correlation distance. For the plasma, the correlation distance may be estimated to be of the order of the Debye length; hence, we have

$$\tau_2 \sim \lambda_D / \langle v \rangle. \tag{2.24}$$

Comparison between τ_1 and τ_2 may be facilitated if we realize that $1/\tau_1 \cong \nu_m$, given by (1.25) for Coulomb collisions, and that $1/\tau_2 \cong \omega_p$, the plasma frequency. In the light of (1.26), we then find

$$\tau_1 \gg \tau_2 \tag{2.25}$$

as long as the plasma parameter remains a very small number.

The inequality (2.25) implies that the pair correlation functions relax much faster than the single-particle distribution functions. The slower processes associated with the evolution of the latter functions are sometimes referred to as the kinetic processes. In such a kinetic stage of development, the pair correlation functions have already completed their own relaxations; these functions must therefore be expressible as functionals of the single-particle distribution functions. In fact, one of the most important conclusions from these hierarchy considerations is that the pair and all the higher-order correlation functions can be written as functionals of the single-particle distribution functions in the kinetic processes.

2.4 DIELECTRIC PROPAGATOR

The conclusion reached in the previous section also implies that the single-particle distribution function may be regarded as time independent as far as the calculations of the pair correlation function are concerned. Accordingly, we may neglect the time dependence of F in (2.18). We then have a manageable problem in which the coupling between the time evolutions of F and G has been disentangled.

For simplicity, we consider a uniform plasma in the absence of external fields ($\mathbf{E}_{ext}=0$ and $\mathbf{B}_{ext}=0$). Then we may substitute

$$L(1) = \mathbf{v}_1 \cdot \frac{\partial}{\partial \mathbf{r}_1}, \tag{2.26}$$

$$F(1;t) = f(\mathbf{v}_1) \tag{2.27}$$

in (2.18); $f(\mathbf{v})$ is the velocity distribution function with normalization

$$\int f(\mathbf{v})\, d\mathbf{v} = 1. \tag{2.28}$$

Taking account of the macroscopic neutrality of the system, we have

$$\left(\frac{\partial}{\partial t}+\mathbf{v}_1\cdot\frac{\partial}{\partial \mathbf{r}_1}\right)U(1,1';t)-\int v(1,2)f(\mathbf{v}_1)U(2,1';t)\,d2=0. \qquad (2.29)$$

With the initial condition (2.19), this equation may be solved through the technique of Fourier–Laplace transformations.

For a translationally invariant system, the propagator $U(1,1';t)$ depends only on $\mathbf{r}_1-\mathbf{r}_1'$ in space; we carry out a spatial Fourier transformation of the propagator according to

$$U_{\mathbf{k}}(\mathbf{v}_1,\mathbf{v}_1';t)=\int_V d(\mathbf{r}_1-\mathbf{r}_1')\,U(1,1';t)\exp[-i\mathbf{k}\cdot(\mathbf{r}_1-\mathbf{r}_1')]. \qquad (2.30a)$$

Its inverse transformation is given by

$$U(1,1';t)=\frac{1}{V}\sum_{\mathbf{k}}U_{\mathbf{k}}(\mathbf{v}_1,\mathbf{v}_1';t)\exp[i\mathbf{k}\cdot(\mathbf{r}_1-\mathbf{r}_1')] \qquad (2.30b)$$

where the $\mathbf{k}$ summation extends over those discrete values of the wave vectors which are determined from the periodic boundary conditions appropriate to the volume V. Since the assumption of translational invariance would imply an infinite plasma (without boundary), the choice of the volume is practically unrestricted; for simplicity, we may sometimes choose $V=1$, as we did in Section 1.4. Passage to an infinite-volume case can be achieved by simply going over from the Fourier series to the Fourier integral via

$$\sum_{\mathbf{k}}\rightarrow V\int\frac{d\mathbf{k}}{(2\pi)^3}\rightarrow\int\frac{d\mathbf{k}}{(2\pi)^3}. \qquad (2.31)$$

We next introduce a one-sided Fourier transformation of $U_{\mathbf{k}}(\mathbf{v}_1,\mathbf{v}_1';t)$ with respect to time

$$\mathsf{U}_{\mathbf{k}}(\mathbf{v}_1,\mathbf{v}_1';\omega)=\int_0^\infty U_{\mathbf{k}}(\mathbf{v}_1,\mathbf{v}_1';t)\exp(i\omega t)\,dt. \qquad (2.32a)$$

It follows that the function $\mathsf{U}_{\mathbf{k}}$ is analytic in the upper half of the complex ω plane; a positive imaginary part of ω may be sufficient to guarantee convergence of the integration (2.32a). $\mathsf{U}_{\mathbf{k}}(\mathbf{v}_1,\mathbf{v}_1';\omega)$ is then continued analytically into the lower half of the ω plane; there, it generally has singularities, such as poles. Inverse transformation of (2.32a) is given by

$$U_{\mathbf{k}}(\mathbf{v}_1,\mathbf{v}_1';t)=\frac{1}{2\pi}\int_C\mathsf{U}_{\mathbf{k}}(\mathbf{v}_1,\mathbf{v}_1';\omega)\exp(-i\omega t)\,d\omega, \qquad (2.32b)$$

where the contour C extends from $-\infty$ to $+\infty$ along a path in the upper half of the ω plane in such a way that all the singularities lie below it. For $t>0$, we can close the contour with an infinite semicircle in the lower half-plane; Cauchy's theorem may thus be used for the inverse transformation. For $t<0$, we close the contour with an infinite semicircle in the upper half-plane; we then have

$$U_{\mathbf{k}}(\mathbf{v}_1,\mathbf{v}_1';t)=0 \qquad (t<0). \tag{2.33}$$

As we see from the description above, the one-sided Fourier transformation is equivalent to the Laplace transformation; this transformation makes it possible to take explicit account of the initial conditions in a solution of a differential equation.

Let us now carry out the Fourier–Laplace transformations to (2.29) with the initial condition (2.19). Since the operator $\mathcal{V}(1,2)$ may be expanded as†

$$\mathcal{V}(1,2)=\omega_p^2\sum_{\mathbf{k}}\frac{i\mathbf{k}}{k^2}\cdot\frac{\partial}{\partial\mathbf{v}_1}\exp[i\mathbf{k}\cdot(\mathbf{r}_1-\mathbf{r}_2)], \tag{2.34}$$

we have

$$\mathcal{U}_{\mathbf{k}}(\mathbf{v}_1,\mathbf{v}_1';\omega)=\frac{i}{\omega-\mathbf{k}\cdot\mathbf{v}_1}\delta(\mathbf{v}_1-\mathbf{v}_1')$$
$$-\frac{\omega_p^2}{k^2}\frac{1}{\omega-\mathbf{k}\cdot\mathbf{v}_1}\mathbf{k}\cdot\frac{\partial}{\partial\mathbf{v}_1}f(\mathbf{v}_1)\int d\mathbf{v}_2\mathcal{U}_{\mathbf{k}}(\mathbf{v}_2,\mathbf{v}_1';\omega). \tag{2.35}$$

Upon integrating both sides with respect to $\mathbf{v}_1$, we find

$$\int d\mathbf{v}\mathcal{U}_{\mathbf{k}}(\mathbf{v},\mathbf{v}_1';\omega)=\frac{i}{(\omega-\mathbf{k}\cdot\mathbf{v}_1')\epsilon(\mathbf{k},\omega)} \tag{2.36}$$

where

$$\epsilon(\mathbf{k},\omega)=1+\frac{\omega_p^2}{k^2}\int d\mathbf{v}\frac{1}{\omega-\mathbf{k}\cdot\mathbf{v}}\mathbf{k}\cdot\frac{\partial}{\partial\mathbf{v}}f(\mathbf{v}). \tag{2.37}$$

This function is called the *dielectric response function*; it plays a very important part in describing the collective behavior of the plasma. We shall study detailed properties of this function and physical consequences arising from them in subsequent chapters. Here, however, we are satisfied to have obtained the dielectric response function through a calculation of the particle propagator for a pair correlation function.

†Here, ω_p is defined as $(4\pi nq^2/m)^{1/2}$.

Substituting (2.36) in (2.35), we finally obtain a solution for the propagator

$$U_{\mathbf{k}}(\mathbf{v}_1,\mathbf{v}_1';\omega)=\frac{i}{\omega-\mathbf{k}\cdot\mathbf{v}_1}\delta(\mathbf{v}_1-\mathbf{v}_1')$$

$$-i\frac{\omega_p^2}{k^2}\frac{1}{(\omega-\mathbf{k}\cdot\mathbf{v}_1)(\omega-\mathbf{k}\cdot\mathbf{v}_1')\epsilon(\mathbf{k},\omega)}\mathbf{k}\cdot\frac{\partial}{\partial\mathbf{v}_1}f(\mathbf{v}_1). \qquad (2.38)$$

Clearly, the first term on the right-hand side represents a free-particle propagator; the second term takes account of the effects of Coulomb interaction.

2.5 BALESCU–LENARD COLLISION TERM

Having thus solved the linearized Vlasov equation for the propagator U, we are now in a position to evaluate the pair correlation function and thereby to calculate the explicit form of the collision term on the right-hand side of (2.14). To begin, we carry out the spatial Fourier transformation for the pair correlation function in the same way as (2.30a) (with $V=1$); with the aid of (2.27) and (2.34), we obtain from (2.20)

$$G_{\mathbf{k}}(\mathbf{v}_1,\mathbf{v}_2)=i\frac{\omega_p^2}{n}\int d\mathbf{v}_1'\int d\mathbf{v}_2'\int_0^{\infty}dt\frac{\mathbf{k}}{k^2}\cdot\left[\left(\frac{\partial}{\partial\mathbf{v}_1'}-\frac{\partial}{\partial\mathbf{v}_2'}\right)f(\mathbf{v}_1')f(\mathbf{v}_2')\right]$$

$$\times U_{\mathbf{k}}(\mathbf{v}_1,\mathbf{v}_1';t)U_{-\mathbf{k}}(\mathbf{v}_2,\mathbf{v}_2';t). \qquad (2.39)$$

In terms of this function, the collision term may be expressed as

$$\left.\frac{\partial f(\mathbf{v})}{\partial t}\right]_c=-i\omega_p^2\sum_{\mathbf{k}}\frac{\mathbf{k}}{k^2}\cdot\frac{\partial}{\partial\mathbf{v}}\int d\mathbf{v}'G_{\mathbf{k}}(\mathbf{v},\mathbf{v}'). \qquad (2.40)$$

We now substitute (2.32b) in (2.39):

$$G_{\mathbf{k}}(\mathbf{v}_1,\mathbf{v}_2)=i\frac{\omega_p^2}{n}\int d\mathbf{v}_1'\int d\mathbf{v}_2'\int_0^{\infty}dt\frac{\mathbf{k}}{k^2}\cdot\left[\left(\frac{\partial}{\partial\mathbf{v}_1'}-\frac{\partial}{\partial\mathbf{v}_2'}\right)f(\mathbf{v}_1')f(\mathbf{v}_2')\right]$$

$$\times\int_C\frac{d\omega}{2\pi}U_{\mathbf{k}}(\mathbf{v}_1,\mathbf{v}_1';\omega)\exp(-i\omega t)\int_{C'}\frac{d\omega'}{2\pi}U_{-\mathbf{k}}(\mathbf{v}_2,\mathbf{v}_2';\omega')\exp(-i\omega' t).$$

To assure the convergence of the t integration, we impose a condition

$$\mathrm{Im}(\omega+\omega')<0. \qquad (2.41)$$

We then have

$$G_{\mathbf{k}}(\mathbf{v}_1,\mathbf{v}_2)=\frac{\omega_p^2}{(2\pi)^2 n}\int d\mathbf{v}_1'\int d\mathbf{v}_2'\int_C d\omega\int_{C'} d\omega'\frac{\mathbf{k}}{k^2}\cdot\left[\left(\frac{\partial}{\partial\mathbf{v}_1'}-\frac{\partial}{\partial\mathbf{v}_2'}\right)f(\mathbf{v}_1')f(\mathbf{v}_2')\right]$$

$$\times\frac{1}{\omega+\omega'}U_{\mathbf{k}}(\mathbf{v}_1,\mathbf{v}_1';\omega)U_{-\mathbf{k}}(\mathbf{v}_2,\mathbf{v}_2';\omega').$$

The ω' integration can be carried out by closing the contour by an infinite semicircle in the upper half-plane; since $U_{-\mathbf{k}}(\mathbf{v}_2,\mathbf{v}_2';\omega')$ vanishes as $|\omega'|\to\infty$, the integration along the semicircle is zero. By virtue of (2.41) and the nature of the contour C', the only contribution to the integration arises from the pole at $\omega'=-\omega$. Hence, we obtain

$$G_{\mathbf{k}}(\mathbf{v}_1,\mathbf{v}_2)=i\frac{\omega_p^2}{2\pi n}\int d\mathbf{v}_1'\int d\mathbf{v}_2'\int_{-\infty}^{\infty} d\omega\frac{\mathbf{k}}{k^2}\cdot\left[\left(\frac{\partial}{\partial\mathbf{v}_1'}-\frac{\partial}{\partial\mathbf{v}_2'}\right)f(\mathbf{v}_1')f(\mathbf{v}_2')\right]$$

$$\times U_{\mathbf{k}}(\mathbf{v}_1,\mathbf{v}_1';\omega)U_{-\mathbf{k}}(\mathbf{v}_2,\mathbf{v}_2';-\omega). \quad (2.42)$$

The contour of ω integration in (2.42) extends from $-\infty$ to $+\infty$; it sees all the singularities of $U_{\mathbf{k}}(\mathbf{v}_1,\mathbf{v}_1';\omega)$ from above. The function $U_{-\mathbf{k}}(\mathbf{v}_2,\mathbf{v}_2';-\omega)$, which is complex conjugate to $U_{\mathbf{k}}(\mathbf{v}_2,\mathbf{v}_2';\omega)$, however, obeys different boundary conditions: It is an analytic function in the lower half of the complex ω plane; it is then continued analytically into the upper half-plane, where it may have singularities. The integration contour, therefore, sees all the singularities of $U_{-\mathbf{k}}(\mathbf{v}_2,\mathbf{v}_2';-\omega)$ from below.

An explicit expression for the collision term may be obtained from (2.40) with the aid of (2.38) and (2.42). The calculations are lengthy and involve extensive use of the analytic properties as described above; some of the intermediate steps in the calculations are explained in Appendix B. The final result is

$$\left.\frac{\partial f(\mathbf{v})}{\partial t}\right]_c=\frac{\pi\omega_p^4}{n}\sum_{\mathbf{k}}\frac{\mathbf{k}}{k^2}\cdot\frac{\partial}{\partial\mathbf{v}}\int d\mathbf{v}'\frac{\mathbf{k}}{k^2}\cdot\left[\left(\frac{\partial}{\partial\mathbf{v}}-\frac{\partial}{\partial\mathbf{v}'}\right)f(\mathbf{v})f(\mathbf{v}')\right]\frac{\delta(\mathbf{k}\cdot\mathbf{v}-\mathbf{k}\cdot\mathbf{v}')}{|\epsilon(\mathbf{k},\mathbf{k}\cdot\mathbf{v})|^2}.$$

(2.43)

This collision term was derived independently by Balescu [7] and Lenard [8]. The collision term of this form with $\epsilon=1$ was first obtained by Landau [9].

For a multicomponent plasma, the *Balescu–Lenard collision term* can likewise be written as

$$\left.\frac{\partial f_\sigma(\mathbf{v})}{\partial t}\right]_c = \frac{\pi\omega_\sigma^2}{n_\sigma}\sum_{\sigma'}\sum_{\mathbf{k}}\omega_{\sigma'}^2\frac{\mathbf{k}}{k^2}\cdot\frac{\partial}{\partial\mathbf{v}}\int d\mathbf{v}'\frac{\mathbf{k}}{k^2}$$

$$\cdot\left[\left(\frac{m_{\sigma'}}{m_\sigma}\frac{\partial}{\partial\mathbf{v}}-\frac{\partial}{\partial\mathbf{v}'}\right)f_\sigma(\mathbf{v})f_{\sigma'}(\mathbf{v}')\right]\frac{\delta(\mathbf{k}\cdot\mathbf{v}-\mathbf{k}\cdot\mathbf{v}')}{|\epsilon(\mathbf{k},\mathbf{k}\cdot\mathbf{v})|^2} \tag{2.44}$$

where

$$\epsilon(\mathbf{k},\omega)=1+\sum_\sigma\frac{\omega_\sigma^2}{k^2}\int d\mathbf{v}\frac{1}{\omega-\mathbf{k}\cdot\mathbf{v}}\mathbf{k}\cdot\frac{\partial}{\partial\mathbf{v}}f_\sigma(\mathbf{v}) \tag{2.45}$$

is the dielectric response function of the multicomponent plasma, and

$$\omega_\sigma^2=4\pi n_\sigma q_\sigma^2/m_\sigma \tag{2.46}$$

is the square of the plasma frequency for the particles of σ species.

It is instructive to investigate the k dependence of the terms involved in (2.43): Apart from the contributions arising from $|\epsilon(\mathbf{k},\mathbf{k}\cdot\mathbf{v})|^2$, the terms under the $\mathbf{k}$ summation are proportional to k^{-3}. Transforming the summation into integration via (2.31), we then find that (2.43) would diverge logarithmically both at the lower and upper limits of the k integration if the k dependence of $\epsilon(\mathbf{k},\mathbf{k}\cdot\mathbf{v})$ has been ignored. The physical origin of the divergence in the small k domain is the same as that discussed in Section 1.2A; it involves scattering events with small momentum transfers. The divergence in this domain has been cured in the Balescu–Lenard collision term, since $\epsilon(\mathbf{k},\mathbf{k}\cdot\mathbf{v})$ is in fact proportional to k^{-2} in the limit of small k; physically, this amounts to taking the screening action of the plasma into account.

The divergence in the large k domain, however, still remains in the Balescu–Lenard collision term; $\epsilon(\mathbf{k},\mathbf{k}\cdot\mathbf{v})$ approaches unity in the limit of large k. The origin of this divergence may be traced to our original assumption, (2.12) and (2.13). At short distances, where the momentum transfer between the particles is large, the major effects come from the discreteness of the particles; expansion in powers of the plasma parameter does not lead to a valid description of the phenomena in this domain.

We have thus observed that the Balescu–Lenard theory gives a convergent result for collisions with small momentum transfers. On the other hand, as we saw in Section 1.2A, a theory based on a binary-collision model provides a correct description for collisions with large momentum transfers. The ranges of validity for these two theories, in fact, greatly overlap when $g\ll 1$. Here, therefore, we may discover a possibility of unifying these two approaches and thereby constructing a theory which contains no divergences associated with the Coulomb interactions; the theory would then enable us to determine the argument of the Coulomb logarithm exactly without making use of free parameters. Such a possibility was first recognized by Hubbard [10]; subsequently, it has been extensively investigated and developed into a unified theory by Kihara and his coworkers [11].

There are a number of different ways to derive the Balescu–Lenard collision term; the methods used by Balescu and Lenard are not the same as the one described here. Wyld and Pines [12] have obtained a quantum-mechanical collision term through a calculation of the transition probability via a Coulomb interaction screened by the dielectric response function. Passing over to the classical limit ($\hbar \to 0$), they have then shown on the one hand that the Balescu–Lenard term may be recovered if we keep the wave vector $\mathbf{k}$ finite (and hence let the momentum transfer $\hbar\mathbf{k}$ vanish), and on the other hand that the ordinary Boltzmann collision term may be obtained if the momentum transfer is kept finite. This finding gives us further physical insight into the nature of the system with Coulomb interaction; in particular, it provides another suggestion for the possibility of a unified theory between distant and close collisions.

Another approach of interest is the one based on the Fokker–Planck equation. Hubbard and Thompson [13] have calculated the Fokker–Planck coefficients through an explicit evaluation of the fluctuation spectrum of the microscopic electric fields; they have thereby reproduced the Balescu–Lenard collision term. We shall return to a discussion of these topics in Chapter 10.

2.6 PROPERTIES OF THE BALESCU–LENARD COLLISION TERM

The collision term (2.43) exhibits a number of remarkable properties; the investigations of these properties are mostly due to Lenard [8]:

(a) If f is positive for $t=0$, it is positive for $t>0$.
(b) The particle density is independent of time.
(c) The mean velocity is independent of time.
(d) The mean kinetic energy is independent of time.
(e) Any Maxwellian distribution is a stationary solution.
(f) As $t \to \infty$, any solution approaches a Maxwellian distribution. Therefore, these are the only stationary solutions.

To prove these properties, we find it convenient to rewrite (2.43) so that

$$\left.\frac{\partial f(\mathbf{v})}{\partial t}\right]_c = -\frac{\partial}{\partial \mathbf{v}}\cdot\mathbf{J}(\mathbf{v}) \tag{2.47}$$

with

$$\mathbf{J}(\mathbf{v}) = \int d\mathbf{v}'\mathsf{Q}(\mathbf{v},\mathbf{v}')\cdot\left(\frac{\partial}{\partial \mathbf{v}} - \frac{\partial}{\partial \mathbf{v}'}\right)f(\mathbf{v})f(\mathbf{v}'), \tag{2.48}$$

$$\mathsf{Q}(\mathbf{v},\mathbf{v}') = -\frac{\pi\omega_p^4}{n}\sum_{\mathbf{k}}\frac{\mathbf{kk}}{k^4}\frac{\delta(\mathbf{k}\cdot\mathbf{v}-\mathbf{k}\cdot\mathbf{v}')}{|\epsilon(\mathbf{k},\mathbf{k}\cdot\mathbf{v})|^2}. \tag{2.49}$$

A proof of (a) proceeds as follows. If f is positive everywhere initially and becomes negative at some later time, there must be an instant at which its minimum value first becomes negative. At such a point we have these four conditions: (i) $f=0$; (ii) $\partial f/\partial \mathbf{v}=0$; (iii) $\partial^2 f/\partial \mathbf{v}\partial \mathbf{v}$ is a nonnegative definite tensor; and (iv) $\partial f/\partial t<0$. Taking account of the first two conditions, we have

$$\frac{\partial f(\mathbf{v})}{\partial t} = -\int d\mathbf{v}' f(\mathbf{v}')\mathsf{Q}(\mathbf{v},\mathbf{v}'):\frac{\partial^2 f(\mathbf{v})}{\partial \mathbf{v}\partial \mathbf{v}}.$$

Since Q is a negative definite tensor, we thus find that conditions (iii) and (iv) are incompatible with the kinetic equation.

Properties (b)–(d) are related to the conservation properties. Property (b) means

$$\frac{d}{dt}\int f(\mathbf{v})\,d\mathbf{v}=0,$$

which may be shown through a direct integration of (2.47). Similarly, properties (c) and (d) can be proved through calculations of $(d/dt)\int d\mathbf{v}\mathbf{v}f(\mathbf{v})$ and $(d/dt)\int d\mathbf{v}(\frac{1}{2})v^2 f(\mathbf{v})$, which involve partial integrations; in the process we note the symmetry $\mathsf{Q}(\mathbf{v},\mathbf{v}')=\mathsf{Q}(\mathbf{v}',\mathbf{v})$ and an identity $(\mathbf{v}-\mathbf{v}')\cdot\mathsf{Q}(\mathbf{v},\mathbf{v}')=0$.

Property (f) is related to the H theorem of a standard kinetic theory. Proof of properties (e) and (f) is left to the reader as an exercise (Problem 2.5).

It is worth noticing that the Balescu–Lenard collision term conserves the mean kinetic energy only and does not concern itself with the interaction energy. This stems from the fact that the theory has taken account of only the first-order effects in the plasma parameter. As we noted in Section 1.2B, the average interaction energy is smaller by a factor of the plasma parameter than the average kinetic energy. To construct a theory which would conserve the sum of both kinetic and potential energies through the collision term, we must consider some of the higher-order effects which are not included in the Balescu–Lenard theory. An attempt to improve the theory in these directions has been made by Klimontovich [14].

References

1. N. N. Bogoliubov, in *Studies in Statistical Mechanics* (E. K. Gora, transl.; J. de Boer and G. E. Uhlenbeck, eds.), Vol. I, p. 1 (North-Holland, Amsterdam, 1962).
2. N. Rostoker and M. N. Rosenbluth, *Phys. Fluids* 3, 1 (1960).
3. Yu. L. Klimontovich, *Zhur. Eksptl. Teoret. Fiz.* **33**, 982 (1957) [*Soviet Phys. JETP* **6**, 753 (1958)]; *The Statistical Theory of Non-Equilibrium Processes in a Plasma* (H. S. H. Massey and O. M. Blunn, transls.; D. ter Haar, ed.) (MIT Press, Cambridge, Mass., 1967).
4. T. H. Dupree, *Phys. Fluids* **6**, 1714 (1963).
5. A. A. Vlasov, *Zhur. Eksptl. Teoret. Fiz.* **8**, 291 (1938); *Usp. Fiz. Nauk* **93**, 444 (1967) [*Soviet Phys. Usp.* **10**, 721 (1968)].
6. J. E. Mayer and M. G. Mayer, *Statistical Mechanics* (Wiley, New York, 1940).
7. R. Balescu, *Phys. Fluids* 3, 52 (1960).
8. A. Lenard, *Ann. Phys. (N.Y.)* **10**, 390 (1960).
9. L. D. Landau, *Phys. Z. Sowjetunion* **10**, 154 (1936); *Zhur. Eksptl. Teoret. Fiz.* 7, 203 (1937).
10. J. Hubbard, *Proc. Roy. Soc. (London)* **A261**, 371 (1961).
11. T. Kihara and O. Aono, *J. Phys. Soc. Japan* **18**, 837 (1963); in *Kinetic Equations* (R. L. Liboff and N. Rostoker, eds.), p. 201 (Gordon & Breach, New York, 1971).
12. H. W. Wyld and D. Pines, *Phys. Rev.* **127**, 1851 (1962).
13. W. B. Thompson and J. Hubbard, *Revs. Mod. Phys.* **32**, 714 (1960); J. Hubbard, *Proc. Roy. Soc. (London)* **A260**, 114 (1961).
14. Yu. L. Klimontovich, *Zhur. Eksptl. Teoret. Fiz.* **60**, 1352 (1971) [*Soviet Phys. JETP* **33**, 732 (1971)].

Problems

2.1. Following the treatments in Appendix A, derive the second ($s=2$) and third ($s=3$) equations of the BBGKY hierarchy (2.1) from the Klimontovich equation.

2.2. When a single-particle distribution f_σ in a constant uniform magnetic field is a function of only $v_\parallel$ and $v_\perp$, the components of the velocity parallel and perpendicular to the magnetic field, show that it is a stationary solution of the Vlasov equation in the absence of external perturbations.

2.3. Derive (2.15).

2.4. Derive (2.35).

2.5. Prove properties (e) and (f) in Section 2.6.

CHAPTER 3

PLASMAS AS DIELECTRIC MEDIA

The plasma has been defined as a statistical ensemble of mobile charged particles; these charged particles move randomly in the system, interact with each other through their own electromagnetic forces, and respond to the electromagnetic disturbances which may be applied from external sources. A plasma is therefore inherently capable of sustaining rich classes of electromagnetic phenomena.

A proper description of such electromagnetic phenomena may be obtained if we know how a plasma will respond macroscopically to a given electromagnetic disturbance. For with this knowledge, we can devise a complete scheme for analyzing its electromagnetic properties with the aid of the macroscopic equations of electrodynamics. The functions which characterize those responses therefore play the central part in such a scheme of describing the plasma; all the properties of the plasma as a macroscopic medium are contained in those response functions.

The microscopic features of the particle interactions in the plasma are not altogether lost in such a macroscopic description. They may be incorporated in the calculations of the response functions through the ways in which the particles, mutually interacting, adjust themselves to the external disturbance. This approach may therefore resemble that of a "black box"; those properties of the plasma that depend on the nature of the microscopic particle interactions may be revealed to the external observer through the course in which the plasma responds and adjusts itself to the external disturbance.

Since our interest lies in the electromagnetic phenomena, it may be appropriate to select an electric field as an external disturbance to the plasma; typically, then, the response appears in the form of an induced current density in the plasma. The relationship between these two vectorial quantities must generally be expressed in terms of the conductivity tensor. If we further take into account the displacement current arising from the time variation of the electric field, the field–current relationship is more compactly expressed by the dielectric tensor; this quantity therefore contains the same physical information about the plasma as the conductivity tensor.

The main objective of this chapter is to lay a theoretical foundation for the electrodynamic description of plasmas as continuous dielectric media: We shall first give a mathematical definition of the dielectric tensor; we shall apply this dielectric tensor to a set of macroscopic electrodynamic equations and find a formal solution in the form of a frequency–wave number dispersion relation. Such a dispersion relation characterizes the basic properties of both the longitudinal and the transverse excitations in the dielectric medium. An important subclass of the dielectric tensor, the dielectric response function, will then be singled out through a consideration of the density response against an external test charge; this function is essential in describing the longitudinal properties of the plasma. Finally, we shall explicitly calculate the dielectric tensor for plasmas, both in the absence and in the presence of a uniform external magnetic field. The analytic properties of the dielectric response function are summarized in Appendix C.

3.1 THE DIELECTRIC TENSOR AND THE DISPERSION RELATION

The dielectric tensor contains essentially all the information about the electromagnetic properties of the plasma; it is determined from a calculation of the plasma response to an electric field disturbance [1–4]. A number of fundamental assumptions are usually involved in such a calculation; we begin with a discussion of the implications of those assumptions.

A. Linearity, Translational Invariance, and Causality

In order to establish a response relationship in the plasma, let us introduce an electric field disturbance $\mathbf{E}(\mathbf{r},t)$, which varies in space and time. This electric field then induces electric current density $\mathbf{J}(\mathbf{r},t)$ in the medium. If the system is stable against such a disturbance and if the strength of the disturbance is sufficiently weak, then the induced current may be represented by that part of the response which is *linear* in the disturbing field. In circumstances where such a linear response relationship provides a valid description of the system behavior, the induced current can be calculated by superposing the effects of the disturbances applied at various points in space and time; thus we may write

$$\mathbf{J}(\mathbf{r},t) = \int_V d\mathbf{r}' \int_{-\infty}^{t} dt' \sigma(\mathbf{r},\mathbf{r}';t,t')\cdot\mathbf{E}(\mathbf{r}',t'). \tag{3.1}$$

Here, the tensor $\sigma(\mathbf{r},\mathbf{r}';t,t')$ describes the propagation characteristics of a disturbance applied at $(\mathbf{r}',t')$ as an electric field and observed at $(\mathbf{r},t)$ as an induced current; insofar as Eq. (3.1) represents a linear response relation, the tensor is determined solely by the properties of the system *without* perturbation.

In addition to linearity, Eq. (3.1) contains the concept of *causality*. This is reflected in the choice of the domain $[-\infty, t]$ for the range of time integration in (3.1). Starting from an unperturbed state at $t=-\infty$, the current is gradually induced in the system owing to application of the perturbation field; in these circumstances, the effect $\mathbf{J}(\mathbf{r},t)$ cannot precede the cause $\mathbf{E}(\mathbf{r}',t')$.

We now assume that the unperturbed state of the medium is homogeneous in space and stationary in time; the kernel $\boldsymbol{\sigma}$ then becomes a function of $\mathbf{r}-\mathbf{r}'$ and $t-t'$ only, that is,

$$\boldsymbol{\sigma}(\mathbf{r},\mathbf{r}';t,t') \to \boldsymbol{\sigma}(\mathbf{r}-\mathbf{r}',t-t'). \tag{3.2}$$

This expression implies the *translational invariance* of the system in space and time. For validity of this assumption, the characteristic dimensions associated with the inhomogeneities of the system must be substantially greater than those of the electromagnetic phenomena under consideration.

For such a translationally invariant system, the natural representations of physical variables are in terms of the plane-wave states. We therefore substitute (3.2) in (3.1) and carry out Fourier transformations of the resulting convolution integral in both space and time. We then have

$$\mathbf{J}(\mathbf{k},\omega) = \boldsymbol{\sigma}(\mathbf{k},\omega)\cdot\mathbf{E}(\mathbf{k},\omega) \tag{3.3}$$

where

$$\mathbf{E}(\mathbf{k},\omega) = \int_V d\mathbf{r} \int_{-\infty}^{\infty} dt \mathbf{E}(\mathbf{r},t) \exp[-i(\mathbf{k}\cdot\mathbf{r}-\omega t)], \tag{3.4}$$

$$\boldsymbol{\sigma}(\mathbf{k},\omega) = \int_V d\mathbf{r} \int_{0}^{\infty} dt \boldsymbol{\sigma}(\mathbf{r},t) \exp[-i(\mathbf{k}\cdot\mathbf{r}-\omega t)], \tag{3.5}$$

and $\mathbf{J}(\mathbf{k},\omega)$ is calculated from $\mathbf{J}(\mathbf{r},t)$ in the same way as $\mathbf{E}(\mathbf{k},\omega)$ in (3.4). It follows from the reality of $\mathbf{E}(\mathbf{r},t)$, $\mathbf{J}(\mathbf{r},t)$, and thus $\boldsymbol{\sigma}(\mathbf{r},t)$ that their Fourier components have the symmetry properties

$$\mathbf{E}^*(\mathbf{k},\omega) = \mathbf{E}(-\mathbf{k},-\omega), \tag{3.6a}$$

$$\boldsymbol{\sigma}^*(\mathbf{k},\omega) = \boldsymbol{\sigma}(-\mathbf{k},-\omega) \tag{3.6b}$$

for real values of ω. (The asterisk means complex conjugate.)

The tensor $\boldsymbol{\sigma}(\mathbf{k},\omega)$ is the *conductivity tensor* of the medium; as is clear from (3.3), it measures the ratio between the induced current density and the electric field in a plane-wave state, $\exp[i(\mathbf{k}\cdot\mathbf{r}-\omega t)]$. It is important to note in (3.5) that $\boldsymbol{\sigma}(\mathbf{k},\omega)$ has been defined in terms of a one-sided Fourier transformation with respect to time [compare Eq. (2.32a)]; this stems from the principle of causality imposed on Eq. (3.1). It thus follows that $\sigma(\mathbf{k},\omega)$ is analytic in the upper half of the complex ω plane. Tensor $\sigma(\mathbf{k},\omega)$ is then continued analytically into the lower half ω plane; there, $\sigma(\mathbf{k},\omega)$ generally has singularities. Inverse transformations of $\sigma(\mathbf{k},\omega)$ are carried out according to

$$\sigma(\mathbf{r},t)=(2\pi V)^{-1}\sum_{\mathbf{k}}\int_C d\omega\sigma(\mathbf{k},\omega)\exp[i(\mathbf{k}\cdot\mathbf{r}-\omega t)] \tag{3.7}$$

where the contour C extends from $-\infty$ to $+\infty$ along a path in the upper half of the ω plane in such a way as to see the singularities of $\sigma(\mathbf{k},\omega)$ from above. All the properties of the inverse transformations as described after (2.32b) are applicable to (3.7) as well. In particular, we have for $t<0$

$$\sigma(\mathbf{r},t)=0,$$

another manifestation of the causality. Finally, for the $\mathbf{k}$ summation, we may recall the comments made after (2.30b); in most cases, we choose $V=1$. Thus, the inverse transformations of (3.4) may be performed as

$$\begin{aligned}\mathbf{E}(\mathbf{r},t)&=(2\pi)^{-1}\sum_{\mathbf{k}}\int_{-\infty}^{\infty}d\omega\mathbf{E}(\mathbf{k},\omega)\exp[i(\mathbf{k}\cdot\mathbf{r}-\omega t)]\\&=[(2\pi)^4]^{-1}\int d\mathbf{k}\int_{-\infty}^{\infty}d\omega\mathbf{E}(\mathbf{k},\omega)\exp[i(\mathbf{k}\cdot\mathbf{r}-\omega t)].\end{aligned}$$

B. Definition of the Dielectric Tensor

We now apply the linear response relationship (3.3) to the macroscopic electromagnetic equations. We start from the Maxwell equations

$$\begin{aligned}&\nabla\times\mathbf{E}+\frac{1}{c}\frac{\partial\mathbf{B}}{\partial t}=0,\\&\nabla\times\mathbf{B}-\frac{1}{c}\frac{\partial\mathbf{E}}{\partial t}=\frac{4\pi}{c}(\mathbf{J}+\mathbf{J}_{\text{ext}}),\\&\nabla\cdot\mathbf{E}=4\pi(\rho+\rho_{\text{ext}}),\\&\nabla\cdot\mathbf{B}=0.\end{aligned} \tag{3.8}$$

Here, ρ is the induced space-charge field connected with $\mathbf{J}$ through the continuity equation†

$$\frac{\partial \rho}{\partial t}+\nabla\cdot\mathbf{J}=0. \tag{3.9}$$

$\mathbf{J}_{\text{ext}}$ and ρ_{ext} are the current and charge densities introduced from external sources; they satisfy a continuity equation similar to (3.9).

Expressing every quantity in terms of its complex Fourier amplitude,§ we find that Eqs. (3.8) reduce to

$$\mathbf{k}\times\mathbf{E}-\frac{\omega}{c}\mathbf{B}=0, \tag{3.10a}$$

$$\mathbf{k}\times\mathbf{B}+\frac{\omega}{c}\boldsymbol{\epsilon}\cdot\mathbf{E}=-\frac{4\pi i}{c}\mathbf{J}_{\text{ext}}, \tag{3.10b}$$

$$\mathbf{k}\cdot\boldsymbol{\epsilon}\cdot\mathbf{E}=-4\pi i\rho_{\text{ext}}, \tag{3.10c}$$

$$\mathbf{k}\cdot\mathbf{B}=0, \tag{3.10d}$$

where Eq. (3.10c) has been obtained with the aid of (3.9). The tensor $\boldsymbol{\epsilon}(\mathbf{k},\omega)$ in Eqs. (3.10b) and (3.10c), defined as

$$\boldsymbol{\epsilon}(\mathbf{k},\omega)=\mathbf{I}-(4\pi/i\omega)\boldsymbol{\sigma}(\mathbf{k},\omega), \tag{3.11}$$

is the *dielectric tensor* of the medium; $\mathbf{I}$ is the unit tensor represented as

$$\mathbf{I}=\begin{pmatrix}1&0&0\\0&1&0\\0&0&1\end{pmatrix}. \tag{3.12}$$

It is instructive to compare the set of equations (3.10) with another standard set of electromagnetic equations in continuous media: For a description of the electromagnetic fields, we usually introduce the electric induction $\mathbf{D}$ and the magnetic field $\mathbf{H}$ in addition to the electric field $\mathbf{E}$ and the magnetic induction $\mathbf{B}$. The latter two quantities satisfy Eqs. (3.10a) and (3.10d); the former pair is described by

$$\mathbf{k}\times\mathbf{H}+\frac{\omega}{c}\mathbf{D}=-\frac{4\pi i}{c}\mathbf{J}_{\text{ext}}, \tag{3.13a}$$

$$\mathbf{k}\cdot\mathbf{D}=-4\pi i\rho_{\text{ext}}. \tag{3.13b}$$

†In some other cases (e.g., in Section 1.4), ρ may denote the number-density fluctuations; this should not cause confusion, as we clearly specify the meaning in each case.

§Unless otherwise stated, we use this representation throughout the remainder of this section.

To close the equations, we must find relations between **E** and **B**, on the one hand, and **D** and **H** on the other; such relationships should reflect the medium properties of the system.

A comparison between (3.10b) and (3.13a) suggests

$$\mathbf{H}=\mathbf{B}, \tag{3.14}$$

$$\mathbf{D}=\boldsymbol{\epsilon}\cdot\mathbf{E} \tag{3.15}$$

for the required relationships; Eq. (3.15) also guarantees compatibility between (3.10c) and (3.13b). The term "dielectric tensor" for $\boldsymbol{\epsilon}(\mathbf{k},\omega)$ derives from Eq. (3.15).

It should be noted, however, that Eqs. (3.14) and (3.15) are not the unique consequence of the comparison between (3.10b) and (3.13a). In fact, defining the transverse projection tensor with respect to the wave vector **k** as

$$\mathsf{I}_T=\mathsf{I}-\frac{\mathbf{kk}}{k^2}, \tag{3.16}$$

we find that substitutions of

$$\mathbf{D}=\left[\mathsf{I}-\frac{4\pi}{i\omega}(\mathsf{I}-\lambda\mathsf{I}_T)\cdot\boldsymbol{\sigma}\right]\cdot\mathbf{E}, \tag{3.17}$$

$$\mathbf{H}=\mathbf{B}+\lambda\frac{4\pi}{ick^2}(\mathbf{k}\times\boldsymbol{\sigma})\cdot\mathbf{E} \tag{3.18}$$

in (3.13a) and (3.13b) yield (3.10b) and (3.10c) for any value of the parameter λ. Obviously, Eqs. (3.14) and (3.15) correspond to a special case of (3.17) and (3.18) in that $\lambda=0$.

Another case of interest appears, for example, when we choose $\lambda=1$; the relation between **E** and **D** now becomes longitudinal in the sense that $\mathbf{E}-\mathbf{D}$ is parallel to **k**. The difference between **B** and **H** now becomes significant; Eq. (3.18) describes the transverse properties of the medium in these circumstances.

In order to avoid such ambiguity in specifying the relationships among the four field quantities, we regard **E** and **B** as primary quantities; physically, they represent the total electric and magnetic fields in the medium. Those quantities satisfy the field equations, Eqs. (3.10). The dielectric tensor, which is the sole agency characterizing the linear electromagnetic properties of the medium in those equations, is determined from the relation between the electric field and the induced current density as

$$\mathbf{J}(\mathbf{k},\omega)=-\frac{i\omega}{4\pi}[\boldsymbol{\epsilon}(\mathbf{k},\omega)-\mathsf{I}]\cdot\mathbf{E}(\mathbf{k},\omega). \tag{3.19}$$

In this way, we may secure a simple and unambiguous basis for formulating an electromagnetic problem of a plasma.

C. Dispersion Relation

We now proceed to find a formal solution to the basic set of electromagnetic equations established above. Upon substituting (3.10a) in (3.10b), we obtain

$$\left[\epsilon-\left(\frac{kc}{\omega}\right)^2 \mathbf{1}_\mathrm{T}\right]\cdot\mathbf{E}=\frac{4\pi}{i\omega}\mathbf{J}_{\mathrm{ext}}, \tag{3.20}$$

or, defining a new tensor $\Delta(\mathbf{k},\omega)$ by

$$\Delta(\mathbf{k},\omega)\equiv\epsilon(\mathbf{k},\omega)-\left(\frac{kc}{\omega}\right)^2\mathbf{1}_\mathrm{T}, \tag{3.21}$$

we have

$$\Delta\cdot\mathbf{E}=\frac{4\pi}{i\omega}\mathbf{J}_{\mathrm{ext}}. \tag{3.22}$$

Equation (3.21) is called the dispersion tensor.

When no external disturbance is applied to the plasma, $\mathbf{J}_{\mathrm{ext}}=0$; that is,

$$\Delta\cdot\mathbf{E}=0. \tag{3.23}$$

Equation (3.23) possesses nontrivial solutions only if the determinant constructed from the elements of the tensor $\Delta(\mathbf{k},\omega)$ vanishes:

$$\det|\Delta(\mathbf{k},\omega)|=0. \tag{3.24}$$

This equation thereby determines the frequency-wave vector dispersion relation for the electromagnetic waves in the medium.

D. Isotropic Medium

The foregoing dispersion relation generally applies to the linear electromagnetic phenomena in a homogeneous medium. If we may further assume that the medium is isotropic, that is, invariant under rotation of the coordinate axes, then a substantial simplification takes place in the dispersion relation.

We first note that the dielectric tensor $\epsilon(\mathbf{k},\omega)$ in these circumstances contains only one element of directional dependence through that of the wave vector $\mathbf{k}$; the tensor character of $\epsilon(\mathbf{k},\omega)$ must be determined from a general consideration so as to reflect this dependence. The tensors of rank two, which can be constructed from the vector $\mathbf{k}$ and which are mutually independent, are the longitudinal projection tensor

$$\mathsf{I}_L \equiv \mathbf{k}\mathbf{k}/k^2 \tag{3.25}$$

and the transverse projection tensor I_T defined by (3.16). The dielectric tensor must therefore be expressed as a linear combination of these two tensors; hence,

$$\boldsymbol{\epsilon}(\mathbf{k},\omega) = \epsilon_L(k,\omega)\mathsf{I}_L + \epsilon_T(k,\omega)\mathsf{I}_T. \tag{3.26}$$

The coefficients $\epsilon_L(k,\omega)$ and $\epsilon_T(k,\omega)$ introduced here depend only on the magnitude of $\mathbf{k}$ and not its direction. Inversely, when the dielectric tensor is given for an isotropic medium, the functions $\epsilon_L(k,\omega)$ and $\epsilon_T(k,\omega)$ are calculated as

$$\epsilon_L(k,\omega) = \frac{\mathbf{k}\cdot\boldsymbol{\epsilon}(\mathbf{k},\omega)\cdot\mathbf{k}}{k^2}, \tag{3.27}$$

$$\epsilon_T(k,\omega) = \tfrac{1}{2}[\mathrm{Tr}\,\boldsymbol{\epsilon}(\mathbf{k},\omega) - \epsilon_L(k,\omega)] \tag{3.28}$$

where Tr means the summation over the diagonal elements.

The dispersion tensor Eq. (3.21) now reads

$$\boldsymbol{\Delta}(\mathbf{k},\omega) = \epsilon_L(k,\omega)\mathsf{I}_L + \left[\epsilon_T(k,\omega) - (kc/\omega)^2\right]\mathsf{I}_T. \tag{3.29}$$

Thus, it is also diagonalized into purely longitudinal and transverse parts. The dispersion relation Eq. (3.24) splits itself into two equations:

$$\epsilon_L(k,\omega) = 0, \tag{3.30}$$

$$\epsilon_T(k,\omega) = (kc/\omega)^2. \tag{3.31}$$

The physical meaning of these equations may become clear if we examine (3.23) by substituting (3.29) and the identity $\mathbf{E} \equiv \mathsf{I}_L\cdot\mathbf{E} + \mathsf{I}_T\cdot\mathbf{E}$. We thus find $\mathsf{I}_L\cdot\mathbf{E} \neq 0$ when Eq. (3.30) is satisfied; Eq. (3.30) therefore represents the dispersion relation for the longitudinal mode. Similarly, $\mathsf{I}_T\cdot\mathbf{E} \neq 0$ when Eq. (3.31) is satisfied; this equation thus gives the dispersion relation for the transverse mode in the medium.

3.2 THE DIELECTRIC RESPONSE FUNCTION

The dielectric tensor, considered in section 3.1, is complete in that it contains all the information about the linear electromagnetic properties of the plasma.

It is sometimes quite useful and significant, however, to single out certain elements of this tensor and construct another important response function, the dielectric response function [5, 6]; this function describes essentially the longitudinal properties of the plasma. We may recall that we have already seen an example of such a function in Section 2.4. In this section, we give a general definition of this response function through a consideration of a test-charge problem and establish its relationship with the dielectric tensor; the dielectric response function provides the dispersion relation for the density-fluctuation excitations in the plasma.

A. Test-Charge Problem

Consider a situation in which an external test-charge field

$$\rho_{\text{ext}}(\mathbf{k},\omega)\exp[i(\mathbf{k}\cdot\mathbf{r}-\omega t)]+\text{cc}$$

is introduced into the plasma (cc stands for complex conjugate); it will disturb the plasma and create density fluctuations. Since we are here concerned only with a linear response, the space-charge field so induced may also be written as

$$\rho(\mathbf{k},\omega)\exp[i(\mathbf{k}\cdot\mathbf{r}-\omega t)]+cc$$

The complex amplitude $\rho_{\text{tot}}(\mathbf{k},\omega)$ of the total space-charge field is then calculated as

$$\rho_{\text{tot}}(\mathbf{k},\omega)=\rho_{\text{ext}}(\mathbf{k},\omega)+\rho(\mathbf{k},\omega). \tag{3.32}$$

This quantity must be linearly related to $\rho_{\text{ext}}(\mathbf{k},\omega)$, so that we may write

$$\rho_{\text{tot}}(\mathbf{k},\omega)=\frac{\rho_{\text{ext}}(\mathbf{k},\omega)}{\epsilon(\mathbf{k},\omega)}. \tag{3.33}$$

The *dielectric response function* $\epsilon(\mathbf{k},\omega)$ is defined through this equation; we can use (3.33) for a calculation of $\epsilon(\mathbf{k},\omega)$. Clearly, the dielectric function measures the extent to which an external test charge is screened and its effectiveness modified by the induced space charge in the plasma.

The dielectric response function may also be calculated simply from the dielectric tensor. To establish the relationship, we first note that the electric field is determined from the Poisson equation as

$$\mathbf{E}(\mathbf{k},\omega)=-i\frac{4\pi\mathbf{k}}{k^2}\rho_{\text{tot}}(\mathbf{k},\omega).$$

Substituting this in (3.3), we obtain

$$\mathbf{J}(\mathbf{k},\omega) = -i\frac{4\pi}{k^2}\boldsymbol{\sigma}(\mathbf{k},\omega)\cdot\mathbf{k}\rho_{\text{tot}}(\mathbf{k},\omega).$$

This induced current is connected with the induced space-charge field via the continuity equation (3.9). Hence, we have

$$\rho(\mathbf{k},\omega) = \frac{1}{\omega}\mathbf{k}\cdot\mathbf{J}(\mathbf{k},\omega)$$

$$= \frac{4\pi}{i\omega}\frac{\mathbf{k}\cdot\boldsymbol{\sigma}(\mathbf{k},\omega)\cdot\mathbf{k}}{k^2}\rho_{\text{tot}}(\mathbf{k},\omega). \tag{3.34}$$

Comparison between (3.33) and (3.34) with the aid of (3.32) yields

$$\epsilon(\mathbf{k},\omega) = 1 - \frac{4\pi}{i\omega}\frac{\mathbf{k}\cdot\boldsymbol{\sigma}(\mathbf{k},\omega)\cdot\mathbf{k}}{k^2}, \tag{3.35}$$

or in terms of the dielectric tensor (3.11), we find

$$\epsilon(\mathbf{k},\omega) = \frac{\mathbf{k}\cdot\boldsymbol{\epsilon}(\mathbf{k},\omega)\cdot\mathbf{k}}{k^2}. \tag{3.36}$$

Consequently, the knowledge of the dielectric tensor is sufficient to determine the dielectric response function.

B. Density-Fluctuation Excitations

Equation (3.36) is quite similar to Eq. (3.27); the only difference arises from the fact that (3.27) is independent of the directions of **k** because of isotropy. The dielectric response function, as defined in terms of the density response to an external test-charge field, leads us to a dispersion relation of longitudinal density-fluctuation excitations which in fact commands a much wider range of applicability than simply the cases of isotropic plasmas.

To see this, we rearrange (3.32) with the aid of (3.33) so that the induced space-charge field is explicitly written

$$\rho(\mathbf{k},\omega) = -\{1-[1/\epsilon(\mathbf{k},\omega)]\}\rho_{\text{ext}}(\mathbf{k},\omega). \tag{3.37}$$

This equation would imply that no space-charge field is induced when $\rho_{\text{ext}}=0$. However, if the situation is such that the frequency and the wave vector coincide with one of the poles of $[1-(1/\epsilon)]$, or equivalently, if **k** and ω

satisfy the equation

$$\epsilon(\mathbf{k},\omega)=0, \tag{3.38}$$

then such a space-charge field could be spontaneously excited in the plasma without an application of external disturbances. Equation (3.38), therefore, provides the dispersion relation for density-fluctuation excitations in the plasma.

3.3 KINETIC-THEORETICAL CALCULATION OF THE DIELECTRIC TENSOR WITHOUT MAGNETIC FIELD

We now calculate the dielectric tensor of a plasma within a kinetic-theoretical description. We therefore start from the Vlasov equation coupled with Maxwell equations.

A. Vlasov–Maxwell Equations

We defined the single-particle distribution function $f_\sigma(\mathbf{r},\mathbf{v};t)$ for the σ species in Section 2.1; its normalization has been specified by Eq. (2.4), so that the local number density, for example, is given by

$$n_\sigma(\mathbf{r},t)=n_\sigma\int f_\sigma(\mathbf{r},\mathbf{v};t)\,d\mathbf{v}, \tag{3.39a}$$

where $n_\sigma\equiv N_\sigma/V$ is the average number density of the particles. These distribution functions obey the Vlasov equation (2.9) so that

$$\frac{\partial f_\sigma}{\partial t}+\mathbf{v}\cdot\frac{\partial f_\sigma}{\partial \mathbf{r}}+\frac{q_\sigma}{m_\sigma}\left(\mathbf{E}+\frac{\mathbf{v}}{c}\times\mathbf{B}\right)\cdot\frac{\partial f_\sigma}{\partial \mathbf{v}}=0. \tag{3.39b}$$

The electromagnetic fields $\mathbf{E}$ and $\mathbf{B}$ in this equation are conspicuously the sum of those applied from outside sources and those induced from the internal particle distributions. Individually they may be determined from

$$\nabla\times\mathbf{E}_{\text{ext}}+\frac{1}{c}\frac{\partial\mathbf{B}_{\text{ext}}}{\partial t}=0, \tag{3.40a}$$

$$\nabla\times\mathbf{B}_{\text{ext}}-\frac{1}{c}\frac{\partial\mathbf{E}_{\text{ext}}}{\partial t}=\frac{4\pi}{c}\mathbf{J}_{\text{ext}}, \tag{3.40b}$$

$$\nabla\cdot\mathbf{E}_{\text{ext}}=4\pi\rho_{\text{ext}}, \tag{3.40c}$$

$$\nabla\cdot\mathbf{B}_{\text{ext}}=0 \tag{3.40d}$$

and

$$\nabla\times\mathbf{E}_{\text{ind}}+\frac{1}{c}\frac{\partial\mathbf{B}_{\text{ind}}}{\partial t}=0 \tag{3.41a}$$

$$\nabla\times\mathbf{B}_{\text{ind}}-\frac{1}{c}\frac{\partial\mathbf{E}_{\text{ind}}}{\partial t}=\frac{4\pi}{c}\sum_{\sigma}\int q_\sigma n_\sigma \mathbf{v} f_\sigma(\mathbf{r},\mathbf{v};t)\,d\mathbf{v} \tag{3.41b}$$

$$\nabla\cdot\mathbf{E}_{\text{ind}}=4\pi\sum_{\sigma}\int q_\sigma n_\sigma f_\sigma(\mathbf{r},\mathbf{v};t)\,d\mathbf{v} \tag{3.41c}$$

$$\nabla\cdot\mathbf{B}_{\text{ind}}=0. \tag{3.41d}$$

The total fields

$$\mathbf{E}=\mathbf{E}_{\text{ext}}+\mathbf{E}_{\text{ind}},\qquad \mathbf{B}=\mathbf{B}_{\text{ext}}+\mathbf{B}_{\text{ind}} \tag{3.42}$$

satisfy Eqs. (3.8).

The set of Vlasov–Maxwell equations, Eqs. (3.8) and Eq. (3.39b), therefore provides a complete kinetic description of a plasma in the fluid limit. In this section we use this set of equations to calculate the dielectric tensor of a plasma in the absence of an external magnetic field; the cases of a plasma in a uniform magnetic field will be treated in the next section.

B. Perturbations

We assume the plasma to be homogeneous and in a stationary state at $t=-\infty$ with the velocity distribution functions $f_\sigma(\mathbf{v})$ for the σ species. We may recall that any distribution function which is independent of spatial coordinates can be a solution of the Vlasov equation (3.39b) in the absence of external perturbations because $\mathbf{E}$ and $\mathbf{B}$ identically vanish in such a system due to the overall charge neutrality.

The electric field disturbance is then turned on *adiabatically* as

$$\delta\mathbf{E}\exp[i(\mathbf{k}\cdot\mathbf{r}-\omega t)+\eta t]+\text{cc} \tag{3.43}$$

The positive infinitesimal η in (3.43) serves to assure the adiabatic turning on of the disturbance and thereby to guarantee a causal response of the system; we let $\eta\rightarrow+0$ eventually. Associated with the electric field (3.43), there is a magnetic field disturbance determined from (3.10a) as

$$\delta\mathbf{B}\exp[i(\mathbf{k}\cdot\mathbf{r}-\omega t)+\eta t]+\text{cc}=(c/\omega)(\mathbf{k}\times\delta\mathbf{E})\exp[i(\mathbf{k}\cdot\mathbf{r}-\omega t)+\eta t]+\text{cc}. \tag{3.44}$$

In response to these disturbances, the distribution functions depart from their stationary values; within the confines of the linear response theory, they may also be written as

$$f_\sigma(\mathbf{v})+\{\delta f_\sigma(\mathbf{v})\exp[i(\mathbf{k}\cdot\mathbf{r}-\omega t)+\eta t]+\mathrm{cc}\}. \tag{3.45}$$

Substituting (3.43), (3.44), and (3.45) in (3.39b) and retaining only those terms linear in $\delta\mathbf{E}$ and $\delta f_\sigma(\mathbf{v})$, we find

$$\left(\frac{\partial}{\partial t}+\mathbf{v}\cdot\frac{\partial}{\partial \mathbf{r}}\right)\delta f_\sigma(\mathbf{v})\exp[i(\mathbf{k}\cdot\mathbf{r}-\omega t)+\eta t]$$

$$=-\frac{q_\sigma}{m_\sigma}\left(\frac{\partial f_\sigma}{\partial \mathbf{v}}\right)\cdot\left[\left(1-\frac{\mathbf{k}\cdot\mathbf{v}}{\omega}\right)\mathsf{I}+\frac{\mathbf{k}\mathbf{v}}{\omega}\right]\cdot\delta\mathbf{E}\exp[i(\mathbf{k}\cdot\mathbf{r}-\omega t)+\eta t]. \tag{3.46}$$

C. The Dielectric Tensor

The amplitude $\delta\mathbf{J}$ of the induced current is calculated from $\delta f_\sigma(\mathbf{v})$; with the aid of (3.46), we have

$$\delta\mathbf{J}=\sum_\sigma q_\sigma n_\sigma\int \mathbf{v}\,\delta f_\sigma(\mathbf{v})\,d\mathbf{v}$$

$$=-\sum_\sigma\frac{q_\sigma^2 n_\sigma}{im_\sigma}\int d\mathbf{v}\frac{\mathbf{v}}{\mathbf{k}\cdot\mathbf{v}-\omega-i\eta}\frac{\partial f_\sigma}{\partial \mathbf{v}}\cdot\left[\left(1-\frac{\mathbf{k}\cdot\mathbf{v}}{\omega}\right)\mathsf{I}+\frac{\mathbf{k}\mathbf{v}}{\omega}\right]\cdot\delta\mathbf{E}. \tag{3.47}$$

A comparison between (3.19) and (3.47) yields

$$\boldsymbol{\epsilon}(\mathbf{k},\omega)=\mathsf{I}-\sum_\sigma\frac{\omega_\sigma^2}{\omega}\int d\mathbf{v}\frac{\mathbf{v}}{\mathbf{k}\cdot\mathbf{v}-\omega-i\eta}\frac{\partial f_\sigma}{\partial \mathbf{v}}\cdot\left[\left(1-\frac{\mathbf{k}\cdot\mathbf{v}}{\omega}\right)\mathsf{I}+\frac{\mathbf{k}\mathbf{v}}{\omega}\right]$$

$$=\left(1-\frac{\omega_p^2}{\omega^2}\right)\mathsf{I}-\sum_\sigma\frac{\omega_\sigma^2}{\omega^2}\int d\mathbf{v}\frac{\mathbf{v}\mathbf{v}}{\mathbf{k}\cdot\mathbf{v}-\omega-i\eta}\mathbf{k}\cdot\frac{\partial f_\sigma}{\partial \mathbf{v}} \tag{3.48}$$

where we have carried out a partial integration with respect to $\mathbf{v}$ in the final step; ω_p has been defined by (1.19).

Let us note that Eq. (3.48) involves a singular integral arising from the pole at $\mathbf{k}\cdot\mathbf{v}=\omega$ when the imaginary part of $\omega+i\eta$ vanishes. The mathematical realization of the adiabatic turning on of the disturbance by choosing a positive infinitesimal η then clearly indicates how to treat this integration: Eq. (3.48) should be calculated with $\mathrm{Im}\,\omega>0$; it is then continued analytically into the lower half of the complex ω plane. In this way, the physical principle of causality has been embodied in the calculation of (3.48).

In the limit of very high frequencies, the last term of (3.48) decreases as ω^{-3}; we thus find that the principal contributions in this limit are

$$\boldsymbol{\epsilon}(\mathbf{k},\omega)\rightarrow\left[1-\left(\omega_{\mathrm{p}}^{2}/\omega^{2}\right)\right]\mathbf{I} \qquad (\omega\rightarrow\infty). \tag{3.49}$$

Equation (3.49) implies that the plasma behaves like a collection of independent charged particles at very high frequencies [7], a property which any correct calculation of the dielectric tensor must satisfy.

D. The Dielectric Response Function

With the aid of (3.36), we can calculate the dielectric response function from (3.48):

$$\epsilon(\mathbf{k},\omega)=1-\sum_{\sigma}\frac{\omega_{\sigma}^{2}}{k^{2}}\int d\mathbf{v}\frac{1}{\mathbf{k}\cdot\mathbf{v}-\omega-i\eta}\mathbf{k}\cdot\frac{\partial f_{\sigma}(\mathbf{v})}{\partial\mathbf{v}}. \tag{3.50}$$

This function likewise has a correct asymptotic behavior

$$\epsilon(\mathbf{k},\omega)\rightarrow1-\left(\omega_{\mathrm{p}}^{2}/\omega^{2}\right) \qquad (\omega\rightarrow\infty). \tag{3.51}$$

Some of the analytic properties of the dielectric response function related to this asymptotic behavior are described in Appendix C.

3.4 KINETIC-THEORETICAL CALCULATION OF THE DIELECTRIC TENSOR IN A UNIFORM MAGNETIC FIELD

We now extend the calculations of the previous section to the cases of a plasma in a uniform magnetic field applied externally [8, 9]. The basic principle of calculation leading to the dielectric tensor is of course unchanged; the actual calculations in this case, however, involve much complicated mathematics because of the spiraling orbits of the particles in the magnetic field. We must therefore find a method to "unwind" the spirals in order to follow the time development of the response in the plasma.

A. Integration along the Unperturbed Trajectory in the Phase Space

In the presence of a uniform magnetic field $\mathbf{B}=B\hat{z}$ in the z direction ($\hat{z}$ is a unit vector in that direction), Eq. (3.46) is modified as

$$\left[\frac{\partial}{\partial t}+\mathbf{v}\cdot\frac{\partial}{\partial\mathbf{r}}+\Omega_{\sigma}(\mathbf{v}\times\hat{z})\cdot\frac{\partial}{\partial\mathbf{v}}\right]\delta f_{\sigma}(\mathbf{v})\exp[i(\mathbf{k}\cdot\mathbf{r}-\omega t)+\eta t]$$

$$=-\frac{q_{\sigma}}{m_{\sigma}}\left(\frac{\partial f_{\sigma}}{\partial\mathbf{v}}\right)\cdot\left[\left(1-\frac{\mathbf{k}\cdot\mathbf{v}}{\omega}\right)\mathbf{I}+\frac{\mathbf{k}\mathbf{v}}{\omega}\right]\cdot\delta\mathbf{E}\exp[i(\mathbf{k}\cdot\mathbf{r}-\omega t)+\eta t], \tag{3.52}$$

where

$$\Omega_\sigma = q_\sigma B / m_\sigma c \tag{3.53}$$

is the cyclotron frequency including the sign of the charge q_σ. We want to solve (3.52) for $\delta f_\sigma(\mathbf{v})$, as we did in the previous section. To do so, we must investigate the nature of the differential operators appearing on the left-hand side so that we can find an appropriate integral operator to invert those differentiations.

For this purpose, it is helpful to consider a differential operator generally defined as

$$\frac{d}{dt} \equiv \frac{\partial}{\partial t} + \frac{d\mathbf{r}}{dt}\cdot\frac{\partial}{\partial \mathbf{r}} + \frac{d\mathbf{v}}{dt}\cdot\frac{\partial}{\partial \mathbf{v}} \tag{3.54}$$

where $\mathbf{r}=\mathbf{r}(t)$ and $\mathbf{v}=\mathbf{v}(t)$ describe a given trajectory of a particle in the phase space $(\mathbf{r},\mathbf{v})$. Clearly, Eq. (3.54) represents a differentiation with respect to time along the prescribed particle orbit in the phase space.

The $d\mathbf{r}/dt$ and $d\mathbf{v}/dt$ are the velocity and the acceleration of the particle; let us choose

$$\frac{d\mathbf{r}}{dt} = \mathbf{v}, \qquad \frac{d\mathbf{v}}{dt} = \Omega_\sigma(\mathbf{v}\times\hat{z}). \tag{3.55}$$

This set of equations then determines an unperturbed free orbit of a charged particle in a uniform magnetic field. With the knowledge that the particle position and velocity at time t are

$$\mathbf{r} = x\hat{x} + y\hat{y} + z\hat{z}, \qquad \mathbf{v} = v_x\hat{x} + v_y\hat{y} + v_z\hat{z}, \tag{3.56}$$

respectively, those at t' are calculated from a solution of (3.55) as

$$\mathbf{r}' = x'\hat{x} + y'\hat{y} + z'\hat{z}, \qquad \mathbf{v}' = v_x'\hat{x} + v_y'\hat{y} + v_z'\hat{z}. \tag{3.57}$$

The solution may be compactly expressed with the aid of tensors (represented in the Cartesian coordinates)

$$\mathsf{H}_\sigma(t) = \begin{bmatrix} \sin\Omega_\sigma t & 1-\cos\Omega_\sigma t & 0 \\ -(1-\cos\Omega_\sigma t) & \sin\Omega_\sigma t & 0 \\ 0 & 0 & \Omega_\sigma t \end{bmatrix}, \tag{3.58}$$

$$\mathsf{B}_\sigma(t)=\frac{1}{\Omega_\sigma}\frac{d}{dt}\mathsf{H}_\sigma(t)$$

$$=\begin{bmatrix} \cos\Omega_\sigma t & \sin\Omega_\sigma t & 0 \\ -\sin\Omega_\sigma t & \cos\Omega_\sigma t & 0 \\ 0 & 0 & 1 \end{bmatrix} \tag{3.59}$$

as

$$\mathbf{v}'=\mathsf{B}_\sigma(t'-t)\cdot\mathbf{v}, \tag{3.60}$$

$$\mathbf{r}'=\mathbf{r}+(1/\Omega_\sigma)\mathsf{H}_\sigma(t'-t)\cdot\mathbf{v}. \tag{3.61}$$

As may be clear from a combination of (3.54) and (3.55), the differential operators on the left-hand side of (3.52) together describe a time differentiation along the unperturbed orbit, Eqs. (3.60) and (3.61). We can thus invert this differentiation by integrating (3.52) with respect to time along the unperturbed particle trajectory as

$$\delta f_\sigma(\mathbf{v})\exp[i(\mathbf{k}\cdot\mathbf{r}-\omega t)+\eta t]$$

$$=-\frac{q_\sigma}{m_\sigma}\int_{-\infty}^{t}\left[\frac{\partial f_\sigma(\mathbf{v}')}{\partial\mathbf{v}'}\right]\cdot\left[\left(1-\frac{\mathbf{k}\cdot\mathbf{v}'}{\omega}\right)\mathsf{I}+\frac{\mathbf{k}\mathbf{v}'}{\omega}\right]$$

$$\cdot\delta\mathbf{E}\exp[i(\mathbf{k}\cdot\mathbf{r}'-\omega t)+\eta t']\,dt'$$

or

$$\delta f_\sigma(\mathbf{v})=-\frac{q_\sigma}{m_\sigma}\int_{0}^{\infty}\left[\frac{\partial f_\sigma(\mathbf{v}')}{\partial\mathbf{v}'}\right]\cdot\left[\left(1-\frac{\mathbf{k}\cdot\mathbf{v}'}{\omega}\right)\mathsf{I}+\frac{\mathbf{k}\mathbf{v}'}{\omega}\right]$$

$$\cdot\delta\mathbf{E}\exp[-i\phi(\tau)-\eta\tau]\,d\tau \tag{3.62}$$

where

$$\phi(\tau)=\mathbf{k}\cdot(\mathbf{r}-\mathbf{r}')-\omega\tau,\qquad \tau=t-t'. \tag{3.63}$$

We may then calculate the induced current and the dielectric tensor from an explicit evaluation of the integration in Eq. (3.62).

B. The Dielectric Tensor

The first expression of the induced current in Eq. (3.47) applies in this case also. The unperturbed velocity distribution functions $f_\sigma(\mathbf{v})$ may be written as a function of the parallel and perpendicular components of the velocity, $v_\parallel$ and $v_\perp$. (See Problem 2.2.) We thus have

$$\mathbf{v}=v_\perp\cos\theta\hat{x}+v_\perp\sin\theta\hat{y}+v_\parallel\hat{z} \tag{3.64}$$

in place of the second equation of (3.56); Eq. (3.60) becomes

$$v'_x=v_\perp\cos(\Omega_\sigma\tau+\theta),\qquad v'_y=v_\perp\sin(\Omega_\sigma\tau+\theta),\qquad v'_z=v_\parallel. \tag{3.65}$$

The induced current is now calculated as

$$\delta\mathbf{J}=-\sum_\sigma\frac{q_\sigma^2n_\sigma}{m_\sigma}\int_0^{2\pi}d\theta\int_0^\infty v_\perp dv_\perp\int_{-\infty}^\infty dv_\parallel$$

$$\times\int_0^\infty d\tau\,\mathbf{v}\left[\frac{\partial f_\sigma(\mathbf{v}')}{\partial\mathbf{v}'}\right]\cdot\left[\left(1-\frac{\mathbf{k}\cdot\mathbf{v}'}{\omega}\right)\mathsf{I}+\frac{\mathbf{k}\mathbf{v}'}{\omega}\right]\cdot\delta\mathbf{E}\exp[-i\phi(\tau)-\eta\tau],$$

whence the dielectric tensor is obtained as

$$\epsilon(\mathbf{k},\omega)=\mathsf{I}+\sum_\sigma\frac{\omega_\sigma^2}{i\omega}\int_0^{2\pi}d\theta\int_0^\infty v_\perp dv_\perp\int_{-\infty}^\infty dv_\parallel$$

$$\times\int_0^\infty d\tau\,\mathbf{v}\left[\frac{\partial f_\sigma(\mathbf{v}')}{\partial\mathbf{v}'}\right]\cdot\left[\left(1-\frac{\mathbf{k}\cdot\mathbf{v}'}{\omega}\right)\mathsf{I}+\frac{\mathbf{k}\mathbf{v}'}{\omega}\right]\exp[-i\phi(\tau)-\eta\tau].$$

Since $(d/d\tau)\exp[-i\phi(\tau)-\eta\tau]=-i(\mathbf{k}\cdot\mathbf{v}'-\omega)\exp[-i\phi(\tau)-\eta\tau]$, we may carry out a partial integration with respect to τ for the first half of the integral term. Thus, we obtain

$$\epsilon(\mathbf{k},\omega)=\left(1-\frac{\omega_p^2}{\omega^2}\right)\mathsf{I}+\sum_\sigma\frac{\omega_\sigma^2}{\omega^2}\int_0^{2\pi}d\theta\int_0^\infty v_\perp dv_\perp\int_{-\infty}^\infty dv_\parallel$$

$$\times\int_0^\infty d\tau\,\mathbf{v}\left(\frac{\partial}{\partial\tau}\frac{\partial f_\sigma(\mathbf{v}')}{\partial\mathbf{v}'}-i\mathbf{k}\cdot\frac{\partial f_\sigma(\mathbf{v}')}{\partial\mathbf{v}'}\mathbf{v}'\right)\exp[-i\phi(\tau)-\eta\tau]. \tag{3.66}$$

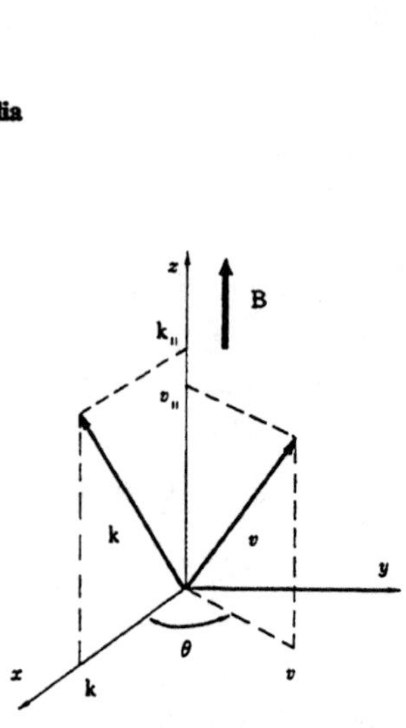

Fig. 3.1 Configuration of various vectors.

In carrying out those integrations, we specifically choose the wave vector **k** on the x–z plane (see Fig. 3.1) and write

$$\mathbf{k}=k_\perp\hat{x}+k_\|\hat{z}. \tag{3.67}$$

We also note that

$$\frac{\partial f_\sigma(\mathbf{v}')}{\partial\mathbf{v}'}=\frac{\partial f_\sigma}{\partial v_\perp}\left[\cos(\Omega_\sigma\tau+\theta)\hat{x}+\sin(\Omega_\sigma\tau+\theta)\hat{y}\right]+\frac{\partial f_\sigma}{\partial v_\|}\hat{z}. \tag{3.68}$$

Since

$$\phi(\tau)=\mathfrak{z}\left[\sin(\Omega_\sigma\tau+\theta)-\sin\theta\right]+k_\|v_\|\tau-\omega\tau$$

where

$$\mathfrak{z}\equiv k_\perp v_\perp/\Omega_\sigma, \tag{3.69}$$

we have†

$$\exp[-i\phi(\tau)-\eta\tau]=\sum_{n=-\infty}^{\infty}\sum_{n'=-\infty}^{\infty}J_n(\mathfrak{z})J_{n'}(\mathfrak{z})$$

$$\times\exp\{-i[n(\Omega_\sigma\tau+\theta)-n'\theta+k_\|v_\|\tau-\omega\tau]-\eta\tau\}. \tag{3.70}$$

We substitute (3.64), (3.65), (3.68), and (3.70) in Eq. (3.66); integration with respect to θ then yields selection rules for the integers n and n' in the summations of Eq. (3.70). After a substantial amount of algebraic manipula-

† Note that $\exp(-iz\sin\psi)=\sum_{n=-\infty}^{\infty}J_n(z)\exp(-in\psi)$ where J_n is the Bessel function of the nth order.

tion, which we leave to the reader as an exercise (Problem 3.7), we find

$$\epsilon(\mathbf{k},\omega)=\left(1-\frac{\omega_p^2}{\omega^2}\right)\mathbf{I}-\sum_\sigma\frac{\omega_\sigma^2}{\omega^2}\sum_{n=-\infty}^{\infty}\int d\mathbf{v}\left(\frac{n\Omega_\sigma}{v_\perp}\frac{\partial f_\sigma}{\partial v_\perp}+k_\parallel\frac{\partial f_\sigma}{\partial v_\parallel}\right)$$

$$\times\frac{\Pi_\sigma(v_\perp,v_\parallel;n)}{n\Omega_\sigma+k_\parallel v_\parallel-\omega-i\eta} \qquad (3.71)$$

where

$$\Pi_\sigma(v_\perp,v_\parallel;n)=\begin{bmatrix} \dfrac{n^2\Omega_\sigma^2}{k_\perp^2}J_n^2 & iv_\perp\dfrac{n\Omega_\sigma}{k_\perp}J_nJ'_n & v_\parallel\dfrac{n\Omega_\sigma}{k_\perp}J_n^2 \\ -iv_\perp\dfrac{n\Omega_\sigma}{k_\perp}J_nJ'_n & v_\perp^2(J'_n)^2 & -iv_\parallel v_\perp J_nJ'_n \\ v_\parallel\dfrac{n\Omega_\sigma}{k_\perp}J_n^2 & iv_\parallel v_\perp J_nJ'_n & v_\parallel^2J_n^2 \end{bmatrix},$$

$$(3.72)$$

$$\int d\mathbf{v}\equiv2\pi\int_0^\infty v_\perp dv_\perp\int_{-\infty}^\infty dv_\parallel,\qquad J_n=J_n(\mathfrak{z}),\qquad J'_n=\frac{dJ_n(\mathfrak{z})}{d\mathfrak{z}}, \qquad (3.73)$$

and $\mathfrak{z}$ has been defined by (3.69). We remark in passing that the last term in Eq. (3.71) decreases as ω^{-3} in the limit of very high frequencies; the calculation Eq. (3.71) is therefore in accord with general asymptotic behavior, Eq. (3.49).

C. The Dielectric Response Function

With the aid of (3.36) again, we may calculate from (3.71) an explicit expression for the dielectric response function in a magnetic field; the result is

$$\epsilon(\mathbf{k},\omega)=1-\sum_\sigma\frac{\omega_\sigma^2}{k^2}\sum_{n=-\infty}^{\infty}\int d\mathbf{v}\left(\frac{n\Omega_\sigma}{v_\perp}\frac{\partial f_\sigma}{\partial v_\perp}+k_\parallel\frac{\partial f_\sigma}{\partial v_\parallel}\right)\frac{J_n^2(\mathfrak{z})}{n\Omega_\sigma+k_\parallel v_\parallel-\omega-i\eta}.$$

$$(3.74)$$

This function is useful in describing the longitudinal fluctuations in the magnetic field.

References

1. A. A. Rukhadze and V. P. Silin, *Usp. Fiz. Nauk* **74**, 223 (1961) [*Soviet Phys. Usp.* **4**, 459 (1961)]; *Usp. Fiz. Nauk* **76**, 79 (1962) [*Soviet Phys. Usp.* **5**, 37 (1962)].
2. T. H. Stix, *The Theory of Plasma Waves* (McGraw-Hill, New York, 1962).
3. A. I. Akhiezer, I. A. Akhiezer, R. V. Polovin, A. G. Sitenko, and K. N. Stepanov, *Collective Oscillations in a Plasma* (H. S. H. Massey and R. J. Tayler, transls.) (MIT Press, Cambridge, Mass., 1967).
4. V. L. Ginzburg, *The Propagation of Electromagnetic Waves in Plasmas*, 2nd ed. (J. B. Sykes and R. J. Tayler, transls.) (Pergamon, Oxford, 1970).
5. J. Lindhard, *Kgl. Danske Videnskab Selskab Mat.-Fys. Medd.* **28**, No. 8 (1954).
6. P. Nozières and D. Pines, *Nuovo Cimento* **9**, 470 (1958); D. Pines and P. Nozières, *The Theory of Quantum Liquids* (W. A. Benjamin, New York, 1966).
7. L. D. Landau and E. M. Lifshitz, *Electrodynamics of Continuous Media* (J. B. Sykes and J. S. Bell, transls.), Section 59 (Addison-Wesley, Reading, Mass., 1960).
8. A. G. Sitenko and K. N. Stepanov, *Zhur. Eksptl. Teoret. Fiz.* **31**, 642 (1956) [*Soviet Phys. JETP* **4**, 512 (1957)].
9. I. B. Bernstein, *Phys. Rev.* **109**, 10 (1958).

Problems

3.1. The total energy adsorbed by the medium in the presence of an electric field disturbance $\mathbf{E}(\mathbf{r},t)$ is given by

$$\mathcal{E}_{\text{abs}}=\int_V d\mathbf{r}\int_{-\infty}^{\infty} dt\mathbf{E}(\mathbf{r},t)\cdot\mathbf{J}(\mathbf{r},t).$$

Show that this can also be written in terms of the spectral components as

$$\mathcal{E}_{\text{abs}}=\frac{1}{2\pi V}\sum_{\mathbf{k}}\int_{-\infty}^{\infty} d\omega\mathbf{E}^*(\mathbf{k},\omega)\cdot\frac{1}{2}[\boldsymbol{\sigma}(\mathbf{k},\omega)+\boldsymbol{\sigma}^+(\mathbf{k},\omega)]\cdot\mathbf{E}(\mathbf{k},\omega)$$

where $\boldsymbol{\sigma}^+(\mathbf{k},\omega)$ is the Hermitian conjugate tensor of $\boldsymbol{\sigma}(\mathbf{k},\omega)$, that is,

$$[\boldsymbol{\sigma}^+(\mathbf{k},\omega)]_{ij}=[\boldsymbol{\sigma}(\mathbf{k},\omega)]^*_{ji}.$$

[It follows from this calculation that the conductivity is anti-Hermitian ($\boldsymbol{\sigma}=-\boldsymbol{\sigma}^+$) when there is no adsorption.]

3.2. The inverse permeability tensor $\boldsymbol{\mu}^{-1}(\mathbf{k},\omega)$ may be defined through

$$\mathbf{H}(\mathbf{k},\omega)=\boldsymbol{\mu}^{-1}(\mathbf{k},\omega)\cdot\mathbf{B}(\mathbf{k},\omega).$$

Find the relationship between $\boldsymbol{\mu}^{-1}(\mathbf{k},\omega)$ and $\boldsymbol{\epsilon}(\mathbf{k},\omega)$ of Eq. (3.11) when $\lambda=1$ is chosen in Eq. (3.18).

3.3. For an isotropic medium, instead of $\epsilon_{\mathrm{L}}(k,\omega)$ and $\epsilon_{\mathrm{T}}(k,\omega)$ of Section 3.1D, it is customary to use another set of scalar functions, $\epsilon(k,\omega)$ and $\mu(k,\omega)$, to set relations

$$\mathbf{D}(k,\omega)=\epsilon(k,\omega)\mathbf{E}(\mathbf{k},\omega), \qquad \mathbf{H}(k,\omega)=\frac{1}{\mu(k,\omega)}\mathbf{B}(\mathbf{k},\omega).$$

In fact, this may be the approach that is most frequently followed in the electrodynamics. Choose the value of λ in Eqs. (3.17) and (3.18) properly and show that the two sets of scalar functions have the relations

$$\epsilon(k,\omega)=\epsilon_{\mathrm{L}}(k,\omega)$$

$$\frac{1}{\mu(k,\omega)}=1+\left(\frac{\omega}{ck}\right)^2[\epsilon_{\mathrm{L}}(k,\omega)-\epsilon_{\mathrm{T}}(k,\omega)].$$

3.4. When the unperturbed velocity distribution functions are isotropic and are written as

$$f_\sigma(\mathbf{v})=F_\sigma(w)$$

where $w=\frac{1}{2}m_\sigma v^2$ is the particle energy, show that in the absence of an external magnetic field the functions Eqs. (3.27) and (3.28) are given by

$$\epsilon_{\mathrm{L}}(k,\omega)=1-\sum_\sigma \frac{8\pi^2 n_\sigma q_\sigma^2}{k^2}\int_0^\infty dv v^2\left[2+\frac{\omega}{kv}\ln\frac{\omega-kv}{\omega+kv}\right]F'_\sigma,$$

$$\epsilon_{\mathrm{T}}(k,\omega)=1-\frac{3}{2}\frac{\omega_{\mathrm{p}}^2}{\omega^2}$$

$$+\sum_\sigma \frac{4\pi^2 n_\sigma q_\sigma^2}{\omega^2}\int_0^\infty dv v^4\left(\frac{\omega^2}{k^2v^2}-1\right)\left[2+\frac{\omega}{kv}\ln\frac{\omega-kv}{\omega+kv}\right]F'_\sigma$$

where

$$F'_\sigma=\frac{dF_\sigma}{dw}.$$

Discuss how to treat the singularities (branch points) at $\omega=\pm kv$ in the integration.

3.5. In a fluid-dynamical approach, the behavior of various components in the plasma may be described by the continuity equation

$$\frac{\partial n_\sigma}{\partial t}+\frac{\partial}{\partial \mathbf{r}}\cdot(n_\sigma \mathbf{u}_\sigma)=0$$

and the equation of motion

$$n_\sigma\left(\frac{\partial}{\partial t}+\mathbf{u}_\sigma\cdot\frac{\partial}{\partial \mathbf{r}}\right)\mathbf{u}_\sigma=-\gamma_\sigma\frac{T_\sigma}{m_\sigma}\frac{\partial n_\sigma}{\partial \mathbf{r}}+\frac{q_\sigma}{m_\sigma}n_\sigma\left(\mathbf{E}+\frac{\mathbf{u}_\sigma}{c}\times\mathbf{B}\right)-\frac{n_\sigma\mathbf{u}_\sigma}{\tau_\sigma}.$$

Here, **u** is the flow velocity, τ the relaxation time, and γ a constant of order unity. For the unperturbed state, we take a homogeneous plasma with a constant uniform magnetic field **B** in the z direction; we assume no net flows of the particles (i.e., $\mathbf{u}_\sigma=0$) in this state. Calculate the dielectric tensor for this system and show that it may be expressed as

$$\epsilon(\mathbf{k},\omega)=\mathsf{I}+\sum_\sigma\frac{\omega_\sigma^2}{\omega^2(1-i\omega\tau_\sigma)}\frac{i\tau_\sigma\omega\mathbf{b}}{\omega(1-i\omega\tau_\sigma)+i(\gamma_\sigma T_\sigma\tau_\sigma/m_\sigma)\mathbf{k}\cdot\mathbf{b}\cdot\mathbf{k}}$$

$$\cdot\left\{\left[\omega(1-i\omega\tau_\sigma)+i\frac{\gamma_\sigma T_\sigma\tau_\sigma}{m_\sigma}\mathbf{k}\cdot\mathbf{b}\cdot\mathbf{k}\right]\mathsf{I}-i\frac{\gamma_\sigma T_\sigma\tau_\sigma}{m_\sigma}\mathbf{k}\mathbf{k}\cdot\mathbf{b}\right\}$$

where

$$\mathbf{b}=\left[(1-i\omega\tau_\sigma)^2+(\Omega_\sigma\tau_\sigma)^2\right]^{-1}$$

$$\times\begin{bmatrix}(1-i\omega\tau_\sigma)^2 & \Omega_\sigma\tau_\sigma(1-i\omega\tau_\sigma) & 0\\ -\Omega_\sigma\tau_\sigma(1-i\omega\tau_\sigma) & (1-i\omega\tau_\sigma)^2 & 0\\ 0 & 0 & (1-i\omega\tau_\sigma)^2+(\Omega_\sigma\tau_\sigma)^2\end{bmatrix}.$$

Check the asymptotic property (3.49) for this dielectric tensor.

3.6. Derive Eq. (3.50).

3.7. Carry out calculations leading from Eq. (3.66) to Eq. (3.71). The following formulas may be used in this derivation:

$$(2n/\mathfrak{z})J_n(\mathfrak{z})=J_{n-1}(\mathfrak{z})+J_{n+1}(\mathfrak{z}),\qquad 2J_n'(\mathfrak{z})=J_{n-1}(\mathfrak{z})-J_{n+1}(\mathfrak{z}),$$

$$\sum_{n=-\infty}^{\infty}J_n^2(\mathfrak{z})=1,\qquad \sum_{n=-\infty}^{\infty}[nJ_n(\mathfrak{z})]^2=\mathfrak{z}^2/2.$$

3.8. Derive Eq. (3.74).

3.9. Show directly from (3.74) that this function has the asymptotic property (3.51).

CHAPTER 4

LONGITUDINAL PROPERTIES OF A PLASMA IN THERMODYNAMIC EQUILIBRIUM

In Chapters 2 and 3, we have studied the basic theoretical tools for investigating various plasma phenomena. In particular, we have learned that the dielectric tensor is capable of describing both the longitudinal and transverse properties of a plasma. In this chapter, we are concerned with the longitudinal properties of a plasma in thermodynamic equilibrium. We shall, therefore, be dealing with the dielectric response function evaluated with a Maxwellian velocity distribution; a function of a complex variable, the plasma dispersion function, will be introduced in this context.

Our first topic of study will be the properties of the collective modes in such a plasma; the plasma oscillation and its damping will be discussed. We shall next turn to those phenomena related to the screening properties of the dielectric response function: these include the Debye screening, the stopping power of the plasma, and the Cherenkov emission of the plasma oscillations. We shall then extend our consideration to those cases of a two-component plasma in which the positive ions as well as the electrons participate in the dynamical processes. In such a system, we shall find that a new mode of wave propagation, the ion-acoustic wave, is possible in addition to the ordinary plasma oscillation; the properties of the ion-acoustic wave will be investigated. Finally, we shall consider the cases in which a uniform magnetic field is applied to the plasma. The propagation characteristics of both the plasma oscillation and the ion-acoustic wave will then be modified; an additional mode of propagation associated with the cyclotron motion of the charged particles will also appear. The latter is called the Bernstein mode.

4.1 THE PLASMA DISPERSION FUNCTION

Since we are concerned here with a plasma in thermodynamic equilibrium, we evaluate the dielectric response function with the aid of a Maxwellian velocity distribution (1.24). The system is isotropic; we may arbitrarily choose the x axis in the direction of the wave vector. For the electron-gas model of the plasma, Eq. (3.50) may then be calculated as

$$\epsilon(\mathbf{k},\omega)=1+\frac{k_{\mathrm{D}}^2}{k^2}\left(\frac{m}{2\pi T}\right)^{1/2}\int_{-\infty}^{\infty}dv_x\frac{kv_x}{kv_x-\omega-i\eta}\exp\left(-\frac{mv_x^2}{2T}\right)$$

$$=1+\frac{k_{\mathrm{D}}^2}{k^2}W\left(\frac{\omega}{k(T/m)^{1/2}}\right) \tag{4.1}$$

where k_{D} is the Debye wave number defined by (1.35). In the last step, we have introduced a new function, the *W function*, according to

$$W(Z)=(2\pi)^{-1/2}\int_{-\infty}^{\infty}\frac{x}{x-Z-i\eta}\exp\left(-\frac{x^2}{2}\right)dx.$$

This is a well-defined and analytic function in the upper half of the complex Z plane; it is then continued analytically into the lower half-plane. Such an analytical continuation may be realized if we deform the contour of x integration so that the point Z always stays above it. We thus write

$$W(Z)\doteq(2\pi)^{-1/2}\int_{\bar{C}}\frac{x}{x-Z}\exp\left(-\frac{x^2}{2}\right)dx \tag{4.2}$$

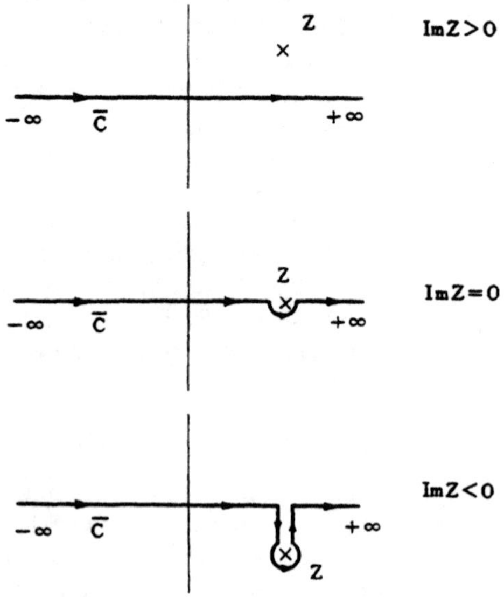

Fig. 4.1 Integration contour $\bar{C}$.

where $\bar{C}$ is the integration contour as specified in Fig. 4.1. The W function satisfies the differential equation

$$\frac{dW}{dZ}=\left(\frac{1}{Z}-Z\right)W-\frac{1}{Z}$$

with $W(0)=1$.

For actual calculation of (4.2), it is most convenient to start with the case where $\operatorname{Im} Z>0$; for real values of x, we then have

$$(x-Z)^{-1}=i\int_0^{\infty} dt \exp[-i(x-Z)t].$$

We substitute this identity in (4.2), and carry out the x integration first. The remaining t integration may then be carried out by splitting the integrand into two parts via

$$\exp(iZt)=\cos(Zt)+i\sin(Zt).$$

We thus find

$$W(Z)=1-Z\exp(-Z^2/2)\int_0^Z dy \exp(y^2/2)+i(\pi/2)^{1/2}Z\exp(-Z^2/2). \tag{4.3}$$

Finally, we extend the applicability of this expression into the lower half of the Z plane to complete the required analytical continuation. As we see in (4.3), the W function is closely related to the error function of a complex argument.

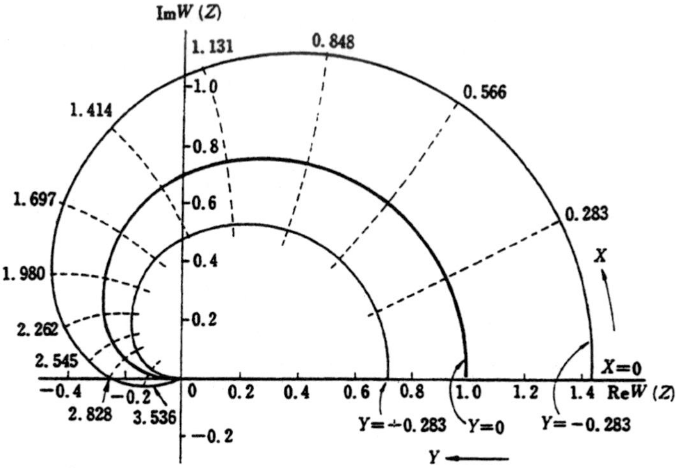

Fig. 4.2 Behavior of the function $W(Z)$ on the complex plane ($Z=X+iY$).

Fried and Conte [1] have computed the numerical values of this function for complex arguments. (See Problem 4.2 also.) In Fig. 4.2, we illustrate the typical behavior of the W function as a function of the real part X of Z; the imaginary part Y is varied as a parameter. It is important to note that the curve does not go across the negative real axis when $Y \geqslant 0$; one or more intersections start to appear when $Y<0$.

For the purpose of mathematical analysis, it is useful to find series expansions of $W(Z)$ for small and large arguments. For $|Z|<1$, $W(Z)$ can be expressed in a convergent series

$$W(Z)=i\left(\frac{\pi}{2}\right)^{1/2} Z \exp\left(-\frac{Z^2}{2}\right)+1-Z^2+\frac{Z^4}{3}-\cdots+\frac{(-1)^{n+1}Z^{2n+2}}{(2n+1)!!}\cdots \tag{4.4}$$

where

$$(2n+1)!!=(2n+1)(2n-1)\cdots 3\cdot 1. \tag{4.5}$$

For large Z, we have an asymptotic series

$$W(Z)=i\left(\frac{\pi}{2}\right)^{1/2} Z \exp\left(-\frac{Z^2}{2}\right)-\frac{1}{Z^2}-\frac{3}{Z^4}-\cdots-\frac{(2n-1)!!}{Z^{2n}}-\cdots. \tag{4.6}$$

4.2 PLASMA OSCILLATION AND LANDAU DAMPING

In Section 3.2B, we considered the dispersion relation for the density fluctuation excitations and obtained Eq. (3.38). We now study the consequence of this equation for an electron plasma in thermodynamic equilibrium.

A. Collective Mode

Since the dielectric function has been calculated as (4.1), the dispersion relation (3.38) becomes

$$1+\frac{k_{\mathrm{D}}^2}{k^2} W\left(\frac{\omega}{k(T/m)^{1/2}}\right)=0. \tag{4.7}$$

This equation can be solved analytically if we may assume

$$\left|\frac{\omega}{k(T/m)^{1/2}}\right| \gg 1, \tag{4.8}$$

for, in these circumstances, we can use the expansion (4.6) for W; Eq. (4.7) becomes

$$1-\frac{\omega_{\rm p}^2}{\omega^2}-3\frac{\omega_{\rm p}^2k^2(T/m)}{\omega^4}-\cdots+i\left(\frac{\pi}{2}\right)^{1/2}\frac{\omega k_{\rm D}^2}{k^3(T/m)^{1/2}}\exp\left[-\frac{\omega^2}{2k^2(T/m)}\right]=0. \tag{4.9}$$

We seek a solution of this equation in the form of a complex frequency

$$\omega=\omega_k+i\gamma_k. \tag{4.10}$$

Introducing one more assumption

$$|\gamma_k/\omega_k|\ll 1, \tag{4.11}$$

we find that Eq. (4.9) may be solved as

$$\omega_k^2=\omega_{\rm p}^2+3(T/m)k^2+\cdots, \tag{4.12}$$

$$\gamma_k=-\left(\frac{\pi}{8}\right)^{1/2}\omega_{\rm p}\frac{k_{\rm D}^3}{k^3}\exp\left[-\frac{\omega_k^2}{2k^2(T/m)}\right]. \tag{4.13}$$

In order that the solution (4.12) be consistent with assumption (4.8), the values of k must be restricted to

$$k\ll k_{\rm D}. \tag{4.14}$$

In the light of (4.13), this condition in turn guarantees the validity of another assumption, (4.11). Consequently, Eqs. (4.12) and (4.13) represent a correct solution of (4.7) in the long-wavelength domain. As k increases, these expressions for ω_k and γ_k become inaccurate; we must use the numerical table of the W function to find a solution for (4.7) in such a short-wavelength domain.

Equation (4.12) obviously corresponds to the collective mode (the plasma oscillation) considered in Section 1.2C. For $k\neq 0$, the frequency is slightly modified from $\omega_{\rm p}$ due to the thermal motion of the electrons.

The appearance of the negative imaginary part (4.13) implies that the plasma oscillation cannot live forever. This phenomenon is called the *Landau damping*, as the existence of such a damping was first predicted by Landau [2].

A question that would immediately come to mind concerns the relationship between this damping phenomenon and the apparent time reversibility of the Vlasov equation, upon which our calculation of the dielectric response function is based. According to Landau's original treatment, macroscopic quantities such as the electric field and charge density are damped exponentially, but perturbations in the electron phase-space distribution $f(\mathbf{r}, \mathbf{v}; t)$ oscillate indefinitely. Since the electron density is given by the integration (3.39a), we may think of the damping as arising from the phase mixing of various parts of the distribution function. A macroscopic physical quantity might therefore reappear in the plasma if we could reverse the direction of phase evolution of the microscopic elements. This aspect has been clearly demonstrated through investigation of the plasma-wave echoes; we shall consider these problems in Chapter 6.

In a quantum-theoretical approach, we can treat the time evolution of the wave intensity through calculations of emission and absorption probabilities of the wave quanta (plasmons) by scattering of single particles [3]. Thus, the Landau damping may be interpreted as arising from the difference between the rate of absorption and that of *induced* emission. The *spontaneous* emission, on the other hand, represents an effect of the first order in the discreteness parameters; it is outside the scope of the Vlasov equation.

A classical counterpart to the foregoing explanation may be given in terms of the resonant coupling between the waves and those particles which are moving with approximately the same velocities as the phase velocity $v_p = \omega/k$ of the wave [4]. If the amplitude of the wave is small but finite, particles moving slightly faster than the wave will decrease their average velocity to v_p through the resonant interaction, transferring their extra kinetic energy to the wave; particles moving slightly slower than the wave will be accelerated and absorb energy from the wave. Therefore, if the distribution function decreases with increasing velocity, as is the case with the Maxwellian, damping takes place, since there will be more slow particles to absorb energy than fast particles to transfer energy to the wave. If, on the contrary, the distribution function increases with velocity near v_p, it is possible that the wave will grow in amplitude, an indication of a plasma-wave instability which we shall study in detail in Chapter 7.

B. Collective versus Individual-Particle Aspects

In the long-wavelength domain of (4.14), the plasma oscillations with frequencies determined from (4.12) are well-defined elementary excitations; their lifetimes are relatively long, so that (4.11) is satisfied. Let us now split the dielectric response function explicitly into the real and imaginary parts and write

$$\epsilon(\mathbf{k},\omega) = \epsilon_1(\mathbf{k},\omega) + i\epsilon_2(\mathbf{k},\omega). \tag{4.15}$$

Generally, as long as (4.11) is satisfied, we may express (4.15) in the vicinity of the collective modes $\omega=\pm\omega_k$ as

$$\epsilon(\mathbf{k},\omega)\cong\left[\frac{\partial\epsilon_1}{\partial\omega}\right]_{\omega=\pm\omega_k}[\omega\mp\omega_k-i\gamma_k]. \tag{4.16}$$

Since γ_k is a negative quantity satisfying (4.11) and

$$\epsilon_1(\mathbf{k},\omega)=1-\frac{\omega_p^2}{\omega^2}-\frac{\omega_p^2}{\omega^4}(\omega_k^2-\omega_p^2)-\cdots, \tag{4.17}$$

we obtain

$$\text{Im}[\epsilon(\mathbf{k},\omega)]^{-1}\cong\frac{\omega_k^5}{2\omega_p^2(3\omega_k^2-2\omega_p^2)}[\delta(\omega-\omega_k)-\delta(\omega+\omega_k)] \tag{4.18}$$

with the aid of the Dirac formula (B.4).

We now wish to examine the approximate expression (4.18) in the light of the sum rule (C. 10). A calculation directly from (4.18) gives

$$\int_{-\infty}^{\infty}\omega\,\text{Im}[\epsilon(\mathbf{k},\omega)]^{-1}d\omega=-\frac{\pi\omega_k^6}{\omega_p^2(3\omega_k^2-2\omega_p^2)}.$$

Although ω_k^2 exceeds ω_p^2 by an additional term proportional to k^2, we here find that Eq. (4.18) exhausts the sum rule (C. 10) up to the contributions proportional to k^2. A simple approximation, which is in accord with the sum rule, may be obtained by adjusting the coefficients of the δ functions and writing, instead of (4.18),

$$\text{Im}[\epsilon(\mathbf{k},\omega)]^{-1}\cong-\frac{\pi}{2}\frac{\omega_p^2}{\omega_k}[\delta(\omega-\omega_k)-\delta(\omega+\omega_k)]\qquad(k\ll k_D). \tag{4.19}$$

Another important observation to be made in this connection is that the contributions from the vicinity of the collective modes $\omega=\pm\omega_p$ completely exhaust the entire sum rule of (C. 10) for $k\ll k_D$; the values of the integrand $\omega\,\text{Im}[\epsilon(\mathbf{k},\omega)]^{-1}$ would be practically zero outside the domain $\omega\cong\pm\omega_p$. Hence, we are led to envision a two-peak structure, centered at $\omega=\pm\omega_p$, for the functional shape of $-\omega\,\text{Im}[\epsilon(\mathbf{k},\omega)]^{-1}$ when $k\ll k_D$. (See the solid lines in Fig. 4.3.) As we shall study later in Chapter 9, $-\omega\,\text{Im}[\epsilon(\mathbf{k},\omega)]^{-1}$ is proportional to the density-fluctuation spectrum in the plasma. A qualitative conclusion that we reached in Section 1.4 on the behavior of the density fluctuations is here seen to be sustained through a sum-rule analysis of the dielectric response function [5].

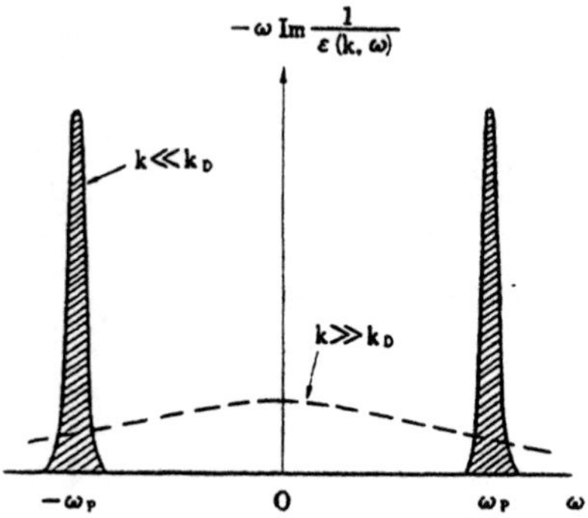

Fig. 4.3 Shapes of $-\omega \mathrm{Im}[\epsilon(\mathbf{k},\omega)]^{-1}$ for $k \ll k_D$ and $k \gg k_D$.

To complete comparison with the results of Section 1.4, we must extend our sum-rule investigation also into the domain $k \gg k_D$. Here we have $\epsilon_1 = 1$, and

$$\mathrm{Im}[\epsilon(\mathbf{k},\omega)]^{-1} = -\left(\frac{\pi}{2}\right)^{1/2} \frac{\omega k_D^2}{k^3 (T/m)^{1/2}} \exp\left(-\frac{\omega^2}{2k^2 (T/m)}\right) \qquad (k \gg k_D). \tag{4.20}$$

Clearly, this function represents the contribution arising from the thermal motion of the individual electrons; the density-fluctuation spectrum has the shape described by the dashed line in Fig. 4.3 with the width $\sqrt{\langle\omega^2\rangle} \cong k(T/m)^{1/2}$. Again this result confirms our previous conclusion reached in Section 1.4 on the behavior of the density fluctuations for $k \gg k_D$. We can easily show that Eq. (4.20) satisfies the sum rule (C. 10).

4.3 DYNAMIC SCREENING

As we remarked in Section 3.2A, the dielectric response function provides a direct measure of the extent to which an external test charge is screened by the induced space charge in the plasma. The screening is dynamic in the sense that it depends on the frequency variable as well as on the wave vector. In this section we wish to study a number of consequences arising from such a screening action as described by the dielectric response function (4.1).

A. Debye Screening and Effective Mass Correction

Consider a test particle with charge q_0, mass m_0, and velocity $\mathbf{v}_0$ introduced in the electron plasma. The space-charge density associated with it is

$$\rho_{\mathrm{ext}}(\mathbf{r},t) = q_0 \delta(\mathbf{r} - \mathbf{v}_0 t), \tag{4.21a}$$

so that the Fourier components are

$$\rho_{ext}(\mathbf{k},\omega)=2\pi q_0\delta(\omega-\mathbf{k}\cdot\mathbf{v}_0). \tag{4.21b}$$

The induced space-charge field is then obtained from (3.37) as

$$\rho(\mathbf{k},\omega)=-2\pi q_0\left\{1-[\epsilon(\mathbf{k},\mathbf{k}\cdot\mathbf{v}_0)]^{-1}\right\}\delta(\omega-\mathbf{k}\cdot\mathbf{v}_0), \tag{4.22}$$

whence the induced potential field is calculated as

$$\varphi_{ind}(\mathbf{k},\omega)=-\frac{8\pi^2 q_0}{k^2}\left\{1-[\epsilon(\mathbf{k},\mathbf{k}\cdot\mathbf{v}_0)]^{-1}\right\}\delta(\omega-\mathbf{k}\cdot\mathbf{v}_0). \tag{4.23a}$$

Carrying out the inverse transformations, we find

$$\varphi_{ind}(\mathbf{r},t)=-4\pi q_0\sum_{\mathbf{k}}(1/k^2)\left\{1-[\epsilon(\mathbf{k},\mathbf{k}\cdot\mathbf{v}_0)]^{-1}\right\}\exp[i\mathbf{k}\cdot(\mathbf{r}-\mathbf{v}_0 t)]. \tag{4.23b}$$

We now substitute (4.1) into (4.23b) to obtain

$$\varphi_{ind}(\mathbf{r},t)=-4\pi q_0\sum_{\mathbf{k}}\left\{(1/k^2)-[k^2+k_D^2 W(\mu\beta_0)]^{-1}\right\}\exp[i\mathbf{k}\cdot(\mathbf{r}-\mathbf{v}_0 t)], \tag{4.24}$$

where μ is the direction cosine between $\mathbf{k}$ and $\mathbf{v}_0$; and

$$\beta_0 \equiv v_0/(T/m)^{1/2}. \tag{4.25}$$

Let us in particular consider the case when the test charge is stationary at the origin (i.e., $v_0=0$). The induced potential field (4.24), which is now time independent, can be calculated as

$$\varphi_{ind}(\mathbf{r})=(-q_0/r)[1-\exp(-k_D r)]. \tag{4.26}$$

The field produced by the test charge is of course

$$\varphi_{ext}(\mathbf{r})=q_0/r.$$

Addition of this and Eq. (4.26) thus reproduces the Debye–Hückel result Eq. (1.10). We have therefore seen that the notion of Debye screening is contained in the dielectric response function (4.1).

The interaction energy $\Delta E(v_0)$ between the test charge and the plasma may be calculated from (4.24) as

$$\Delta E(v_0) = q_0 \varphi_{\text{ind}}(\mathbf{v}_0 t, t)$$

$$= -\frac{q_0^2}{\pi} \int_{-1}^{1} d\mu \int_0^\infty dk \frac{k_D^2 W(\mu\beta_0)}{k^2 + k_D^2 W(\mu\beta_0)}. \tag{4.27}$$

For a slowly moving test charge such that $\beta_0 \ll 1$, we may expand (4.27) with respect to β_0; we thus have

$$\Delta E(v_0) = -q_0^2 k_D + \left(1 - \frac{\pi}{8}\right) \frac{q_0^2 k_D}{6} \frac{v_0^2}{(T/m)} \cdots. \tag{4.28}$$

An effective mass correction Δm_0 of the test charge due to its interaction with the plasma is then defined and calculated as

$$\Delta m_0 = \left[\frac{\partial^2 (\Delta E)}{\partial v_0^2} \right]_{v_0 = 0} = \frac{1}{3}\left(1 - \frac{\pi}{8}\right) \frac{q_0^2 k_D}{(T/m)}. \tag{4.29}$$

This increase in the effective mass arises from deformation of the screening cloud from a spherical shape (4.26) to an ellipsoidal one. As may be clear from (4.24), we can regard $k_D[W(\mu\beta_0)]^{1/2}$ as an effective Debye wave number for screening; when $\beta_0 \neq 0$, this quantity, being a function of μ, gives rise to a directionally dependent screening length.

B. Stopping Power

The effect of finite v_0 is not limited only to the deformation of the screening cloud, however; more importantly, it makes the position of the test charge off the center of the induced potential. This may be seen clearly through an examination of (4.23b) under the inversion $\mathbf{r} - \mathbf{v}_0 t \to -(\mathbf{r} - \mathbf{v}_0 t)$; the presence of $\operatorname{Im}\epsilon(\mathbf{k}, \mathbf{k}\cdot\mathbf{v}_0)$, which is antisymmetric with respect to $\mathbf{k}$, makes (4.23b) asymmetric under such an inversion. As a consequence, the test charge suffers a retarding force from the plasma [6–8]. Let us therefore calculate such a stopping power of the plasma against a moving test charge.

The electric field due to the induced space charge is obtained from (4.23b) as

$$\mathbf{E}_{\text{ind}}(\mathbf{r}, t) = 4\pi q_0 \sum_{\mathbf{k}} (i\mathbf{k}/k^2) \left\{ 1 - [\epsilon(\mathbf{k}, \mathbf{k}\cdot\mathbf{v}_0)]^{-1} \right\} \exp[i\mathbf{k}\cdot(\mathbf{r} - \mathbf{v}_0 t)],$$

whence the force acting on the test charge is

$$\mathbf{F}(\mathbf{v}_0)=q_0\mathbf{E}_{\rm ind}(\mathbf{v}_0t,t)=4\pi q_0^2\sum_{\mathbf{k}}(\mathbf{k}/k^2)\,\mathrm{Im}[\epsilon(\mathbf{k},\mathbf{k}\cdot\mathbf{v}_0)]^{-1}. \tag{4.30}$$

The symmetry properties of the dielectric response function [see, e.g., Eq. (B.3)] have been used in this calculation.

The loss rate of the kinetic energy, $w=\frac{1}{2}m_0v_0{}^2$, per unit time is then calculated as

$$\frac{dw}{dt}=\mathbf{v}_0\cdot\mathbf{F}(\mathbf{v}_0)=4\pi q_0^2\sum_{\mathbf{k}}\frac{\mathbf{k}\cdot\mathbf{v}_0}{k^2}\,\mathrm{Im}[\epsilon(\mathbf{k},\mathbf{k}\cdot\mathbf{v}_0)]^{-1}. \tag{4.31}$$

The stopping power of the plasma is usually expressed in terms of the loss rate of the test-particle energy per unit flight length along its trajectory; denoting such a length by x, we have

$$-\frac{dw}{dx}=-\frac{1}{v_0}\frac{dw}{dt}=-\frac{q_0^2}{\pi v_0}\int_{-1}^{1}d\mu\int dk\,kv_0\mu\,\mathrm{Im}[\epsilon(\mathbf{k},kv_0\mu)]^{-1},$$

where we have transformed the $\mathbf{k}$ summation into integration. Changing an integration variable from μ to ω via $\omega=kv_0\mu$, we finally obtain

$$-\frac{dw}{dx}=-\frac{q_0^2}{\pi v_0^2}\int\frac{dk}{k}\int_{-kv_0}^{kv_0}d\omega\,\omega\,\mathrm{Im}[\epsilon(\mathbf{k},\omega)]^{-1}. \tag{4.32}$$

In the form of Eq. (4.32), we find an important application of the f-sum rule (C. 10) discussed in Appendix C. In fact, when the velocity of the test charge is so large that

$$v_0\gg(T/m)^{1/2}, \tag{4.33}$$

we may replace the upper and lower limits of the ω integration in (4.32) by $+\infty$ and $-\infty$, respectively. With the aid of (C.10), then Eq. (4.32) becomes

$$-\frac{dw}{dx}=\frac{q_0^2\omega_p^2}{v_0^2}\int\frac{dk}{k}. \tag{4.34}$$

We thus have a sum rule, due to Bethe, on the stopping power; Bethe's sum rule states that the total energy loss of a fast particle depends only on the total number of scatterers.

Let us further examine the contents of (4.32) in the light of our investigation in Section 4.2B of the collective versus individual-particle aspects of the plasma. In the long-wavelength domain where $k<k_D$, the collective mode represents the major contribution to $-\omega \operatorname{Im}[\epsilon(\mathbf{k},\omega)]^{-1}$; approximately, we can use Eq. (4.19) in this domain. In view of this equation or Fig. 4.3, we then find that the ω integration in (4.32) vanishes for $k<\omega_p/v_0$. Hence, the contribution of the collective mode to the stopping power may be expressed as

$$-\left(\frac{dw}{dx}\right)_{\text{coll}} \cong \frac{q_0^2\omega_p^2}{v_0^2}\int_{\omega_p/v_0}^{k_D}\frac{dk}{k}=\frac{q_0^2\omega_p^2}{v_0^2}\ln\frac{k_D v_0}{\omega_p}. \tag{4.35a}$$

For $k>k_D$, the individual-particle mode represents the major effects; we may use Eq. (4.20) there. With the aid of (4.33), the ω integration is carried out in (4.32). We thus write the contribution of the individual-particle mode to the stopping power as

$$-\left(\frac{dw}{dx}\right)_{\text{ind}} \cong \frac{q_0^2\omega_p^2}{v_0^2}\int_{k_D}^{k_{\max}}\frac{dk}{k}=\frac{q_0^2\omega_p^2}{v_0^2}\ln\frac{k_{\max}}{k_D} \tag{4.35b}$$

where $k_{\max}$ is an upper limit of the k integration to be determined later from physical considerations. Collecting (4.35a) and (4.35b), we find

$$-\left(\frac{dw}{dx}\right)=-\left(\frac{dw}{dx}\right)_{\text{coll}}-\left(\frac{dw}{dx}\right)_{\text{ind}}=\frac{q_0^2\omega_p^2}{v_0^2}\ln\frac{v_0 k_{\max}}{\omega_p}. \tag{4.36}$$

Note that k_D, which was introduced in the foregoing treatment as the boundary between the collective and individual-particle effects, has disappeared in the final expression of Eq. (4.36).

Classically, the maximum wave number may be chosen in the vicinity of the inverse of a distance of closest approach between the test particle and an electron in the plasma. Hence, we write

$$k_{\max}=\alpha m v_0^2/|q_0 e| \tag{4.37}$$

where m is the reduced mass and α is a constant of order unity. The exact value of α has been determined to be $2/\gamma=1.123$ [7], where $\ln\gamma=0.57721\ldots$ (Euler's constant) or $\gamma=1.78107\ldots$.

When the velocity of the test charge becomes extremely large, so that

$$v_0>|q_0 e|/\hbar, \tag{4.38}$$

then the de Broglie wavelength begins to exceed the classical distance of closest approach. In these circumstances, we choose [9]

$$k_{\max}=2mv_0/\hbar \tag{4.39}$$

for the maximum wave number in (4.36).

C. Cherenkov Emission of Plasma Waves

Let us now pay special attention to that part of the stopping power which arises from the interaction of the test charge with the collective mode of the plasma. As we did in obtaining (4.35a), we substitute (4.19) in (4.31):†

$$-\left(\frac{dw}{dt}\right)_{\text{coll}}=2\pi^2q_0^2\omega_p^2\sum_{\mathbf{k}(k<k_D)}\frac{1}{k^2}[\delta(\omega_p-\mathbf{k}\cdot\mathbf{v}_0)+\delta(\omega_p+\mathbf{k}\cdot\mathbf{v}_0)]. \tag{4.40}$$

This is the rate at which the test particle loses its energy to the collective mode of the plasma. Apparently, then, this must be the rate of energy at which plasma waves are emitted as a result of the passage through the plasma of a fast test particle. The condition for the emission is given from the δ function as

$$\omega_p=|\mathbf{k}\cdot\mathbf{v}_0|, \tag{4.41}$$

which is the well-known *Cherenkov condition* for wave emission by a fast particle. To satisfy both (4.41) and $k<k_D$, the particle velocity must exceed the electron thermal velocity in the plasma.

It is instructive to note that this emission mechanism does not depend on the amplitude of the plasma wave already existent in the plasma; the situation is therefore in sharp contrast with the induced emission processes, which we argued in Section 4.2A to constitute a part of Landau damping term. We may in fact interpret the emission processes described by (4.40) as those corresponding to the *spontaneous emission* of plasma waves, obtained originally through a quantum-theoretical treatment.

We thus calculate the rate of spontaneous emission from (4.40) by regarding each electron in the plasma as a test charge. Summing up such a contribution for all the electrons (in a unit volume) with the aid of their velocity distribution function $f(\mathbf{v})$, we obtain

$$\left(\frac{\partial \mathcal{E}_{\mathbf{k}}}{\partial t}\right)_{\text{sp}}=\frac{\pi m\omega_p^4}{2k^2}\int d\mathbf{v}f(\mathbf{v})\delta(\omega_p-\mathbf{k}\cdot\mathbf{v}) \tag{4.42}$$

†In (4.40), we cannot lump the two δ functions together by using symmetry arguments in $\mathbf{k}$ space. Generally, for a collective mode, $\omega_{-\mathbf{k}}=-\omega_{\mathbf{k}}$ from a reality consideration of physical variables; ω_p, which originates from (4.12), must satisfy the same symmetry requirement.

for the energy rate of spontaneous emission of plasma waves with wave vector $\mathbf{k}$ such that $k<k_D$. This rate is of the first order in the discreteness parameters; the discreteness obviously appears from that of the test charges.

4.4 COLLECTIVE MODES IN A TWO-COMPONENT PLASMA

We now extend our considerations to cases of two-component plasmas, which involve both electrons and ions. We shall use the subscripts $\sigma=\mathrm{e}$ for the electrons and $\sigma=\mathrm{i}$ for the ions; for simplicity, we shall assume $q_i=e$, the unit electric charge, and $n_e=n_i=n$. In this chapter we have been concerned with a situation close to thermodynamic equilibrium; hence, we may write

$$f_\sigma(\mathbf{v})=\left(\frac{m_\sigma}{2\pi T_\sigma}\right)^{1/2}\exp\left(-\frac{m_\sigma v^2}{2T_\sigma}\right) \qquad (\sigma=\mathrm{e},\mathrm{i}) \tag{4.43}$$

for the unperturbed distributions. In so doing we are assuming the possibility that the temperatures of the electrons and the ions may be different, while each species may be described by its own Maxwellian distribution. We begin with a brief discussion of the plausibility of such an assumption.

A. Two-Temperature Plasmas

Approach to equilibrium or Maxwellization of the particle distribution is achieved through collisional processes which exchange energy between the particles. The relaxation times for Maxwellization of electrons and of ions and for temperature equality are in the approximate ratios

$$\tau_{ee}:\tau_{ii}:\tau_{ei}\sim 1:(m_i/m_e)^{1/2}:(m_i/m_e). \tag{4.44}$$

Because

$$m_i/m_e \gg 1, \tag{4.45}$$

the use of Maxwellians with unequal temperatures is reasonable.

The relationship (4.44) can be understood simply from a consideration of binary elastic collisions. Suppose that two particles with masses m_1 and m_2 and velocities $\mathbf{v}_1$ and $\mathbf{v}_2$ collide each other and exchange momentum $\Delta\mathbf{p}$ between them. The amount of energy exchanged, then, is approximately given by

$$\Delta w \cong \mathbf{P}\cdot\Delta\mathbf{p}/(m_1+m_2) \tag{4.46}$$

where $\mathbf{P}=m_1\mathbf{v}_1+m_2\mathbf{v}_2$ is the total momentum; we obtain (4.46) through a Galilean transformation between the center-of-mass frame and the laboratory frame. The cross sections of momentum transfer and energy transfer Q_m and Q_E are defined by

$$Q_m = \int \frac{(\mathbf{v}_1 - \mathbf{v}_2)\cdot\Delta\mathbf{p}}{m|\mathbf{v}_1 - \mathbf{v}_2|^2}\, dQ, \tag{4.47a}$$

$$Q_E = \int \frac{\Delta w}{(m/2)|\mathbf{v}_1 - \mathbf{v}_2|^2}\, dQ \tag{4.47b}$$

where m is the reduced mass and dQ is the differential cross section for scattering [cf. Eqs. (1.2) and (1.3)].

When $m_1 = m_2$, we find $\Delta w \sim (\mathbf{v}_1 - \mathbf{v}_2)\cdot\Delta\mathbf{p}$, so that $Q_m \cong Q_E$. Since the collision frequency for momentum transfer as calculated in (1.25) is proportional to $m^{-1/2}$, the relation

$$\tau_{ee}/\tau_{ii} \cong (m_e/m_i)^{1/2}$$

follows.

When $m_1 \ll m_2$, the transfer of energy becomes very inefficient, so that $\Delta w \sim (m_1/m_2)(\mathbf{v}_1 - \mathbf{v}_2)\cdot\Delta\mathbf{p}$; we thus find $Q_E \cong (m_1/m_2)Q_m$. (See Problem 4.6.) Since the particle mass of the lighter species enters in the collision frequency for momentum transfer (1.25) in these circumstances, we obtain the remaining relation

$$\tau_{ee}/\tau_{ei} \cong m_e/m_i.$$

B. The Dielectric Response Function

We may now substitute (4.43) in (3.50) to obtain

$$\epsilon(\mathbf{k},\omega) = 1 + \frac{k_e^2}{k^2} W\left(\frac{\omega}{k(T_e/m_e)^{1/2}}\right) + \frac{k_i^2}{k^2} W\left(\frac{\omega}{k(T_i/m_i)^{1/2}}\right) \tag{4.48}$$

where

$$k_\sigma^2 = \frac{4\pi n q_\sigma^2}{T_\sigma} \qquad (\sigma = e, i). \tag{4.49}$$

In the limit of low frequencies such that

$$|\omega| \ll k(T_i/m_i)^{1/2} < k(T_e/m_e)^{1/2}, \tag{4.50}$$

we may expand the W function in (4.48) with the aid of (4.4); the lowest-order terms of such an expansion are

$$\epsilon(\mathbf{k},0) = 1 + (k_D^2/k^2) \tag{4.51}$$

with

$$k_D^2 = k_e^2 + k_i^2. \tag{4.52}$$

Equation (4.51) represents the static screening constant for a two-component plasma; the Debye length $\lambda_D \equiv 1/k_D$ is somewhat reduced here, because the ions also participate in the act of screening.

C. Electron Plasma Oscillation

The properties of the collective modes may be investigated through the dispersion relation (3.38); with the aid of (4.48), we have

$$1+\frac{k_e^2}{k^2}W\left(\frac{\omega}{k(T_e/m_e)^{1/2}}\right)+\frac{k_i^2}{k^2}W\left(\frac{\omega}{k(T_i/m_i)^{1/2}}\right)=0. \tag{4.53}$$

The high-frequency collective mode corresponding to the plasma oscillation of Eq. (4.12) may be obtained from (4.53) by expanding the two W functions in asymptotic series (4.6). Assuming thus

$$|\omega| \gg k(T_e/m_e)^{1/2} > k(T_i/m_i),^{1/2} \tag{4.54a}$$

we have

$$1=\frac{\omega_e^2+\omega_i^2}{\omega^2}+3\left(\omega_e^2\frac{T_e}{m_e}+\omega_i^2\frac{T_i}{m_i}\right)\frac{k^2}{\omega^4}$$

$$-i\left(\frac{\pi}{2}\right)^{1/2}\left[\frac{\omega}{k(T_e/m_e)^{1/2}}\frac{k_e^2}{k^2}\exp\left(-\frac{\omega^2}{2k^2(T_e/m_e)}\right)\right.$$

$$\left.+\frac{\omega}{k(T_i/m_i)^{1/2}}\frac{k_i^2}{k^2}\exp\left(-\frac{\omega^2}{2k^2(T_i/m_i)}\right)\right]. \tag{4.54b}$$

Setting $\omega=\omega_k+i\gamma_k$ and again assuming

$$|\gamma_k/\omega_k| \ll 1, \tag{4.54c}$$

we find that the solution to Eq. (4.54b) is

$$\omega_k^2=\omega_p^2+3\frac{[\omega_e^2(T_e/m_e)+\omega_i^2(T_i/m_i)]}{\omega_p^2}k^2, \tag{4.55}$$

$$\gamma_k=-\left(\frac{\pi}{8}\right)^{1/2}\frac{\omega_k^2}{k^3}\left[\frac{k_e^2}{(T_e/m_e)^{1/2}}\exp\left(-\frac{\omega_k^2}{2k^2(T_e/m_e)}\right)\right.$$

$$\left.+\frac{k_i^2}{(T_i/m_i)^{1/2}}\exp\left(-\frac{\omega_k^2}{2k^2(T_i/m_i)}\right)\right] \tag{4.56}$$

where

$$\omega_p^2 = \omega_e^2 + \omega_i^2. \tag{4.57}$$

The solution (4.55) and (4.56) is consistent with the assumptions (4.54a) and (4.54c) as long as we stay in the long-wavelength domain:

$$k^2 \ll k_e^2. \tag{4.58}$$

The nature of the collective mode obtained here is essentially the same as that discussed in Section 4.2A. Both ω_k and γ_k are slightly modified, however, from the values in (4.12) and (4.13) because of the presence of the ions. The last term in (4.56) represents Landau damping due to the ions.

D. The Ion-Acoustic Wave

Thus far we have not discovered anything particularly new from our study of a two-component system; Debye screening and plasma oscillation are basic features associated with a single-component system as well. The presence of the ions, however, makes it possible to look for an additional frequency domain, other than (4.50) and (4.54a), in which the imaginary part of the dielectric response function may take on small values. This possibility is offered especially because of the large mass ratio, (4.45), between the ion and the electron. We may thus investigate an intermediate-frequency domain such that

$$k(T_e/m_e)^{1/2} \gg |\omega| \gg k(T_i/m_i)^{1/2}. \tag{4.59}$$

The validity of this assumption will be examined later.

In the frequency domain of (4.59), we may use the small argument expansion (4.4) for the electrons and the large argument expansion (4.6) for the ions in Eq. (4.53); we have

$$1 = \frac{\omega_i^2}{\omega^2}\left(1 + 3\frac{k^2 T_i}{\omega^2 m_i} + \cdots\right) - \frac{k_e^2}{k^2}$$

$$-i\left(\frac{\pi}{2}\right)^{1/2}\left[\frac{\omega}{k(T_e/m_e)^{1/2}}\frac{k_e^2}{k^2} + \frac{\omega}{k(T_i/m_i)^{1/2}}\frac{k_i^2}{k^2}\exp\left(-\frac{\omega^2}{2k^2(T_i/m_i)}\right)\right].$$

This equation may be solved by following the procedure of the previous section with assumption (4.54c); in the long-wavelength domain such that $k^2 \ll k_e^2$, we find

$$\omega = \omega_k\left\{\pm 1 - i\left(\frac{\pi}{8}\right)^{1/2}\left[\left(\frac{m_e}{m_i}\right)^{1/2} + \left(\frac{T_e}{T_i}\right)^{3/2}\exp\left(-\frac{T_e}{2T_i} - \frac{3}{2}\right)\right]\right\} \tag{4.60a}$$

where

$$\omega_k = \left(\frac{T_e}{m_i} + \frac{3T_i}{m_i} \right)^{1/2} k. \tag{4.60b}$$

In the light of (4.60), we see that (4.54c) is valid when both (4.45) and

$$T_e/T_i \gg 1 \tag{4.61}$$

are satisfied; these conditions also guarantee the validity of (4.59).

The collective mode described by (4.60) is called the *ion-acoustic wave*; in the long-wavelength limit, it has a dispersion relation of sound-wave type:

$$\omega_k = sk. \tag{4.62a}$$

In the temperature range of (4.61), the propagation velocity is

$$s = (T_e/m_i)^{1/2} \tag{4.62b}$$

and its damping rate is

$$|\gamma_k/\omega_k| = (\pi/8)^{1/2}(m_e/m_i)^{1/2}. \tag{4.63}$$

The ion-acoustic wave is essentially the collective mode associated with the ions. In the absence of mobile electrons, such a single-component ion plasma would be characterized by the collective mode with the frequencies $\pm\omega_i$, the ion plasma frequency. This frequency is, however, so low that the mobile electrons can easily follow the motion of the ions. The electrons respond adiabatically to the ionic motion and thereby act to screen their electrostatic fields; the screening constant in these adiabatic circumstances is clearly

$$\epsilon_e(\mathbf{k}) = 1 + (k_e^2/k^2). \tag{4.64}$$

As a result, the collective mode of the ions is described by the screened plasma frequency

$$\omega_k^2 = \omega_i^2/\epsilon_e(\mathbf{k}). \tag{4.65}$$

In the limit of long wavelengths, Eq. (4.65) reproduces (4.62). As the wave number increases, the frequency deviates from the linear relationship of Eq. (4.62) and approaches the ion plasma frequency; there the ion Landau damping becomes substantial. [The physical considerations presented in this paragraph have in fact been embodied in earlier mathematical treatment through a high-frequency expansion for the ions and a low-frequency expansion for the electrons based on (4.59).]

4.5 DIELECTRIC RESPONSE FUNCTION WITH EXTERNAL MAGNETIC FIELD

We now proceed to consider the longitudinal plasma properties in the presence of a uniform stationary magnetic field. The dielectric response function has been obtained in this case as Eq. (3.74), in which we substitute a Maxwellian:†

$$f(v_\perp, v_\parallel) = \left(\frac{m}{2\pi T}\right)^{3/2} \exp\left[-\frac{v_\perp^2 + v_\parallel^2}{2(T/m)}\right]. \tag{4.66}$$

The calculation is facilitated by noticing a relationship among the Bessel functions

$$\int_0^\infty \exp(-a^2x^2)xJ_n(px)J_n(qx)\,dx = \frac{1}{2a^2}\exp\left(-\frac{p^2+q^2}{4a^2}\right)I_n\left(\frac{pq}{2a^2}\right) \tag{4.67}$$

where I_n is the modified Bessel function of the nth order. We thus obtain

$$\epsilon(\mathbf{k},\omega) = 1 + \frac{k_D^2}{k^2}\left\{1 + \sum_n \frac{\omega}{\omega - n\Omega}\left[W\left(\frac{\omega - n\Omega}{|k_\parallel|(T/m)^{1/2}}\right) - 1\right]\Lambda_n(\beta)\right\} \tag{4.68}$$

where

$$\beta \equiv k_\perp^2 T/m\Omega^2, \tag{4.69}$$

$$\Lambda_n(\beta) \equiv I_n(\beta)\exp(-\beta). \tag{4.70}$$

The function $\Lambda_n(\beta)$ appears quite commonly in a theoretical treatment of Maxwellian plasmas in a magnetic field. We note: (i) $0 \leqslant \Lambda_n(\beta) \leqslant 1$; (ii) $\Lambda_0(\beta)$ decreases monotonically from $\Lambda_0(0) = 1$; and (iii) $\Lambda_n(\beta)$ for $n \neq 0$ starts from $\Lambda_n(0) = 0$, reaches the maximum, and then decreases. For large values of β, it approaches

$$\Lambda_n(\beta) \to (2\pi\beta)^{-1/2}\exp(-n^2/2\beta) \qquad (\beta \to \infty). \tag{4.71}$$

This expression may also be used as an approximation for $\Lambda_n(\beta)$ with a large

† More generally, we could assign different temperatures for the parallel and perpendicular degrees of freedom. In the light of (4.44), such an anisotropic situation is more likely to be realized, if at all, for the ions than for the electrons. We shall consider this possibility in Chapter 7.

harmonic number n. We then find from (4.71) that the maximum of $\Lambda_n(\beta)$ occurs at

$$\beta \cong n^2 \qquad \text{or} \qquad |n\Omega| \cong k_\perp (T/m)^{1/2} \tag{4.72}$$

and that the maximum value of $\Lambda_n(\beta)$ decreases as $1/|n|$.

An important application of (4.71) takes place when the limit of $B \to 0$ is considered. In this limit, the cyclotron frequency Ω approaches zero; we may thus replace the summation over the harmonic numbers by an integration over a continuous variable ξ according to

$$\Lambda_n(\beta) \to \frac{|\Omega|}{(2\pi T/m)^{1/2} k_\perp} \exp\left(-\frac{n^2\Omega^2}{2k_\perp^2 (T/m)}\right), \qquad n\Omega \to \xi, \qquad \sum_{n=-\infty}^{\infty} \to \int_{-\infty}^{\infty} \frac{d\xi}{|\Omega|}. \tag{4.73}$$

This scheme of going over to a continuum limit offers a convenient device by which a connection between a calculation with a magnetic field and one without a magnetic field may be established.

4.6 EXCITATIONS IN PLASMAS WITH A MAGNETIC FIELD

In the presence of a magnetic field, the charged particles move with helical orbits along the magnetic lines of force; freedom of motion in the perpendicular directions is thereby suppressed substantially. If we pass to the limit of an extremely strong magnetic field, then the plasma behaves as if it were a one-dimensional gas. These features are naturally reflected in the properties of the collective modes; the propagation characteristics of the electron plasma oscillation and the ion-acoustic wave with a magnetic field are greatly altered from those without a magnetic field. Another notable feature is the periodicity brought about by the cyclotron motion of the particles. A plasma may thus exhibit a new collective mode, propagating mainly in directions perpendicular to the magnetic field, with a frequency near the fundamental or a higher harmonic frequency of the cyclotron motion [10]. In this section we first study these aspects of longitudinal excitatations for a single-component electron plasma; the results will then be extended to the cases of a two-component electron–ion plasma.

A. Electron Plasma Oscillation

The nature of the collective modes may be investigated by substituting (4.68) into the dispersion relation (3.38). We begin with the case of an electron gas, for which we suppress the subscript e altogether; the magnetic field is assumed to be strong, so that

$$\Omega^2 \gg \omega_p^2 \tag{4.74}$$

where Ω is the electron cyclotron frequency given by (3.53).

In the vicinity of the plasma frequency ($\omega \cong \pm\omega_p$), the dielectric function (4.68) may then be approximated by

$$\epsilon(\mathbf{k},\omega) \cong 1 + \frac{k_D^2}{k^2}\left\{[1-\Lambda_0(\beta)] + W\left(\frac{\omega}{|k_\parallel|(T/m)^{1/2}}\right)\Lambda_0(\beta)\right\}, \tag{4.75}$$

because the neglected terms are smaller by a factor of the order of ω_p^2/Ω^2. As may be clear from (4.75) [and as we shall explicitly show in (4.79)], the plasma oscillation can be a well-defined excitation only in the long-wavelength domain, $k^2 \ll k_D^2$. Combination of this and (4.74) yields

$$\beta \ll 1 \tag{4.76}$$

for the plasma oscillation. We now seek a solution for the dispersion relation in the form

$$\omega = \pm\omega_0(\mathbf{k}) + i\gamma_0(\mathbf{k}). \tag{4.77}$$

With the aid of expansion (4.6) as well as (4.74) and (4.76), we then obtain from Eq. (4.75)

$$\omega_0(\mathbf{k}) = \frac{|k_\parallel|}{k}\omega_p[\Lambda_0(\beta)]^{1/2} \cong \frac{|k_\parallel|}{k}\omega_p, \tag{4.78a}$$

$$\frac{\gamma_0(\mathbf{k})}{\omega_0(\mathbf{k})} = -\left(\frac{\pi}{8}\right)^{1/2}\frac{k_D^3}{k^3}[\Lambda_0(\beta)]^{3/2}\exp\left[-\frac{k_D^2\Lambda_0(\beta)}{2k^2}\right]. \tag{4.78b}$$

Equation (4.78a) clearly indicates a one-dimensional propagation of the plasma oscillation in the direction of the magnetic field. For such a wave to have a relatively long lifetime, the condition

$$k^2 \ll k_D^2\Lambda_0(\beta) \tag{4.79}$$

must then be satisfied.

B. The Electron Bernstein Mode

In addition to the plasma-wave mode (4.78), Eq. (4.68) permits propagation of the collective modes with frequencies in the vicinity of the harmonics of the electron cyclotron frequency. In such a frequency domain, we may write

$$\epsilon(\mathbf{k},\omega)=1+\frac{k_{\mathrm{D}}^2}{k^2}[1-\Lambda_0(\beta)]-\frac{k_{\mathrm{D}}^2}{k^2}\frac{\omega}{\omega-n\Omega}\left[1-W\left(\frac{\omega-n\Omega}{|k_{\parallel}|(T/m)^{1/2}}\right)\right]\Lambda_n(\beta)$$

$$(\omega\cong n\Omega;\quad n=\pm 1,\pm 2,\ldots). \tag{4.80}$$

We again seek a solution for the dispersion relation in the form

$$\begin{aligned}\omega&=\omega_n(\mathbf{k})+i\gamma_n(\mathbf{k})\\&=n\Omega[1+\Delta_n(\mathbf{k})]+i\gamma_n(\mathbf{k})\end{aligned} \tag{4.81a}$$

with an assumption

$$\frac{|k_{\parallel}|(T/m)^{1/2}}{|n\Omega|}\ll|\Delta_n(\mathbf{k})|\ll 1. \tag{4.81b}$$

We thus obtain

$$\Delta_n(\mathbf{k})=\frac{k_{\mathrm{D}}^2\Lambda_n(\beta)}{k^2+k_{\mathrm{D}}^2[1-\Lambda_0(\beta)]}, \tag{4.82a}$$

$$\left|\frac{\gamma_n(\mathbf{k})}{\omega_n(\mathbf{k})}\right|=\left(\frac{\pi}{2}\right)^{1/2}\frac{|n\Omega|[\Delta_n(\mathbf{k})]^2}{|k_{\parallel}|(T/m)^{1/2}}\exp\left\{-\frac{[n\Omega\Delta_n(\mathbf{k})]^2}{2k_{\parallel}^2(T/m)}\right\}. \tag{4.82b}$$

Roughly, $\Lambda_n(\beta)$ takes on the maximum value $\sim 1/|n|$ at (4.72). Hence, we estimate

$$[\Delta_n(\mathbf{k})]_{\max}\sim\frac{1}{|n|^3}\frac{\omega_{\mathrm{p}}^2}{\Omega^2}. \tag{4.83}$$

In the light of (4.74), the latter half of (4.81b) is thus guaranteed.

The first half of (4.81b) amounts to the condition $|\gamma_n(\mathbf{k})/\omega_n(\mathbf{k})|\ll 1$. With the aid of (4.72), we may then rewrite this condition approximately as

$$|k_{\parallel}|\ll k_{\perp}[\Delta_n(\mathbf{k})]. \tag{4.84}$$

In view of (4.83), we find that the propagation directions of this collective mode are confined near the directions perpendicular to the magnetic field. In particular, when $k_{\parallel}=0$, we have $\gamma_n(\mathbf{k})=0$; the wave propagates without damping across the magnetic field.

Physically we can understand these damping properties in terms of the wave–particle resonance picture of Landau damping. When a wave propagates with a finite phase velocity in a direction perpendicular to the magnetic field, no electrons are available to interact resonantly with the wave because the electrons are effectively bound to the magnetic lines of force; no Landau damping is thus expected. When $k_{\|}$ becomes finite, there are some electrons that can satisfy the resonance condition $\omega_n \cong k_{\|} v_{\|}$; damping thereby becomes finite.

C. Ion-Acoustic Wave

We now recover the subscripts e and i to treat plasmas consisting of electrons and ions. Because of the large mass ratio (4.45), the properties of the electron plasma oscillation and the electron Bernstein mode as described earlier are little affected by the presence of the ions. We shall therefore be concerned only with the low-frequency collective modes associated with the ions.

We look for a solution corresponding to the ion-acoustic wave in the frequency domain such that

$$|k_{\|}|(T_i/m_i)^{1/2} \ll |\omega| \ll |k_{\|}|(T_e/m_e)^{1/2} \tag{4.85a}$$

and in the long-wavelength limit such that

$$|\omega| \ll \Omega_i. \tag{4.85b}$$

The dielectric response function is then simplified as

$$\begin{aligned}\epsilon(\mathbf{k},\omega) = 1 &+ \frac{k_e^2}{k^2} + \frac{k_i^2}{k^2}[1-\Lambda_0(\beta_i)] - \frac{k_{\|}^2}{k^2}\frac{\omega_i^2}{\omega^2}\Lambda_0(\beta_i) \\ &+ i\left(\frac{\pi}{2}\right)^{1/2}\left\{\frac{\omega}{|k_{\|}|(T_e/m_e)^{1/2}}\frac{k_e^2}{k^2}\right. \\ &\left. + \frac{\omega}{|k_{\|}|(T_i/m_i)^{1/2}}\frac{k_i^2}{k^2}\Lambda_0(\beta_i)\exp\left[-\frac{\omega^2}{2k_{\|}^2(T_i/m_i)}\right]\right\}.\end{aligned} \tag{4.86}$$

Adopting the form of (4.77) as a solution for the dispersion relation, we find

$$\omega_0(\mathbf{k}) = \frac{|k_{\|}|}{k}\omega_i \frac{[\Lambda_0(\beta_i)]^{1/2}}{\{\epsilon_e(\mathbf{k}) + [1-\Lambda_0(\beta_i)]k_i^2/k^2\}^{1/2}}.$$

In the limit of long wavelengths, this expression reduces to

$$\omega_0(\mathbf{k}) = |k_{\|}|s \tag{4.87a}$$

where $\epsilon_e(\mathbf{k})$ and s have been defined by Eqs. (4.64) and (4.62b), respectively. Again, Eq. (4.87a) shows a one-dimensional propagation of the ion-acoustic wave in the direction of the magnetic field.

The existence of the frequency domain (4.85a) implies condition (4.61). In the long-wavelength limit, we then obtain a simplified expression for the damping rate

$$\frac{\gamma_0(\mathbf{k})}{\omega_0(\mathbf{k})}=-\left(\frac{\pi}{8}\right)^{1/2}\left(\frac{m_e}{m_i}\right)^{1/2}. \tag{4.87b}$$

This formula is essentially the same as (4.63).

D. The Ion Bernstein Mode

Finally let us investigate the dispersion relation in the vicinity of the harmonics $n\Omega_i$ of the ion cyclotron frequency. The dielectric response function may be approximated in this case as

$$\epsilon(\mathbf{k},\omega)=1+\frac{k_e^2}{k^2}+\frac{k_i^2}{k^2}[1-\Lambda_0(\beta_i)]$$
$$-\frac{k_i^2}{k^2}\frac{\omega}{\omega-n\Omega_i}\left[1-W\left(\frac{\omega-n\Omega_i}{|k_\parallel|(T_i/m_i)^{1/2}}\right)\right]\Lambda_n(\beta_i)$$
$$(\omega\cong n\Omega_i;\quad n=\pm1,\pm2,\ldots). \tag{4.88}$$

Equation (4.88) differs from (4.80) by an additional static screening coming from the electrons. Again setting $\omega=n\Omega_i[1+\Delta_n(\mathbf{k})]+i\gamma_n(\mathbf{k})$ for the dispersion relation, we obtain

$$\Delta_n(\mathbf{k})=\frac{k_i^2\Lambda_n(\beta_i)}{k^2\epsilon_e(\mathbf{k})+k_i^2[1-\Lambda_0(\beta_i)]}, \tag{4.89a}$$

$$\left|\frac{\gamma_n(\mathbf{k})}{n\Omega_i}\right|=\left(\frac{\pi}{2}\right)^{1/2}\frac{|n\Omega_i|[\Delta_n(\mathbf{k})]^2}{|k_\parallel|(T_i/m_i)^{1/2}}\exp\left\{-\frac{[n\Omega_i\Delta_n(\mathbf{k})]^2}{2k_\parallel^2(T_i/m_i)}\right\}. \tag{4.89b}$$

In many practical cases, it is quite reasonable to assume†

$$\omega_i^2\gg\Omega_i^2 \tag{4.90}$$

†As we shall study in Chapter 5, the ratio ω_i^2/Ω_i^2 provides an effective transverse dielectric constant responsible for the Alfven wave propagation in the plasma.

even with a substantially strong magnetic field. In these circumstances, we may approximate

$$\Delta_n(\mathbf{k}) \cong \frac{T_e}{T_e + T_i} \Lambda_n(\beta_i). \tag{4.91}$$

The wave vectors are restricted within the domain

$$|k_\parallel| \ll k_\perp [\Delta_n(\mathbf{k})]. \tag{4.92}$$

Propagations nearly perpendicular to the magnetic field are therefore indicated in this case also.

References

1. B. D. Fried and S. D. Conte, *The Plasma Dispersion Function* (Academic, New York, 1961).
2. L. D. Landau, *J. Phys. (USSR)* **10**, 25 (1946).
3. D. Pines and J. R. Schrieffer, *Phys. Rev.* **125**, 804 (1962).
4. D. Bohm and E. P. Gross, *Phys. Rev.* **75**, 1851 (1949); *Phys. Rev.* **75**, 1864 (1949).
5. S. Ichimaru, *Ann. Phys. (N.Y.)* **20**, 78 (1962).
6. E. Fermi, *Z. Phys.* **29**, 315 (1924).
7. N. Bohr, *Kgl. Danske Videnskab. Selskab Mat.-Fys. Medd.* **18**, No. 8 (1948).
8. H. Frölich and R. L. Platzmann, *Phys. Rev.* **92**, 1152 (1953); H. Frölich and H. Pelzer, *Proc. Phys. Soc. (London)* **68A**, 525 (1955).
9. J. Lindhard, *Kgl. Danske Videnskab. Selskab Mat.-Fys. Medd.* **28**, No. 8 (1954).
10. I. B. Bernstein, *Phys. Rev.* **109**, 10 (1958).

Problems

4.1. Carry out the calculation leading to Eq. (4.3).

4.2. Fried and Conte in [1] tabulate the values of the functions

$$Z(\zeta) = \pi^{-1/2} \int_{\bar{C}} dx (x-\zeta)^{-1} \exp(-x^2), \qquad Z'(\zeta) = -2[1+\zeta Z(\zeta)]$$

for complex variable ζ. ($\bar{C}$ is the contour shown in Fig. 4.1.) $W(Z)$, defined by Eq. (4.2), is in fact closely related to $Z'(\zeta)$; find the relationship between the two.

4.3. Derive expansions (4.4) and (4.6).

4.4. Obtain the solution (4.12) and (4.13).

4.5. Derive (4.28) from (4.27).

4.6. Calculate the recoil energy Δw of the scatterer in the collision event of Fig. 1.1, and show

$$\frac{\Delta w}{\frac{1}{2} m v^2} \cong \frac{2m}{M}(1-\cos\chi).$$

4.7. For the ion-acoustic mode (4.62), show that

$$\left.\frac{\partial \epsilon_1(\mathbf{k},\omega)}{\partial \omega}\right]_{\omega=\omega_k} = \frac{2}{\omega_k}\epsilon_e(\mathbf{k})$$

where $\epsilon_e(\mathbf{k})$ is the static screening constant (4.64) of the electrons.

4.8. Following the procedure of Section 4.2B, obtain an approximate expression for $\text{Im}[\epsilon(\mathbf{k},\omega)]^{-1}$ in the vicinity of the ion-acoustic mode (4.60).

4.9. Derive (4.68).

4.10. Show that in the limit of $B \to 0$ Eq. (4.68) reduces to (4.1).

4.11. Consider a test particle with electric charge q_0 and velocity $\mathbf{v}_0$ introduced in an electron plasma with the dielectric response function given by (4.68); we assume that $\mathbf{v}_0$ is parallel to $\mathbf{B}$.
(a) Calculate the interaction energy between the electron plasma and the test charge at $\mathbf{v}_0=0$. Discuss the physical meaning of the result.
(b) Consider the limit of a very strong magnetic field and calculate that part of the interaction energy which is proportional to v_0^2 for a slowly moving test charge.

4.12. Following the procedure of Section 4.2B, obtain an approximate expression for $\text{Im}[\epsilon(\mathbf{k},\omega)]^{-1}$ in the vicinity of the electron Bernstein mode (4.81a).

CHAPTER 5

TRANSVERSE PROPERTIES OF A PLASMA IN THERMODYNAMIC EQUILIBRIUM

We now proceed to consider the propagation of electromagnetic waves in plasmas near thermodynamic equilibrium. As we remarked earlier, a plasma can sustain a rich class of electromagnetic phenomena; this is especially the case for electromagnetic wave propagation in plasmas with an external magnetic field [1–5]. The cyclotron motion of the charged particles around the magnetic lines of force, on the one hand, brings about anisotropy to the medium properties; on the other hand, it introduces an additional frequency, the cyclotron frequency, to characterize the electromagnetic properties of the plasma. For a given electromagnetic phenomenon, we would thus be dealing with a combination of at least three characteristic frequencies: the plasma frequency, the cyclotron frequency, and the frequency of the electromagnetic wave. Frequencies associated with the thermal motion of the particles and collisional frequencies would have to be added to such a list, if any of them were significantly involved in the electromagnetic processes under consideration.

In this chapter, we begin with the simple case of a plasma without an external magnetic field; the medium is isotropic. An important feature of wave propagation in such a plasma is a cutoff at the plasma frequency; an electromagnetic wave with frequency less than the plasma frequency cannot propagate in the plasma. In this connection, we shall also study the anomalous skin effect, a transverse analogue of the Landau damping of the plasma oscillation. We shall then consider the cases of a plasma in an external magnetic field. The medium is now anisotropic; all the elements of the dielectric tensor have to be evaluated with a Maxwellian distribution. With the aid of such a dielectric tensor, we shall first investigate the wave propagation along the magnetic field; we shall find that in these circumstances a representation in terms of circularly polarized waves is more convenient than one based on linearly polarized waves. A remarkable feature is that the plasma with a magnetic field permits propagation of electromagnetic waves with frequencies much lower than the plasma frequency; helicon, Alfvén wave, and whistler are typical examples of such low-frequency modes of propagation. A phenomenon corresponding to Landau damping now appears in the form of Doppler-shifted cyclotron resonances. Finally, we shall consider wave propagation perpendicular to the magnetic field. Two possible modes of propagation, ordinary and extraordinary modes, will be investigated. The effects of nonlocality, or spatial dispersion, brought about by the finite Larmor radii of the particles will be briefly studied.

5.1 WAVE PROPAGATION IN PLASMAS WITHOUT EXTERNAL MAGNETIC FIELD

In this case the plasma is isotropic; we begin with a calculation of the transverse dielectric function according to (3.28).

A. The Transverse Dielectric Function

We substitute the Maxwellian (1.24) into (3.48) and calculate (3.28); the result is

$$\epsilon_{\mathrm{T}}(k,\omega)=1-\frac{\omega_{\mathrm{p}}^2}{\omega^2}\left[1-W\left(\frac{\omega}{k(T/m)^{1/2}}\right)\right] \tag{5.1}$$

where the W function has been defined by (4.2). The dispersion relation of the transverse wave is then determined from (3.31).

The propagation mode we are interested in should not have appreciable damping. We thus look for frequency domains in which the imaginary part of W is small; in view of Fig. 4.2, two such domains exist:

$$|\omega/k|\gg(T/m)^{1/2}; \tag{5.2a}$$

$$|\omega/k|\ll(T/m)^{1/2}. \tag{5.2b}$$

In the domain (5.2a), we may use the expansion (4.6); hence

$$\epsilon_{\mathrm{T}}(k,\omega)=1-\frac{\omega_{\mathrm{p}}^2}{\omega^2}\left(1+\frac{k^2T}{\omega^2m}+\cdots\right)+i\left(\frac{\pi}{2}\right)^{1/2}\frac{\omega_{\mathrm{p}}k_{\mathrm{D}}}{\omega k}\exp\left[-\frac{\omega^2}{2k^2(T/m)}\right].$$

Clearly, $\mathrm{Re}\,\epsilon_{\mathrm{T}}<1$; it follows from (3.31) that $|\omega/k|>c$, the light velocity in vacuum. This result is consistent with the original assumption (5.2a). The imaginary part of $\epsilon_{\mathrm{T}}(k,\omega)$ in the expression above should then be set equal to zero, for it physically represents the effects of resonant coupling between wave and particles; no particles may exist with velocities greater than c. The finite value in the expression above has resulted from our use of (1.24), which is valid only for a nonrelativistic situation. By the same token, we may ignore the effects of thermal motion in the real part of ϵ_{T} for a nonrelativistic plasma.

We may similarly examine the dielectric function and the dispersion relation starting from (5.2b). It will then be found that the results of the solution so obtained are incompatible with the assumption (5.2b); hence, this case must be disregarded as a possible propagation mode in the plasma.

Consequently, the propagation of an electromagnetic wave in the plasma under consideration is described by the dispersion relation

$$(kc/\omega)^2 = 1 - (\omega_p^2/\omega^2) \tag{5.3a}$$

or

$$\omega^2 = \omega_p^2 + k^2c^2. \tag{5.3b}$$

In the long-wavelength domain $k \ll \omega_p/c$, the phase velocity of the wave substantially exceeds the light velocity in vacuum.

Equation (5.3a) or (5.3b) accounts for the dispersion and broadening of radio signals traveling through interstellar space. Since ω_p^2 is proportional to the density of charged particles, the density fluctuations in interstellar space create local variations in the velocity and direction of radio-wave propagation. The effective path length of the signal from a radio source to the observer on the earth would thus be altered by these fluctuations; the signal would be broadened in shape and scintillate in time. Obviously from (5.3a), such effects would be more emphasized for radio signals with lower frequencies.

B. Cutoffs and Resonances

According to (5.3), the electromagnetic wave propagates in the plasma only when its frequency is greater than the plasma frequency. As we decrease the frequency, the transverse dielectric function $\epsilon_T(k,\omega)$ also decreases; it vanishes at $\omega = \omega_p$. The square of the phase velocity would diverge to $+\infty$ at this point. If we further decrease the frequency, both $\epsilon_T(k,\omega)$ and $(\omega/k)^2$ become negative; the wave cannot propagate in the frequency domain $|\omega| < \omega_p$.

As the argument above may have indicated, a point at which $\epsilon_T(k,\omega)$ or $(kc/\omega)^2$ [compare Eq. (3.31)] changes its sign marks a boundary between propagating and nonpropagating frequency domains. Generally, there are two such possibilities, as Fig. 5.1 illustrates: One is the case in which $(kc/\omega)^2$ changes sign through a point

$$(kc/\omega)^2 = 0. \tag{5.4}$$

The frequency at which Eq. (5.4) takes place is called a *cutoff frequency*.

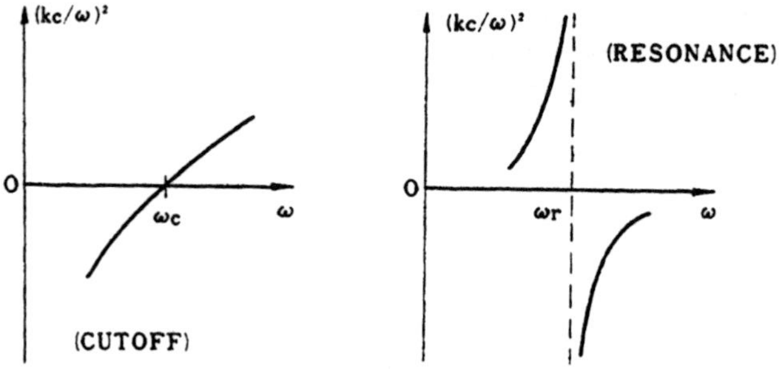

Fig. 5.1 (a) Cutoff; (b) resonance.

Those waves in the propagating domain near the cutoff frequency travel with extremely large phase velocities; dissipation processes in the medium thus become insignificant. The electromagnetic waves will be reflected at those places where the cutoff condition (5.4) is satisfied. Consequently, the plasma frequency represents the cutoff frequency for the electromagnetic waves in the plasma without an external magnetic field.

The transverse dielectric function or $(kc/\omega)^2$ can also change its sign through infinity (see Fig. 5.1b); at the boundary, we thus have

$$(\omega/kc)^2=0. \tag{5.5}$$

This situation is called *resonance*, and the frequency at which Eq. (5.5) takes place determines the *resonance frequency*. In the vicinity of a resonance, the waves travel with very small phase velocities; dissipations in the medium thus become substantial. The electromagnetic waves will be strongly absorbed by the medium under these resonant conditions. We shall see that various resonances take place in plasmas with an external magnetic field.

C. The Anomalous Skin Effect

Let us return to our dispersion relation (5.3) and consider the case where $|\omega|<\omega_p$ in more detail. As the equation indicates, the electromagnetic wave with such a low frequency cannot propagate in the plasma; instead it is reflected by the plasma with a penetration depth

$$d \cong |1/k| = c/\left(\omega_p^2 - \omega^2\right)^{1/2}$$

$$\cong c/\omega_p. \tag{5.6}$$

The electromagnetic fields penetrate into the plasma only within a thin surface layer determined by (5.6).

This may not be the whole story, however, for the fact that the electromagnetic fields are confined within a given (thin) layer implies that the effective wavelengths associated with the spatial variations of the fields are also limited within a similar distance. In the surface layer, a possibility therefore exists that $|\omega/k| \lesssim (T/m)^{1/2}$. If this is the case, then the imaginary part of (5.1), or the transverse conductivity of the plasma, now plays a more important role than its real part. In these circumstances, the conductivity of the medium determines the penetration depth of the fields, as we know from a theory of the ordinary skin effect. Unlike the cases of the ordinary skin effect, however, the effective conductivity in the present case exhibits a strong spatial dispersion (i.e., k dependence) and thus in particular depends critically on the penetration depth; the overall situation must therefore be determined in a self-consistent manner. The imaginary part of (5.1) appears as a result of resonant coupling between the waves and particles; no collisional effects between particles are involved. This is therefore a transverse analogue of the Landau damping discussed in Section 4.2A. It is in this sense that the phenomenon presently under consideration is called the *anomalous skin effect* [6–8].

To be more specific, let us consider the situation depicted in Fig. 5.2: A

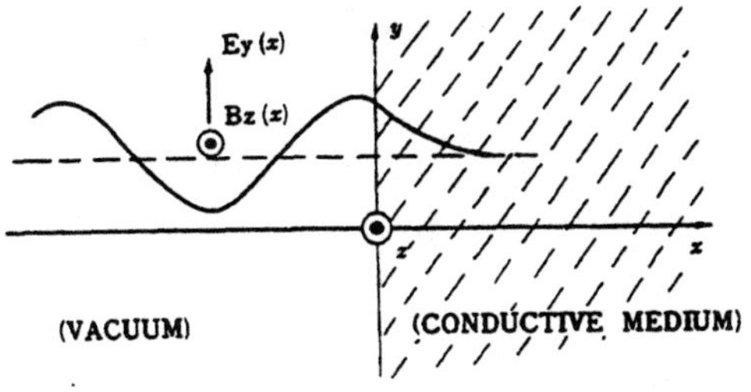

Fig. 5.2 The anomalous skin effect.

conductive medium occupies the semi-infinite space $x>0$; a plane electromagnetic wave is incident onto the medium from the vacuum region $x<0$. We wish to study the anomalous skin effect in these circumstances; when the medium is a plasma, we must assume $|\omega|<\omega_p$.

The transverse conductivity of the plasma is calculated from (5.1) as

$$\sigma_T(k,\omega)=\frac{\omega}{4\pi i}[\epsilon_T(k,\omega)-1]$$

$$=\frac{1}{4\sqrt{2\pi}}\frac{\omega_p k_D}{k}\exp\left[-\frac{\omega^2}{2k^2(T/m)}\right]. \tag{5.7}$$

Clearly, this quantity takes on a substantial magnitude only when

$$|\omega/k|<(T/m)^{1/2}. \tag{5.8}$$

The electric field penetrated into the plasma then induces a current density; this relationship should be described by an equation similar to (3.1).

Spatially, all the physical quantities involved are functions of x alone. The configuration under consideration is thus translationally invariant in the y and z directions. We may, therefore, simply integrate a transverse version of Eq. (3.1) in those two directions; only that part of $\sigma_T(k,\omega)$ in which $k_y=0$ and $k_z=0$ remains.

All the physical quantities are assumed to vary as $\exp(-i\omega t)$ in time; we may thus suppress the time argument in each quantity throughout the problem. Expressing the foregoing ideas in mathematical terms, we find that the induced current in the y direction is written as

$$J_y(x)=\int dx'\Sigma(x-x')E_y(x') \tag{5.9}$$

where

$$\Sigma(x)=(2\pi)^{-1}\int_{-\infty}^{\infty}dk_x\,\sigma_T(|k_x|,\omega)\exp(ik_xx). \tag{5.10}$$

Superficially, it might be suggested from Fig. 5.2 that the domain of x' integration in (5.9) would be from 0 to ∞; actually, the domain does depend on the boundary condition for particle reflection at the plasma–vacuum interface, $x=0$.

Suppose, for example, that the particles are reflected randomly at the surface. The average momentum carried by a particle in the y or z direction after reflection will then be zero; no effects of the electric field will remain there. In these circumstances, we may take Eq. (5.9) and set the lower and upper limits of integration to be 0 and ∞; hence,

$$J_y(x) = \int_0^\infty dx' \Sigma(x-x') E_y(x'). \tag{5.11}$$

Such a situation is called the diffuse boundary condition.

We may go over to the other limiting cases and adopt a perfectly reflecting boundary condition. In this case,

$$J_y(x) = \int_0^\infty dx' [\Sigma(x-x') + \Sigma(x+x')] E_y(x'),$$

or, extending the definition to $E(x)$ into the domain $x<0$ by

$$E_y(-x) = E_y(x), \tag{5.12}$$

we have

$$J_y(x) = \int_{-\infty}^\infty dx' \Sigma(x-x') E(x'). \tag{5.13}$$

For the sake of definiteness, we shall adopt this boundary condition in the following.

We combine Eq. (5.13) with the first two equations of (3.8), which now read

$$\frac{\partial E_y}{\partial x} = \frac{i\omega}{c} B_z, \tag{5.14a}$$

$$-\frac{\partial B_z}{\partial x} = \frac{4\pi}{c} J_y. \tag{5.14b}$$

In Eq. (5.14b), we have neglected the displacement current or the imaginary part of the conductivity. Since $E_y(x)$ is an even function with respect to x [see Eq. (5.12)], Eq. (5.14a) indicates that $B_z(x)$ is an odd function; $B_z(x)$ is discontinuous at $x=0$ as $B_z(+0) = -B_z(-0)$. We differentiate (5.14a) with respect to x and substitute (5.14b) in it; taking account of the foregoing discontinuity and (5.13), we obtain

$$\frac{\partial^2 E_y(x)}{\partial x^2} + i\frac{4\pi\omega}{c^2} \int_{-\infty}^\infty dx' \Sigma(x-x') E_y(x') = \frac{2i\omega}{c} B(+0)\delta(x).$$

This differential equation may be solved through Fourier transformation;

with the aid of (5.10), we find

$$E_y(x) = -i\frac{\omega}{\pi c}B(+0)\int_{-\infty}^{\infty} dk_x \frac{\exp(ik_x x)}{k_x^2 - i(4\pi\omega/c^2)\sigma_T(|k_x|,\omega)}. \tag{5.15}$$

The anomalous skin effect is usually measured through the surface impedance, which is calculated from (5.15) as

$$\begin{aligned} z &\equiv (E_y/B_z)_{x=+0} \\ &= -i(\omega/\pi c)\int_{-\infty}^{\infty} dk_x [k_x^2 - i(4\pi\omega/c^2)\sigma_T(|k_x|,\omega)]^{-1}. \end{aligned} \tag{5.16}$$

As may be clear from the derivation above, the applicability of Eqs. (5.15) and (5.16) is not restricted to cases of classical plasmas. With the knowledge of the transverse conductivity, we can use these formulas for any conductive medium.

For the classical electron gas, the conductivity is given by (5.7). Assuming (5.8), we may approximate the exponential function in (5.7) to be unity. Substitution of this approximate expression into Eq. (5.16) then yields

$$z = \frac{2\omega\delta}{3c}\left(\frac{1}{\sqrt{3}} - i\right) \tag{5.17}$$

where

$$\delta = \left[\left(\frac{2T}{\pi m}\right)^{1/2}\frac{c^2}{\omega_p^2\omega}\right]^{1/3} \tag{5.18}$$

is the anomalous skin depth of the classical plasma (Problem 5.3).

In the light of this specific calculation, let us finally examine the validity of the assumption on (5.8). Within the skin layer, $k \gtrsim 1/\delta$; hence,

$$\frac{\omega}{k(T/m)^{1/2}} \lesssim \frac{\omega\delta}{(T/m)^{1/2}} = \left[\left(\frac{2}{\pi}\right)^{1/2}\frac{\omega^2c^2}{\omega_p^2(T/m)}\right]^{1/3}. \tag{5.19}$$

The finiteness of a laboratory plasma places a lower bound on the frequencies usable in such a skin-effect experiment. For those parameters available to classical plasmas, it then appears quite difficult to make the right-hand side of (5.19) less than unity.

If, however, we move over to the cases of degenerate plasmas in solids, the situation changes significantly. Here, we may approximately replace T by the Fermi energy in expressions such as (5.18) and (5.19). In addition, the plasma frequency increases substantially because of high concentration and difference in the effective mass. Experimentally, therefore, the anomalous skin effect is more easily observed in a solid state plasma than in a gaseous plasma.

5.2 DIELECTRIC TENSOR WITH AN EXTERNAL MAGNETIC FIELD

We proceed to consider the cases of a plasma with a uniform stationary magnetic field applied to it. The system is no longer isotropic; for the analysis of electromagnetic wave propagation in such a system, all the elements of the dielectric tensor must be calculated.

The velocity distribution function under present consideration is the Maxwellian, (4.66). The explicit expression for the dielectric tensor may be obtained by substituting this distribution function into Eq. (3.71). The intermediate steps of the calculation are left to the reader as an exercise (Problem 5.4); the result for a single-component plasma is

$$\epsilon(\mathbf{k},\omega)=1-\frac{\omega_{\mathrm{p}}^2}{\omega^2}\left\{\sum_{n=-\infty}^{\infty}\frac{Z_0}{Z_n}\Pi(\beta,Z_n;n)[1-W(Z_n)]-Z_0^2\hat{\mathbf{z}}\hat{\mathbf{z}}\right\} \qquad (5.20)$$

where

$$\Pi(\beta,Z_n;n)=\begin{bmatrix}\dfrac{n^2}{\beta}\Lambda_n(\beta) & in\Lambda_n'(\beta) & \dfrac{k_\parallel}{|k_\parallel|}\dfrac{n}{\sqrt{\beta}}Z_n\Lambda_n(\beta)\\ -in\Lambda_n'(\beta) & \dfrac{n^2}{\beta}\Lambda_n(\beta)-2\beta\Lambda_n'(\beta) & -i\dfrac{k_\parallel}{|k_\parallel|}\sqrt{\beta}\,Z_n\Lambda_n'(\beta)\\ \dfrac{k_\parallel}{|k_\parallel|}\dfrac{n}{\sqrt{\beta}}Z_n\Lambda_n(\beta) & i\dfrac{k_\parallel}{|k_\parallel|}\sqrt{\beta}\,Z_n\Lambda_n'(\beta) & Z_n^2\Lambda_n(\beta)\end{bmatrix}, \qquad (5.21)$$

$$Z_n=\frac{\omega-n\Omega}{|k_\parallel|(T/m)^{1/2}}. \qquad (5.22)$$

Equations (4.2), (4.70), and (4.69) define the functions W and Λ_n and the variable β, respectively; the configuration as specified in Fig. 3.1 has been used for the vectors $\mathbf{B}$ and $\mathbf{k}$. Equation (5.20) is extended to the case of a multicomponent plasma in a standard way.

The propagation of the electromagnetic waves may be investigated through the dispersion relation (3.24), or

$$\det\left|\epsilon(\mathbf{k},\omega)-\left(\frac{kc}{\omega}\right)^2\left(\mathbf{I}-\frac{\mathbf{kk}}{k^2}\right)\right|=0. \tag{5.23}$$

In the following sections, we shall discuss the consequences of this dispersion relation in two special cases: when **k** is parallel to **B**; and when **k** is perpendicular to **B**. A treatment involving more general directions of propagation is described in Problem 5.5.

5.3 WAVE PROPAGATION PARALLEL TO THE MAGNETIC FIELD

In this case, $k_\perp=0$ and $|k_\parallel|=k$; the tensor $\pi(\beta,Z_n;n)$ given by Eq. (5.21) now takes a simplified form in the limit of $\beta\to0$ as

$$\pi(0,Z_n;n)=\begin{bmatrix} \frac{1}{2}[\delta(n,1)+\delta(n,-1)] & \frac{i}{2}[\delta(n,1)-\delta(n,-1)] & 0 \\ -\frac{i}{2}[\delta(n,1)-\delta(n,-1)] & \frac{1}{2}[\delta(n,1)+\delta(n,-1)] & 0 \\ 0 & 0 & Z_0^2\delta(n,0) \end{bmatrix} \tag{5.24}$$

where $\delta(n,n')$ is Kronecker's delta, defined as

$$\delta(n,n')=\begin{cases} 1 & (n=n'), \\ 0 & (n\neq n'). \end{cases} \tag{5.25}$$

A. Transformation to Circularly Polarized Waves

Thus far we have been describing a vectorial quantity such as the electric field in terms of its Cartesian components; this representation is most convenient when the linearly polarized waves are eigenmodes of propagation in the system. In the present case, however, the circularly polarized waves turn out to be the proper solution to the dispersion relation. It is therefore useful at this time to introduce a scheme of transformation from linearly polarized waves to circularly polarized waves.

The transformation may be achieved with the aid of a unitary matrix

$$\mathbf{U}=\begin{pmatrix} 1/\sqrt{2} & -i/\sqrt{2} & 0 \\ -i/\sqrt{2} & 1/\sqrt{2} & 0 \\ 0 & 0 & 1 \end{pmatrix}. \tag{5.26}$$

It satisfies the condition for unitarity

$$\mathbf{U}\cdot\mathbf{U}^{+}=\mathbf{U}^{+}\cdot\mathbf{U}=\mathbf{I}$$

where $\mathbf{U}^{+}$ is the Hermitian conjugate matrix to $\mathbf{U}$ (compare Problem 3.1). An electric field vector expressed in terms of its Cartesian components is transformed as

$$\begin{pmatrix} E_{\mathrm{r}} \\ E_{\mathrm{l}} \\ E_z \end{pmatrix}=\mathbf{U}\cdot\begin{pmatrix} E_x \\ E_y \\ E_z \end{pmatrix}$$

or

$$E_{\mathrm{r}}=(E_x-iE_y)/\sqrt{2}\,, \tag{5.27a}$$

$$E_{\mathrm{l}}=-i(E_x+iE_y)/\sqrt{2}\,. \tag{5.27b}$$

The new components E_{r} and E_{l} given by Eqs. (5.27) clearly represent the right and left circularly polarized components, respectively.†

A tensorial quantity, such as the dielectric tensor (5.20), is transformed as $\mathbf{U}\cdot\boldsymbol{\epsilon}\cdot\mathbf{U}^{+}$. With the aid of (5.24), we find that the dielectric tensor after such a transformation is now expressed in a diagonal form

$$\mathbf{U}\cdot\boldsymbol{\epsilon}\cdot\mathbf{U}^{+}=\begin{pmatrix} \epsilon_{\mathrm{r}} & 0 & 0 \\ 0 & \epsilon_{\mathrm{l}} & 0 \\ 0 & 0 & \epsilon \end{pmatrix}.$$

The diagonal elements for a multicomponent plasma may thus be calculated as

$$\epsilon_{\mathrm{r}}(k,\omega)=1-\sum_{\sigma}\frac{\omega_{\sigma}^{2}}{\omega(\omega+\Omega_{\sigma})}\left[1-W\left(\frac{\omega+\Omega_{\sigma}}{k(T_{\sigma}/m_{\sigma})^{1/2}}\right)\right], \tag{5.28a}$$

$$\epsilon_{\mathrm{l}}(k,\omega)=1-\sum_{\sigma}\frac{\omega_{\sigma}^{2}}{\omega(\omega-\Omega_{\sigma})}\left[1-W\left(\frac{\omega-\Omega_{\sigma}}{k(T_{\sigma}/m_{\sigma})^{1/2}}\right)\right], \tag{5.28b}$$

$$\epsilon(k,\omega)=1+\sum_{\sigma}\frac{k_{\sigma}^{2}}{k^{2}}W\left(\frac{\omega}{k(T_{\sigma}/m_{\sigma})^{1/2}}\right). \tag{5.28c}$$

Equation (5.28c) is the longitudinal dielectric function with $k_{\perp}=0$, which was investigated in the previous chapter.

† This statement is true only when $\omega>0$; we shall confine ourselves to this frequency domain throughout the remainder of this chapter.

The dispersion tensor (3.21) is also diagonalized after the unitary transformation; the dispersion relation for the right (or left) circularly polarized wave is thus obtained as

$$(kc/\omega)^2 = \epsilon_{r(l)}(k,\omega). \tag{5.29}$$

We now examine these dispersion relations for specific cases.

B. Electron Gas—Helicon

We first take up the case of an electron gas; recalling the definition (3.53) of Ω_σ, which includes the sign of the charge, we have for the right circularly polarized wave

$$\epsilon_r(k,\omega) = 1 - \frac{\omega_e^2}{\omega(\omega - |\Omega_e|)}\left[1 - W\left(\frac{\omega - |\Omega_e|}{k(T_e/m_e)^{1/2}}\right)\right]. \tag{5.30}$$

The imaginary part of this dielectric function will be negligible if

$$||\Omega_e| - \omega| \gg k(T_e/m_e)^{1/2}. \tag{5.31}$$

With this assumption, the dispersion relation (5.29) for the right circularly polarized wave becomes

$$\left(\frac{kc}{\omega}\right)^2 = 1 - \frac{\omega_e^2}{\omega(\omega - |\Omega_e|)}. \tag{5.32}$$

This relationship is illustrated by the solid lines in Fig. 5.3.

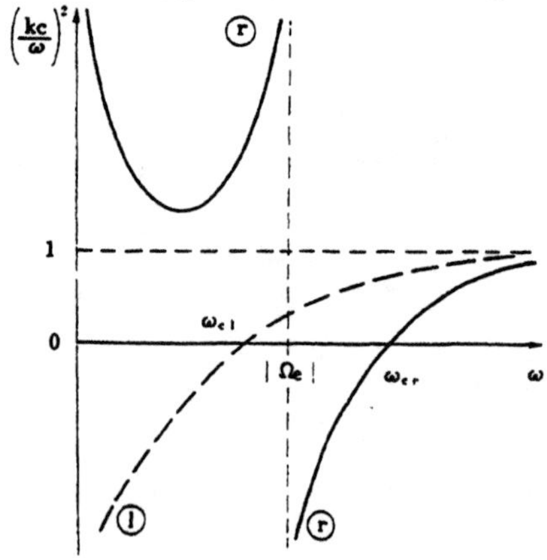

Fig. 5.3 Right and left circularly polarized waves in the electron gas.

The cutoff frequency ω_{cr} is calculated from (5.32) as

$$\omega_{cr}=\tfrac{1}{2}\left[\left(\Omega_e^2+4\omega_e^2\right)^{1/2}+|\Omega_e|\right]. \tag{5.33}$$

The resonance occurs at the frequency

$$\omega_R=|\Omega_e|. \tag{5.34}$$

In the magnetic field, the electrons gyrate in the same direction as the right circularly polarized wave; a resonance is expected when the wave frequency matches the cyclotron frequency.

When the thermal velocities of the electrons are finite, however, condition (5.31) may be violated in the vicinity of the resonance frequency (5.34). Thus, we expect that in the domain

$$|\Omega_e|-k(T_e/m_e)^{1/2}<\omega<|\Omega_e|+k(T_e/m_e)^{1/2} \tag{5.35}$$

the imaginary part of Eq. (5.30) and therefore the damping of the wave will be substantial. The physical origin of this attenuation can be easily understood: When a wave with frequency ω and wave number k propagates along the magnetic field in the plasma, an electron with velocity $v_\parallel$ along the magnetic field experiences, because of the usual Doppler effect, a shifted frequency

$$\omega'=\omega\pm kv_\parallel.$$

If ω' coincides with the electron cyclotron frequency, then the resonant absorption of the wave energy by the electron will take place; Eq. (5.35) represents the condition for availability of many such electrons. This phenomenon, called the *Doppler-shifted cyclotron resonance*, is therefore closely related to the Landau damping of the plasma oscillations. It has been investigated for both solids [9, 10] and plasmas [11]; the use of this Doppler-shifted absorption for a point-by-point mapping of the Fermi surface has been considered by Stern [12].

A most remarkable feature in the dispersion relation (5.32) is the emergence of a new propagation mode at low frequencies. In the low-frequency domain, we may neglect the first term, unity, on the right-hand side of Eq. (5.32); the solution then is

$$\omega=\frac{|\Omega_e|(kc)^2}{\omega_e^2+(kc)^2}. \tag{5.36}$$

In the limit of long wavelengths, this relation becomes

$$\omega = \frac{|\Omega_e| c^2}{\omega_e^2} k^2. \tag{5.37}$$

For short wavelengths, Eq. (5.36) approaches $|\Omega_e|$. Figure 5.4 illustrates such a behavior of Eq. (5.36). In the vicinity of $|\Omega_e|$, the Doppler-shifted cyclotron resonance becomes important; the domain of strong absorption (5.35) is also described in Fig. 5.4 by the shaded area. Such a possibility of electromagnetic wave propagation along the magnetic field in a plasma, at a frequency much lower than both the plasma frequency and the cyclotron frequency, was first predicted by Aigrain [13] and by Konstantinov and Perel' [14]; the word "helicon" was coined by Aigrain to describe the propagation mode (5.37).

The appearance of the helicon mode in the low-frequency limit is in fact a unique feature of an uncompensated plasma like the electron gas. Such a situation occurs frequently in solids, in which the oppositely charged mobile carriers may not exactly compensate each other. For this reason, the helicon has been investigated extensively in solid state plasmas [5, 15, 16].

The left circularly polarized mode may be treated similarly. In place of Eq. (5.32), we now have

$$\left(\frac{kc}{\omega}\right)^2 = 1 - \frac{\omega_e^2}{\omega(\omega + |\Omega_e|)}. \tag{5.38}$$

This relationship is depicted by the dashed line in Fig. 5.3. The cutoff occurs at the frequency

$$\omega_{cl} = \tfrac{1}{2}\left[(\Omega_e^2 + 4\omega_e^2)^{1/2} - |\Omega_e|\right]. \tag{5.39}$$

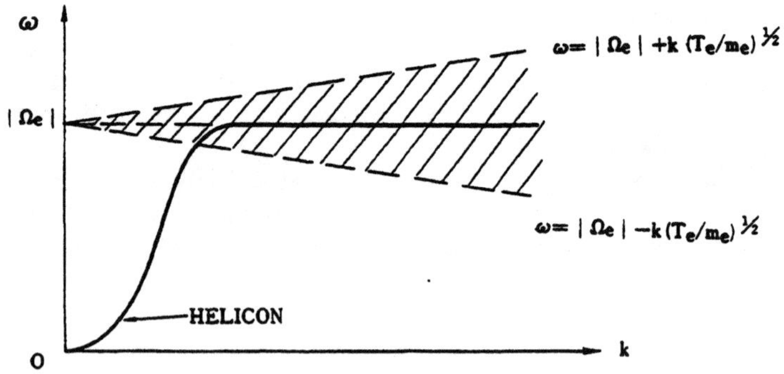

Fig. 5.4 Low-frequency mode in electron plasma.

No resonance takes place for a left circularly polarized wave in an electron plasma.

As Fig. 5.3 clearly illustrates, a substantial difference generally exists in propagation characteristics between the right and left circularly polarized waves in plasmas with a magnetic field. When a linearly polarized wave is injected toward a plasma along the magnetic field, its right and left circularly polarized components thereby propagate differently in the plasma: In those cases in which the propagation velocities are different, the plane of linear polarization will rotate after a passage of a finite distance in the plasma, a phenomenon called *Faraday rotation*. In some other cases, only one of the two circularly polarized components may be able to propagate in the plasma; in these circumstances, we have a converter of a linearly polarized wave into a circularly polarized wave through a plasma.

C. Compensated Two-Component Plasma—Alfvén Wave and Whistler

Let us next consider a two-component plasma consisting of electrons and singly charged ions; their average densities are assumed to be equal.

The dielectric function for the right circularly polarized wave is

$$\epsilon_r(k,\omega)=1-\frac{\omega_e^2}{\omega(\omega-|\Omega_e|)}\left[1-W\left(\frac{\omega-|\Omega_e|}{k(T_e/m_e)^{1/2}}\right)\right]$$

$$-\frac{\omega_i^2}{\omega(\omega+\Omega_i)}\left[1-W\left(\frac{\omega+\Omega_i}{k(T_i/m_i)^{1/2}}\right)\right]. \tag{5.40}$$

Assuming the conditions for little absorption,

$$||\Omega_e|-\omega|\gg k(T_e/m_e)^{1/2}, \qquad \Omega_i+\omega\gg k(T_i/m_i)^{1/2}, \tag{5.41}$$

we find the dispersion relation for the right circularly polarized wave

$$\left(\frac{kc}{\omega}\right)^2=1-\frac{\omega_e^2}{\omega(\omega-|\Omega_e|)}-\frac{\omega_i^2}{\omega(\omega+\Omega_i)}. \tag{5.42}$$

The behavior of this function is illustrated by solid lines in Fig. 5.5. The cutoff frequency is located at

$$\omega_{cr}=\tfrac{1}{2}\left\{\left[(|\Omega_e|+\Omega_i)^2+4(\omega_e^2+\omega_i^2)\right]^{1/2}+|\Omega_e|-\Omega_i\right\} \tag{5.43}$$

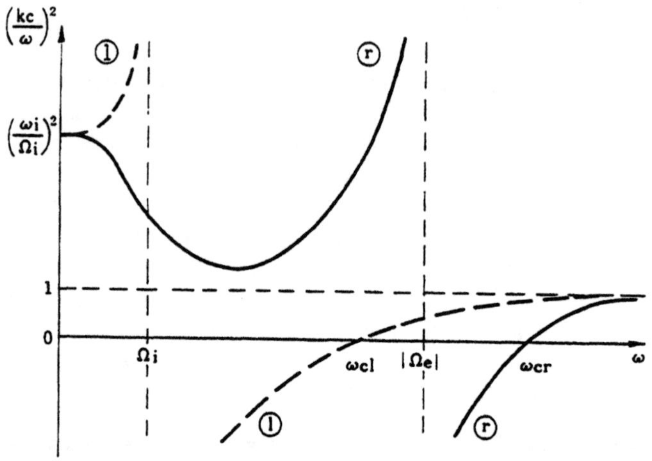

Fig. 5.5 Right and left circularly polarized waves in a compensated two-component plasma.

and the resonance frequency is again given by

$$\omega_R = |\Omega_e|. \tag{5.44}$$

In the vicinity of this resonance, the first condition of (5.41) becomes violated; strong absorption ensues due to the Doppler-shifted cyclotron resonances of the electrons. The second condition of (5.41), on the other hand, is trivially satisfied in ordinary circumstances.

The feature to be remarked here is again the appearance of a low-frequency propagation mode. Neglecting the first term on the right-hand side of Eq. (5.42) and with the aid of the mass relationship (4.45), we find in the low-frequency domain

$$\omega = \frac{|\Omega_e|(kc)^2}{\omega_e^2 + (kc)^2}\left[\frac{1}{2} + \left(\frac{\omega_i^2}{(kc)^2} + \frac{1}{4}\right)^{1/2}\right]. \tag{5.45}$$

In the long-wavelength limit such that

$$(kc)^2 \ll 4\omega_i^2 \ll \omega_e^2, \tag{5.46}$$

Eq. (5.45) reduces to

$$\omega = (\Omega_i/\omega_i)kc, \tag{5.47}$$

where we use an equality, $|\Omega_e|\omega_i^2=\Omega_i\omega_e^2$, for the compensated plasma. The propagation mode described by Eq. (5.47) is the *Alfvén wave* [17]; its propagation velocity, denoted by c_A, is called the Alfvén velocity. With the aid of (2.46) and (3.53), we find from (5.47)

$$c_A=\frac{B}{(4\pi nm_i)^{1/2}}=\frac{B}{(4\pi\rho_m)^{1/2}} \tag{5.48}$$

where $\rho_m=nm_i$ is the mass density of the plasma. The Alfvén wave has been investigated in various substances, including gaseous-discharge plasmas [18, 19], liquid metals [20, 21], and solids [5, 16, 22].

It is instructive at this moment to examine the numerical values involved about an effective dielectric constant ϵ_A for the Alfvén wave propagation. As may be clear from a comparison between Eq. (5.47) and a formula such as Eq. (3.31) or Eq. (5.29), such a dielectric constant in this case may be determined as

$$\epsilon_A=\omega_i^2/\Omega_i^2. \tag{5.49}$$

For a hydrogen plasma, we compute

$$\epsilon_A=1.9\times10^{-2}(n/B^2) \tag{5.50}$$

where n and B are measured in units of cm^{-3} and gauss. When $n=10^{12}$ and $B=10^4$, we have $\epsilon_A=2\times10^2$; hence, it may be reasonable to assume $\epsilon_A\gg1$ even with a substantial strength of the magnetic field.

In the intermediate-wavelength domain such that

$$4\omega_i^2\ll(kc)^2\ll\omega_e^2, \tag{5.51}$$

we find that Eq. (5.45) becomes

$$\omega=\frac{|\Omega_e|c^2}{\omega_e^2}k^2. \tag{5.52}$$

Note that this dispersion relation is identical to Eq. (5.37) for helicon; the only difference appears in the domains of wave numbers in which they are defined. The propagation mode described in Eq. (5.52) is called a *whistler*. It is particularly famous in connection with the ionospheric radio noises with decreasing pitches which originate from disturbances created by lightning in the polar areas of the earth.

The physical origin of the emergence of a heliconlike mode in a compensated plasma is easy to understand: In the wave number domain as given by

(5.51), the corresponding frequencies are too high for the ions to be able to follow dynamically; the ions therefore act as if they are immobile space charges. On the other hand, for the electrons, the frequencies involved are still low, so that a low-frequency analysis is applicable. Hence, the combined result should resemble the low-frequency mode obtained in an uncompensated electron plasma.

Overall behavior of the low-frequency right circularly polarized mode (5.45) together with the domain of strong absorption due to the Doppler-shifted electron cyclotron resonances is shown in Fig. 5.6.

The propagation characteristics of the left circularly polarized wave may be analyzed quite similarly. Assuming

$$|\Omega_i-\omega|\gg k(T_i/m_i)^{1/2}, \qquad |\Omega_e|+\omega\gg k(T_e/m_e)^{1/2}, \tag{5.53}$$

we obtain the dispersion relation

$$\left(\frac{kc}{\omega}\right)^2=1-\frac{\omega_e^2}{\omega(\omega+|\Omega_e|)}-\frac{\omega_i^2}{\omega(\omega-\Omega_i)}. \tag{5.54}$$

This relationship is illustrated by the dashed lines in Fig. 5.5. The cutoff frequency is calculated to be

$$\omega_{cl}=\tfrac{1}{2}\left\{\left[(|\Omega_e|+\Omega_i)^2+4(\omega_e^2+\omega_i^2)\right]^{1/2}-|\Omega_e|+\Omega_i\right\} \tag{5.55}$$

and the resonance frequency is

$$\omega_L=\Omega_i. \tag{5.56}$$

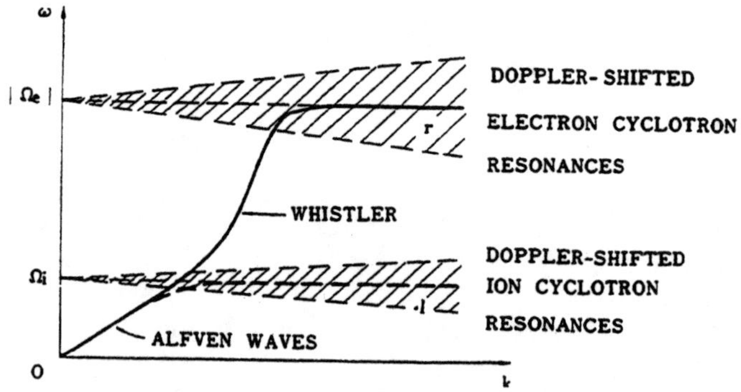

Fig. 5.6 Low-frequency modes in a compensated electron–ion plasma.

In the vicinity of this resonance, the first condition of (5.53) becomes violated; the Doppler-shifted cyclotron resonances of the ions now cause strong attenuation to the wave.

In the low-frequency domain, Eq. (5.54) becomes

$$\omega=\frac{|\Omega_e|(kc)^2}{\omega_e^2+(kc)^2}\left\{-\frac{1}{2}+\left[\frac{\omega_i^2}{(kc)^2}+\frac{1}{4}\right]^{1/2}\right\}. \tag{5.57}$$

This dispersion relation reduces to that of the Alfvén wave, Eq. (5.47), in the limit of long wavelengths $(kc)^2 \ll 4\omega_i^2$; the Alfvén wave can, therefore, propagate as either a right or a left circularly polarized wave, and hence also as a linearly polarized wave, in a compensated plasma with a magnetic field. In the other wavelength domain, $4\omega_i^2 \ll (kc)^2$, Eq. (5.57) approaches Ω_i. The behavior of Eq. (5.57) is shown in Fig. 5.6 together with the domain of strong absorption due to the Doppler-shifted ion cyclotron resonances.

5.4 WAVE PROPAGATION PERPENDICULAR TO THE MAGNETIC FIELD

In this case, $k_\parallel=0$ and $k_\perp=k$; the wave propagates in the x direction as we have been adopting the configuration shown in Fig. 3.1. The dielectric tensor (5.20) may now be written as

$$\epsilon(k,\omega)=\begin{pmatrix}\epsilon_1 & -i\epsilon_x & 0\\ i\epsilon_x & \epsilon_2 & 0\\ 0 & 0 & \epsilon_3\end{pmatrix} \tag{5.58}$$

where for a single-component plasma

$$\epsilon_1(k,\omega)=1-\frac{k_D^2}{k^2}\sum_n\frac{(n\Omega)^2}{\omega(\omega-n\Omega)}\Lambda_n(\beta), \tag{5.59}$$

$$\epsilon_2(k,\omega)=1-\frac{k_D^2}{k^2}\sum_n\frac{(n\Omega)^2}{\omega(\omega-n\Omega)}\left[\Lambda_n(\beta)-\frac{2\beta^2}{n^2}\Lambda_n'(\beta)\right], \tag{5.60}$$

$$\epsilon_3(k,\omega)=1-\frac{\omega_p^2}{\omega^2}\sum_n\frac{\omega}{\omega-n\Omega}\Lambda_n(\beta), \tag{5.61}$$

$$\epsilon_x(k,\omega)=\frac{\omega_p^2}{\omega^2}\sum_n\frac{\omega}{\omega-n\Omega}n\Lambda_n'(\beta), \tag{5.62}$$

and

$$\beta = k^2T/m\Omega^2.$$

The dispersion relation (5.23) splits itself into two equations,

$$\left(\frac{kc}{\omega}\right)^2 = \frac{\epsilon_1\epsilon_2 - \epsilon_x^2}{\epsilon_1}, \tag{5.63}$$

$$\left(\frac{kc}{\omega}\right)^2 = \epsilon_3. \tag{5.64}$$

The former case, Eq. (5.63), is called the *extraordinary wave*. From an examination of (3.23), we find that $E_z = 0$, and that the remaining components satisfy

$$E_x/E_y = i(\epsilon_x/\epsilon_1). \tag{5.65}$$

Since E_x corresponds to the longitudinal component of the electric field while E_y is a transverse one, the extraordinary wave represents a *hybrid mode*. The cutoff frequencies are determined from

$$\epsilon_1\epsilon_2 = \epsilon_x^2. \tag{5.66}$$

The *hybrid resonances* take place at

$$\epsilon_1 = 0. \tag{5.67}$$

Under these resonance conditions, we find from Eq. (5.65) that E_y also vanishes; the extraordinary wave turns into a longitudinal mode at a hybrid resonance. In fact, this longitudinal wave may be identified as a Bernstein mode considered in Section 4.6; we can show with the aid of the identity

$$\sum_n \Lambda_n(\beta) = 1 \tag{5.68}$$

that the dielectric functions (4.68) and (5.59) are identical when $k_\parallel = 0$.

The latter mode, Eq. (5.64), is called the *ordinary wave*. In this case, $E_z \neq 0$; the electric field of the wave is parallel to the external magnetic field. The ordinary wave is obviously a transverse wave.

A. Waves in Cold Plasmas

The elements of the dielectric tensor, Eqs. (5.59)–(5.62), exhibit strong spatial dispersion arising from the cyclotron motion of charged particles with finite Larmor radii. If, however, the wavelength under consideration is much longer than the average Larmor radius of the particles, we may approximately treat the problem by neglecting the effects of thermal motion and thus by going over to the limit $T \to 0$. In such a cold plasma limit, the spatial dispersion is suppressed; Eqs. (5.59)–(5.62) reduce to substantially simplified expressions

$$\epsilon_1 = \epsilon_2 = 1 - \frac{\omega_p^2}{\omega^2 - \Omega^2}, \tag{5.69}$$

$$\epsilon_3 = 1 - \frac{\omega_p^2}{\omega^2}, \tag{5.70}$$

$$\epsilon_x = \frac{\omega_p^2 \Omega}{\omega(\omega^2 - \Omega^2)}. \tag{5.71}$$

It is clear from Eq. (5.70) that the ordinary wave (5.64) is not affected by the presence of the magnetic field in this limit; the results obtained from Eq. (5.3a) are applicable in this case.

The dispersion relation for the extraordinary wave in a compensated electron–ion plasma at $T=0$ is shown in Fig. 5.7. In the limit of low frequencies, we find

$$\epsilon_1 = \epsilon_2 \to 1 + (4\pi\rho_m c^2/B^2), \tag{5.72a}$$

$$\epsilon_x \to 0 \tag{5.72b}$$

where ρ_m is the mass density of the plasma and the condition for charge compensation $\Omega_e \omega_i^2 + \Omega_i \omega_e^2 = 0$ has been used in Eq. (5.72b). The extraordinary wave thus starts in the low-frequency domain as an Alfvén wave. The first

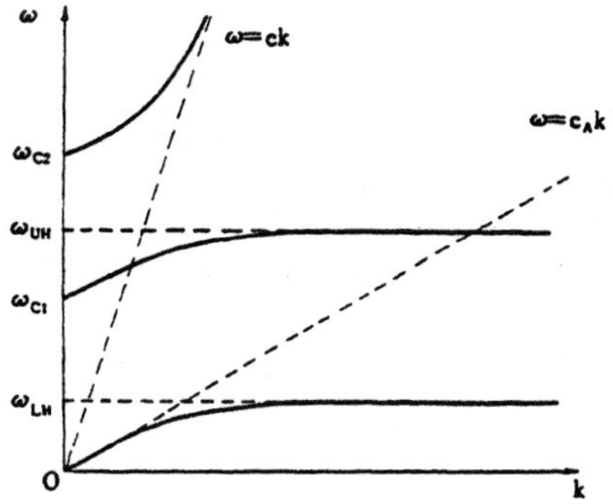

Fig. 5.7 Extraordinary wave in a cold electron–ion plasma.

resonance at ω_{LH} is usually called the *lower hybrid resonance*; the frequency is given by

$$\omega_{LH}^2 = |\Omega_e \Omega_i| \left[\frac{\omega_e^2 + |\Omega_e \Omega_i|}{\omega_e^2 + \Omega_e^2} \right]. \tag{5.73}$$

The first cutoff frequency ω_{c1} appears at

$$\omega_{c1} = \tfrac{1}{2}\left[(\Omega_e^2 + 4\omega_e^2)^{1/2} - |\Omega_e| \right] \tag{5.74}$$

and then the *upper hybrid resonance* takes place at

$$\omega_{UH} = (\Omega_e^2 + \omega_e^2)^{1/2}. \tag{5.75}$$

Finally, the second cutoff frequency ω_{c2} appears at

$$\omega_{c2} = \tfrac{1}{2}\left[(\Omega_e^2 + 4\omega_e^2)^{1/2} + |\Omega_e| \right] \tag{5.76}$$

and this branch approaches the free propagation mode $\omega = ck$.

B. Nonlocal Effects

For a plasma at a finite temperature, nonlocal effects brought about by the thermal motion of charged particles sometimes play quite an important role in determining the propagation characteristics of the electromagnetic wave. The finiteness of the Larmor radii of the charged particles makes it possible to transport, to a certain extent, the disturbances created by the wave in the directions perpendicular to the magnetic field. When the average Larmor radius is comparable to the wavelength, such a nonlocal effect may become particularly substantial.

To illustrate the point by a specific example, let us consider the dispersion relation (5.64) for the ordinary wave; in the absence of a nonlocal effect, this wave would not be affected by the magnetic field. With the entire expression (5.61) for the dielectric function, however, we shall find that the propagation characteristics of the ordinary wave also exhibit certain effects of the magnetic field.

For simplicity, we take up a single-component plasma. In the vicinity of $\omega = |\Omega|$, we may approximate Eq. (5.61) as

$$\epsilon_3(k,\omega) \cong 1 - \frac{\omega_p^2}{\omega^2}\Lambda_0(\beta) - \frac{\omega_p^2}{\omega^2 - \Omega^2} 2\Lambda_1(\beta) \qquad (\omega \cong |\Omega|). \tag{5.77}$$

When the magnetic field is so strong that $\Omega^2 \gg \omega_p^2$, Eq. (5.77) may further be simplified as

$$\epsilon_3(k,\omega) \cong 1 - \frac{\omega_p^2}{\omega^2 - \Omega^2} 2\Lambda_1(\beta) \qquad (\omega \cong |\Omega|). \tag{5.78}$$

The dispersion relation (5.64) is now solved in the vicinity of $\omega = |\Omega|$ by setting $\omega = |\Omega|(1+\Delta_1)$; we find

$$\Delta_1 = \frac{\omega_p^2}{\Omega^2 - (kc)^2} \Lambda_1(\beta). \tag{5.79}$$

This solution is valid only when $|\Delta_1| \ll 1$. For small k, $\Delta_1 \cong (\omega_p^2/\Omega^2)(\beta/2)$; for large k, $\Delta_1 \cong -(\omega_p/kc)^2/(2\pi\beta)^{1/2} \sim -k^{-3}$. In either of those limiting cases, therefore, the foregoing condition is satisfied. The condition becomes violated, however, in the vicinity of $kc = |\Omega|$. Thus, we expect the dispersion curves to behave in this case as shown by the solid lines in Fig. 5.8.

When $\Omega^2 \ll \omega_p^2$, instead of Eq. (5.78) we have

$$\epsilon_3(k,\omega) \cong -\frac{\omega_p^2}{\omega^2} \Lambda_0(\beta) - \frac{\omega_p^2}{\omega^2 - \Omega^2} 2\Lambda_1(\beta). \tag{5.80}$$

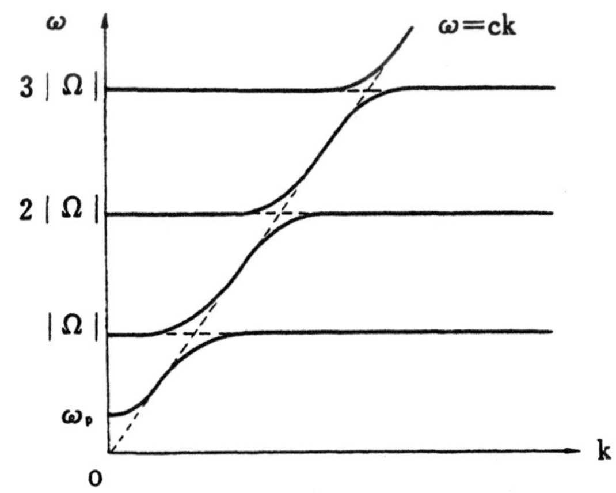

Fig. 5.8 Nonlocal effect in the ordinary wave with $|\Omega| \gg \omega_p$.

Again by setting $\omega=|\Omega|(1+\Delta_1)$, we may solve the dispersion relation (5.64) as

$$\Delta_1=-\frac{\omega_p^2\Lambda_1(\beta)}{(kc)^2+\omega_p^2\Lambda_0(\beta)}. \tag{5.81}$$

The dispersion curves should therefore behave like Fig. 5.9 in these circumstances [5]. For the validity of the treatment above, we must make sure $|\Delta_1|\ll 1$. This condition can no longer be satisfied when the magnetic field becomes so weak that

$$\Omega^2 \lesssim \frac{(T/m)}{c^2}\omega_p^2. \tag{5.82}$$

The calculations described above have illustrated an effect of spatial dispersion in the bulk of the plasma. In a surface layer of a plasma, we discover another important nonlocal effect brought about by the finiteness of the Larmor radii coupled with the anomalous skin effect [23]. Consider a configuration like Fig. 5.2, in which a constant magnetic field is now applied in the z direction, parallel to the surface and perpendicular to the electric-field vector of the wave. The skin depth δ is assumed to be much less than the average Larmor radius $\sim(T/m)^{1/2}/|\Omega|$. In the absence of collisions, an electron that moves in a helical orbit through a number of revolutions will return several times into the layer of thickness δ where the electric field is strong. Thus the motion is similar to that of an electron in a cyclotron with a small accelerating domain; when the frequency of the electromagnetic wave is a multiple of the electron cyclotron frequency, a resonance, called the *Azbel'–Kaner resonance*, will take place.

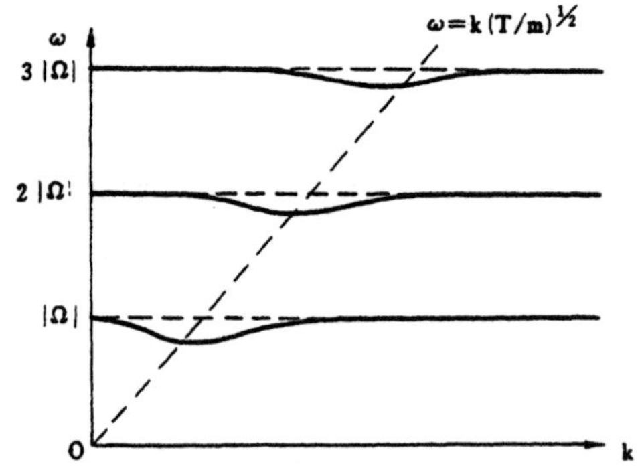

Fig. 5.9 Nonlocal effect in the ordinary wave with $|\Omega|\ll\omega_p$.

References

1. W. P. Allis, S. J. Buchsbaum, and A. Bers, *Waves in Anisotropic Plasmas* (MIT Press, Cambridge, Mass., 1963).
2. V. L. Ginzburg, *The Propagation of Electromagnetic Waves in Plasmas*, 2nd ed. (J. B. Sykes and R. J. Taylor, transls.) (Pergamon, Oxford, 1970).
3. V. D. Shafranov, in *Reviews of Plasma Physics*, (M. A. Leontovich, ed., H. Lashinsky, transl.) Vol. 3, p. 1 (Consultants Bureau, New York, 1967).
4. T. H. Stix, *The Theory of Plasma Waves* (McGraw-Hill, New York, 1962).
5. E. A. Kaner and V. G. Skobov, *Usp. Fiz. Nauk* **89**, 367 (1966) [*Soviet Phys. Usp.* **9**, 480 (1967)]; *Advan. Phys.* **17**, 605 (1968).
6. A. B. Pippard, *Proc. Roy. Soc. (London)* **A191**, 370 and 385 (1947).
7. R. G. Chambers, *Proc. Roy Soc. (London)* **A215**, 481 (1952).
8. J. M. Ziman, *Electrons and Phonons*, Sec. 11.7 (Clarendon Press, Oxford, 1962).
9. R. G. Chambers, *Phil. Mag.* **1**, 459 (1956).
10. P. B. Miller and K. K. Haering, *Phys. Rev.* **128**, 126 (1962).
11. J. E. Drummond, *Phys. Rev.* **110**, 293 (1958); *Phys. Rev.* **112**, 1460 (1958).
12. E. A. Stern, *Phys. Rev. Letters* **10**, 91 (1963).
13. P. Aigrain, in *Proc. Intern. Conf. on Semicond. Phys.*, Prague, 1960, p. 224 (Academic, New York, 1961).
14. O. V. Konstantinov and V. I. Perel', *Zhur. Eksptl. Teoret. Fiz.* **38**, 161 (1960) [*Soviet Phys. JETP* **11**, 117 (1960)].
15. R. Bowers, C. Legendy, and F. Rose, *Phys. Rev. Letters* **7**, 339 (1961).
16. M. Glicksman, in *Solid State Physics* (H. Ehrenreich, F. Seitz, and D. Turnbull, eds.), Vol. 26, p. 275 (Academic, New York, 1971).
17. H. Alfvén, *Nature* **150**, 405 (1942); H. Alfvén and C. G. Fälthammer, *Cosmical Electrodynamics* (Clarendon Press, Oxford, 1963).
18. T. K. Allen, W. R. Baker, R. V. Pyle, and J. M. Wilcox, *Phys. Rev. Letters* **2**, 383 (1959); J. M. Wilcox, F. I. Boley, and A. W. DeSilva, *Phys. Fluids* **3**, 15 (1960).
19. D. F. Jephcott, *Nature* **183**, 1652 (1959).
20. S. Lundquist, *Phys. Rev.* **76**, 1805 (1949).
21. B. Lehnert, *Phys. Rev.* **94**, 815 (1954).
22. S. J. Buchsbaum and J. K. Galt, *Phys. Fluids* **4**, 1514 (1961).
23. M. Ya. Azbel' and E. A. Kaner, *Zhur. Eksptl. Teoret. Fiz.* **30**, 811 (1956) [*Soviet Phys. JETP* **3**, 772 (1956)]; *Zhur. Eksptl. Teoret. Fiz.* **32**, 896 (1956) [*Soviet Phys. JETP* **5**, 730 (1957)]; *Zhur. Eksptl. Teoret. Fiz.* **39**, 80 (1960) [*Soviet Phys. JETP* **12**, 58 (1961)].

Problems

5.1. Carry out the calculation leading to Eq. (5.1).

5.2. Fill in the intermediate steps for deriving Eq. (5.9) from Eq. (3.1).

5.3. Using an approximate expression

$$\sigma_T(k,\omega) = \frac{1}{4\sqrt{2\pi}}\frac{\omega_p k_D}{k},$$

calculate the surface impedance from Eq. (5.16).

5.4. Carry out the calculations leading to Eq. (5.20). In the derivation, it may be helpful to note

$$\sum_{n=-\infty}^{\infty} \pi(v_\perp, v_\parallel; n) = \begin{bmatrix} v_\perp^2/2 & 0 & 0 \\ 0 & v_\perp^2/2 & 0 \\ 0 & 0 & v_\parallel^2 \end{bmatrix},$$

which is derived with the aid of the formulas in Problem 3.7; and

$$\int_0^\infty dx\, x^2 J_n(x) J_n'(x) \exp\left(-\frac{x^2}{2\beta}\right) = \beta^2 \Lambda_n'(\beta),$$

$$\int_0^\infty dx\, x^3 [J_n'(x)]^2 \exp\left(-\frac{x^2}{2\beta}\right) = n^2 \beta \Lambda_n(\beta) - 2\beta^3 \Lambda_n'(\beta),$$

which obtain from differentiations of Eq. (4.67) with respect to p, q, or both.

5.5. When the dielectric tensor is expressed as

$$\epsilon(\mathbf{k}, \omega) = \begin{bmatrix} \epsilon_1 & -i\epsilon_2 & 0 \\ i\epsilon_2 & \epsilon_1 & 0 \\ 0 & 0 & \epsilon_3 \end{bmatrix}$$

in the Cartesian coordinate system, show that the dispersion relation Eq. (5.23) becomes

$$[\epsilon_1 \sin^2\theta + \epsilon_3 \cos^2\theta](kc/\omega)^4 - [(\epsilon_1+\epsilon_2)(\epsilon_1-\epsilon_2)\sin^2\theta + \epsilon_1\epsilon_3(1+\cos^2\theta)](kc/\omega)^2$$
$$+ (\epsilon_1+\epsilon_2)(\epsilon_1-\epsilon_2)\epsilon_3 = 0,$$

where θ is the angle between $\mathbf{k}$ and the z axis. This equation may then be solved for $(kc/\omega)^2$ as

$$\left(\frac{kc}{\omega}\right)^2 = \frac{\begin{array}{l}[(\epsilon_1+\epsilon_2)(\epsilon_1-\epsilon_2)\sin^2\theta + \epsilon_1\epsilon_3(1+\cos^2\theta)] \\ \qquad \pm\{[(\epsilon_1+\epsilon_2)(\epsilon_1-\epsilon_2) - \epsilon_1\epsilon_3]^2 \sin^4\theta + 4\epsilon_2^2\epsilon_3^2\cos^2\theta\}^{1/2}\end{array}}{2[\epsilon_1 \sin^2\theta + \epsilon_3\cos^2\theta]},$$

or for θ as

$$\tan^2\theta = -\frac{\epsilon_3\left[(kc/\omega)^2-(\epsilon_1+\epsilon_2)\right]\left[(kc/\omega)^2-(\epsilon_1-\epsilon_2)\right]}{\left[(kc/\omega)^2\epsilon_1-(\epsilon_1+\epsilon_2)(\epsilon_1-\epsilon_2)\right]\left[(kc/\omega)^2-\epsilon_3\right]}.$$

5.6. Consider the effects of particle collisions on the low-frequency propagation modes of the electromagnetic wave in plasmas along the magnetic field, using the expression for the dielectric tensor obtained in Problem 3.5; we may assume $T_\sigma=0$ in that expression by neglecting the thermal effects.

(a) Show that the helicon mode now suffers attenuation due to the collisions as described by

$$\omega = \frac{|\Omega_e|c^2}{\omega_e^2}k^2\left(1-\frac{i}{\tau_e|\Omega_e|}\right).$$

(b) Show that for the Alfvén wave the dispersion relation becomes

$$\omega = c_A k - \frac{i}{2(m_e+m_i)}\left(\frac{m_e}{\tau_e}+\frac{m_i}{\tau_i}\right).$$

5.7. Derive Eqs. (5.73)–(5.76).

5.8. Consider propagation of an electromagnetic wave in a compensated plasma consisting of electrons and singly charged ions; the effects of the thermal motion and of collisions may be neglected. The frequency of the wave is fixed at a value twice the plasma frequency [i.e., $\omega=2(\omega_e^2+\omega_i^2)^{1/2}$]. Assuming (4.45), we take account of only those terms up to the first order in (m_e/m_i). The direction of the magnetic field is either parallel or perpendicular to the propagation direction of the electromagnetic wave. Only the strength B of the magnetic field is gradually increased from $B=0$ to a value that satisfies $(eB/m_i c\omega)^2 \gg 1$. Discuss the major changes in the propagation characteristics that may take place during these processes.

CHAPTER 6

TRANSIENT PROCESSES

In the preceding two chapters, we studied the fundamental properties of plasmas in or near thermodynamic equilibrium. The subjects treated therein were mostly concerned with description of various plasma properties in a stationary state. Another class of problems, related rather closely to certain experimental aspects of plasma physics, are those involving transient processes in plasmas. It is the purpose of the present chapter to investigate some of the problems in this category, based on the Vlasov description of plasmas.

We begin with a treatment of the problems associated with temporal and spatial propagation of small-amplitude waves in plasmas. We shall consider a situation in which a spatial distribution of potential field is applied to a plasma impulsively at time $t=0$; the potential field will evolve subsequently in the plasma. The evolution is described in terms of the dielectric response function; it involves temporal decay due to the Landau damping. Instead of such a time-dependent situation, we can alternatively formulate the problem in such a way as to single out a spatial evolution of potential disturbance in the plasma. Thus, we shall consider a case in which a time-dependent potential disturbance is applied to the plasma at a plane $x=0$; the potential field will then propagate in space and damp away. A physically important question associated with these phenomena is the relationship between such damping processes and the reversibility of the Vlasov equation. The plasma-wave echo, which demonstrates most clearly the reversibility of microscopic phase evolutions in such Landau damping processes, will be considered in these connections. Finally, we shall treat the problems associated with the behavior of a large-amplitude wave in the plasma. Here, trapping of those particles traveling with almost the same velocities as the phase velocity of the wave plays an essential part; the plasma wave thereby exhibits an amplitude oscillation in its decay processes.

6.1 PROPAGATIONS OF SMALL-AMPLITUDE PLASMA WAVES

Consider the application of a weak external potential field

$$\varphi_{\text{ext}}(\mathbf{r},t)=\sum_{\mathbf{k}}\int\frac{d\omega}{2\pi}\varphi_{\text{ext}}(\mathbf{k},\omega)\exp[i(\mathbf{k}\cdot\mathbf{r}-\omega t)+\eta t]\qquad(\eta\to+0)\quad(6.1)$$

to a single-component plasma. Following the linear response treatments as described in Section 3.2A or 4.3, we find that the Fourier components $\varphi_{\text{ind}}(\mathbf{k},\omega)$ of the induced potential field are given by

$$\varphi_{\text{ind}}(\mathbf{k},\omega) = -[1-(1/\epsilon(\mathbf{k},\omega))]\varphi_{\text{ext}}(\mathbf{k},\omega). \tag{6.2}$$

The distribution function $f(\mathbf{r},\mathbf{v};t)$ deviates from its equilibrium value $f(\mathbf{v})$ so that

$$\begin{aligned} f(\mathbf{r},\mathbf{v};t) &= f(\mathbf{v}) + \delta f(\mathbf{r},\mathbf{v};t) \\ &= f(\mathbf{v}) + \sum_{\mathbf{k}} \int \frac{d\omega}{2\pi} \delta f_{\mathbf{k}\omega}(\mathbf{v}) \exp[i(\mathbf{k}\cdot\mathbf{r}-\omega t)+\eta t] \qquad (\eta \to +0). \end{aligned} \tag{6.3}$$

The Fourier components $\delta f_{\mathbf{k}\omega}(\mathbf{v})$ may be determined from a calculation similar to Eq. (3.46) as

$$\delta f_{\mathbf{k}\omega}(\mathbf{v}) = -\frac{q}{m} \frac{\varphi_{\text{ext}}(\mathbf{k},\omega)}{(\omega-\mathbf{k}\cdot\mathbf{v}+i\eta)\epsilon(\mathbf{k},\omega)} \mathbf{k}\cdot\frac{\partial f(\mathbf{v})}{\partial \mathbf{v}}. \tag{6.4}$$

Equations (6.2) and (6.4) are the basic equations for our study of the temporal and spatial propagation of a small-amplitude wave in a plasma. Before entering into these subjects, however, we wish to examine a little more about the symmetry properties of the dielectric function.

A. Symmetry Properties

The dielectric response function of a plasma without magnetic field has been calculated as Eq. (3.50). We may now choose the x axis arbitrarily in the direction of the wave vector $\mathbf{k}$. Then, for a single-component plasma, Eq. (3.50) can also be written as

$$\epsilon(k,\omega) = 1 - \frac{\omega_p^2}{k^2} \int dv \frac{f'(v)}{v-(\omega/k)-i(\eta/k)} \tag{6.5}$$

where $v \equiv \mathbf{v}\cdot\mathbf{k}/k$ is the velocity component in the direction of the wave vector (i.e., the x axis),

$$f(v) \equiv \int dv_y \int dv_z f(\mathbf{v}) \tag{6.6}$$

is a one-dimensional velocity distribution in the x direction, and the prime denotes a differentiation with respect to v.

Allowing the possibilities that ω and k may be complex variables, we obtain from Eq. (6.5) a function complex conjugate to $\epsilon(k,\omega)$ as

$$\epsilon^*(k,\omega)=1-\frac{\omega_p^2}{(k^*)^2}\int dv\frac{f'(v)}{v-(\omega/k)^*+i(\eta/k^*)}. \tag{6.7}$$

It follows that when k is real,

$$\epsilon(-k,-\omega^*)=\epsilon^*(k,\omega). \tag{6.8}$$

When ω is real, we have

$$\epsilon(-k^*,-\omega)=\epsilon^*(k,\omega). \tag{6.9}$$

For a Maxwellian plasma, the dispersion relation for the plasma oscillation has been given by Eq. (4.9); keeping only up to the lowest-order term with spatial dispersion, we approximate it as†

$$1-\frac{\omega_p^2}{\omega^2}-3\frac{\omega_p^2k^2(T/m)}{\omega^4}+i\left(\frac{\pi}{2}\right)^{1/2}\frac{\omega k_D^2}{|k|^3(T/m)^{1/2}}\exp\left[-\frac{\omega^2}{2k^2(T/m)}\right]=0. \tag{6.10}$$

This equation can likewise be investigated for the two cases when: (1) k is real; and (2) ω is real. These separate cases correspond, respectively, to the temporal and spatial propagations of the plasma oscillation.

When k is real and $k^2 \ll k_D^2$, Eq. (6.10) may be solved for ω as

$$\begin{aligned}\omega &\equiv \pm\omega(k)+i\gamma(k)\\ &=\pm\omega_p\left(1+\frac{3k^2}{2k_D^2}\right)-i\left(\frac{\pi}{8}\right)^{1/2}\omega_p\frac{k_D^3}{|k|^3}\exp\left[-\left(\frac{k_D^2}{2k^2}+\frac{3}{2}\right)\right].\end{aligned} \tag{6.11}$$

Figure 6.1 schematically describes the locations of these solutions on the complex ω plane; this figure applies for both positive and negative values of k.

†In Eq. (4.9), k is the magnitude of the wave vector, and hence, a nonnegative quantity. In the present case, however, k actually represents k_x with $k_y=k_z=0$; the x axis has been fixed along the wave vector. Thus, k takes on a negative value when the x axis and $\mathbf{k}$ are antiparallel. To account for this difference in the definitions of k, the absolute-value sign has been added in Eq. (6.10).

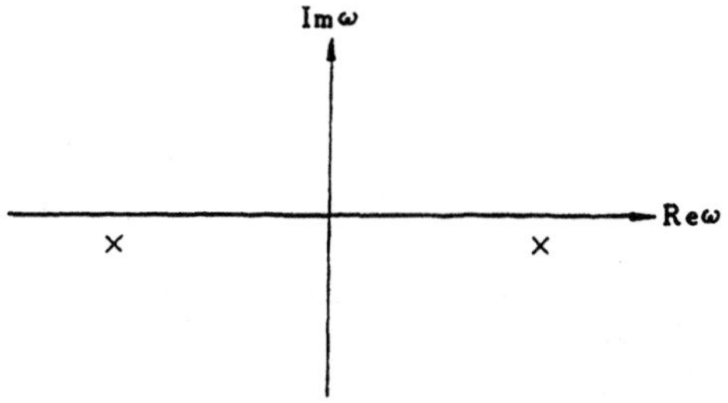

Fig. 6.1 Locations of the plasma-wave solutions Eq. (6.10) for a real value of k.

When ω is real and $\omega^2 \gtrsim \omega_p^2$, Eq. (6.10) may be solved for k as

$$k = \pm \frac{\omega^2}{\omega_p (3T/m)^{1/2}} \left(1 - \frac{\omega_p^2}{\omega^2}\right)^{1/2} \times \left\{ 1 + i\left(\frac{27\pi}{8}\right)^{1/2} \frac{\omega_p^5}{\omega^5} \left(1 - \frac{\omega_p^2}{\omega^2}\right)^{-5/2} \exp\left[-\frac{3\omega_p^2}{2(\omega^2 - \omega_p^2)} \right] \right\}. \tag{6.12}$$

Figure 6.2 shows the locations of these solutions on the complex k plane; in this case, $\omega > 0$ and $\omega < 0$ have to be distinguished.

A remarkable difference exists between the characteristics of the solutions described on the ω plane (Fig. 6.1) and those on the k plane (Fig. 6.2). As a consequence, the plasma wave exhibits a causal response in time, but not in space; we shall study these aspects in the subsequent sections.

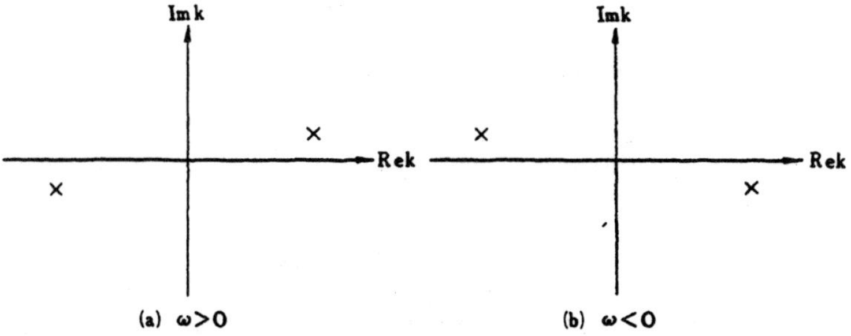

Fig. 6.2 Locations of the plasma-wave solutions Eq. (6.12) for positive and negative values of ω.

B. Temporal Propagation

Let us specify the form of the external disturbance (6.1), and consider

$$\varphi_{\text{ext}}(\mathbf{r},t)=\varphi_0\delta(\omega_{\text{p}}t)[\exp(ik_0x)+\text{cc}] \qquad (k_0>0). \tag{6.13}$$

By adopting Eq. (6.13), we are assuming a situation in which a sinusoid potential distribution with a fixed wavelength $2\pi/k_0$ is impulsively applied to the plasma at $t=0$. The Fourier components of the potential field are

$$\varphi_{\text{ext}}(\mathbf{k},\omega)=(\varphi_0/\omega_{\text{p}})[\delta(k,k_0)+\delta(k,-k_0)]\delta(k_y,0)\delta(k_z,0) \tag{6.14}$$

where $k=k_x$, and Kronecker's delta, defined by Eq. (5.25), is used.

The induced potential field is obtained by substituting (6.14) into (6.2); carrying out the inverse transformations, we find†

$$\varphi_{\text{ind}}(x,t)=-\frac{\varphi_0}{2\pi\omega_{\text{p}}}\int d\omega\left[1-\frac{1}{\epsilon(k_0,\omega)}\right]\exp[i(k_0x-\omega t)]$$
$$-\frac{\varphi_0}{2\pi\omega_{\text{p}}}\int d\omega\left[1-\frac{1}{\epsilon(-k_0,\omega)}\right]\exp[-i(k_0x+\omega t)]. \tag{6.15}$$

For $t>0$, we may close the contour of ω integration by an infinite semicircle in the lower half-plane. The residue at each of the poles of the inverse dielectric response function contributes to the integration; the location of the pole on the complex ω plane determines the frequency and the lifetime of each collective mode. Obviously, those poles lying closest to the real axis represent the collective modes with the longest lifetimes; contributions from such modes eventually predominate. Writing the imaginary part of Eq. (6.11) as $i\gamma(k)[\gamma(k)<0]$, we thus find that the field decays as

$$\varphi_{\text{ind}}(x,t)\sim\exp\{\gamma(k_0)t\} \qquad (t>0), \tag{6.16}$$

an indication of the Landau damping in the plasma. For $t<0$, we may close the contour by an infinite semicicle in the upper half-plane. In view of Fig. 6.1, we have

$$\varphi_{\text{ind}}(x,t)=0 \qquad (t<0), \tag{6.17}$$

another demonstration of causality built in the response function.

†The problem at hand is essentially one dimensional in space; hence, we use x instead of $\mathbf{r}$ to denote the spatial coordinates. Similar simplification arises also in the velocity space for $\delta f(\mathbf{r},\mathbf{v};t)$.

Thus, we have observed that the potential field induced in the plasma by the application of an impulsive perturbation (6.13) will die away exponentially due to the Landau damping processes. The situation differs significantly, however, in the case of an induced perturbation of the phase-space distribution given by Eq. (6.4). To see this, we carry out the inverse transformation

$$\delta f(x,v;t) = -\frac{q\varphi_0 k_0}{2\pi m\omega_p}\int_{-\infty}^{\infty} d\omega \frac{f'(v)}{\omega - k_0 v + i\eta}\frac{1}{\epsilon(k_0,\omega)}\exp[i(k_0 x - \omega t)]$$

$$+\frac{q\varphi_0 k_0}{2\pi m\omega_p}\int_{-\infty}^{\infty} d\omega \frac{f'(v)}{\omega + k_0 v + i\eta}\frac{1}{\epsilon(-k_0,\omega)}\exp[-i(k_0 x + \omega t)]. \quad (6.18)$$

The ω integration can be performed in the same way as before; for $t<0$, Eq. (6.18) vanishes. For $t>0$, various poles of the integrand contribute. Note that the integrand now has poles at $\omega = \pm k_0 v - i\eta$ in addition to the usual poles arising from the inverse dielectric function. We may thus write Eq. (6.18) as

$$\delta f(x,v;t) = -\frac{2q\varphi_0 k_0}{m\omega_p}\,\mathrm{Im}\left\{\frac{f'(v)}{\epsilon(k_0,k_0 v)}\exp[ik_0(x - vt)]\right\}$$

$$+[\text{contributions from the plasma-wave poles}] \qquad (t>0). \quad (6.19)$$

The last term of Eq. (6.19) behaves in the same way as Eq. (6.16); it decays exponentially. The first term on the right-hand side of Eq. (6.19), however, does not show such a decay; it oscillates indefinitely with a frequency $k_0 v$. Thus, in the absence of particle collisions, the phase-space distribution $f(\mathbf{r},\mathbf{v};t)$ will keep the memory of the initial perturbation permanently.

This difference in the time-dependent behavior between (6.16) and (6.19) may be understood in the following way: A macroscopic quantity such as the potential field involves a velocity-space integration of a distribution function like Eq. (6.19). In the process of such an integration, the phases $k_0 vt$ of microscopic elements will now be mixed, because of added contributions from various elements in the velocity space. Consequently, the velocity integration of the first term on the right-hand side of Eq. (6.19) will also decay in time.†

†In actually carrying out the velocity integration for Eq. (6.19), a contour-integration technique may be involved in the complex v plane. When attempting to close the contour by an infinite semicircle, we would ordinarily encounter an essential singularity associated with $f'(v)$ at $v = i\infty$ or $v = -i\infty$. Such a singularity, however, may be canceled in the integrand by a similar singularity occurring in $\epsilon(k_0, k_0 v)$ or $\epsilon(-k_0, -k_0 v)$ (see Problem 6.2). Hence, the integration contour may be closed with the aid of an infinite semicircle in the appropriate half-plane.

On the basis of the foregoing considerations, we can then design the following gedanken experiment to illustrate the reversible nature of the phenomena involved: Apply a pulsed disturbance of the form (6.13) to the plasma and wait for a time interval τ. If $\tau \gg |\gamma(k_0)|^{-1}$ is chosen, the induced potential field (6.16) as well as the last term of (6.19) will have died away; macroscopically, it will appear that the plasma has returned to a state without perturbations. Actually, however, the first term of (6.19) still remains. We now reverse the directions of phase evolution in these microscopic elements. After the passage of another time interval τ, the phase relations of these elements will be restored to their original states at $t=0$; a macroscopic quantity will thereby reappear in the plasma.

Such an experiment, of course, cannot be performed in reality in exactly the same way as just described. We can, however, produce effects physically equivalent to the foregoing through plasma-wave echo experiments; we shall study them after a brief investigation of spatial propagation problems.

C. Spatial Propagation

In spite of its conceptual simplicity and transparency, the external disturbance in the form as specified by Eq. (6.13) may not be easy to create in a given plasma. Instead, we could apply a potential

$$\varphi_{\text{ext}}(\mathbf{r},t)=\varphi_0\delta(k_{\text{D}}x)[\exp(-i\omega_0 t)+\text{cc}] \qquad (\omega_0>0) \tag{6.20}$$

to the plasma without much difficulty through an insertion of a grid at $x=0$ whose potential is sinusoidally modulated by an external source. Let us therefore consider a consequence of such an experiment.

Since the Fourier components of (6.20) are

$$\varphi_{\text{ext}}(\mathbf{k},\omega)=2\pi(\varphi_0/k_D)[\delta(\omega-\omega_0)+\delta(\omega+\omega_0)]\delta(k_y,0)\delta(k_z,0), \tag{6.21}$$

the induced potential field is calculated as

$$\begin{aligned}\varphi_{\text{ind}}(x,t)=&-\frac{\varphi_0}{2\pi k_{\text{D}}}\int_{-\infty}^{\infty}dk\left[1-\frac{1}{\epsilon(k,\omega_0)}\right]\exp[i(kx-\omega_0 t)]\\&-\frac{\varphi_0}{2\pi k_{\text{D}}}\int_{-\infty}^{\infty}dk\left[1-\frac{1}{\epsilon(k,-\omega_0)}\right]\exp[i(kx+\omega_0 t)].\end{aligned} \tag{6.22}$$

For $x>0$, we close the contour of integration by an infinite semicircle in the upper half of the complex k plane. Expressing Eq. (6.12) as

$$k=k_r(\omega)+ik_i(\omega), \tag{6.23}$$

we find from Fig. 6.2 that those poles located in the upper half-plane, $\{k_r(\omega_0)>0,\ k_i(\omega_0)>0\}$ and $\{k_r(-\omega_0)<0,\ k_i(-\omega_0)>0\}$, contribute to Eq. (6.22) through such an integration. Each mode has a phase velocity directed toward the positive x direction (i.e., $\omega/k_r>0$) and decays exponentially as $\exp(-|k_i|x)$. For $x<0$, the contour of the k integration may be closed by an infinite semicircle in the lower half-plane. The poles in this half plane, $\{k_r(\omega_0)<0,\ k_i(\omega_0)<0\}$ and $\{k_r(-\omega_0)>0,\ k_i(-\omega_0)<0\}$, now contribute to Eq. (6.22). Each mode has a phase velocity directed toward the negative x direction (i.e., $\omega/k_r<0$), and decays exponentially as $\exp(|k_i|x)$. Contrary to the case of temporal propagation, the spatial propagation of the plasma wave takes place symmetrically with regard to the positive and negative x directions (unless of course the plasma itself has inherent asymmetry).

The induced perturbation of the phase-space distribution can be calculated similarly from Eq. (6.4):

$$\delta f(x,v;t) = -\frac{q\varphi_0}{2\pi m k_D}\int_{-\infty}^{\infty} dk \frac{kf'(v)}{\omega_0 - kv + i\eta}\,\frac{1}{\epsilon(k,\omega_0)}\exp[i(kx-\omega_0 t)]$$

$$+\frac{q\varphi_0}{2\pi m k_D}\int_{-\infty}^{\infty} dk \frac{kf'(v)}{\omega_0 + kv - i\eta}\,\frac{1}{\epsilon(k,-\omega_0)}\exp[i(kx+\omega_0 t)]. \quad (6.24)$$

The poles at $k=\pm(\omega_0/v)+i(\eta/v)$ contribute to a propagation in the $+x$ direction when $v>0$; naturally, the particles with $v>0$ convey the effects of disturbance imparted at $x=0$ into the region $x>0$. By the same token, the particles with $v<0$ carry information into the negative x region. As in the case of the preceding section, these effects propagate in space without decay; similar comments are applicable here on the relation between spatial evolutions of (6.22) and (6.24).

We can use relationships such as (6.22) to investigate the propagation characteristics of a collective mode in a plasma. The propagation velocity and the spatial decay rate may thus be measured through an interferometric technique; both the electron plasma oscillation [1–3] and the ion-acoustic wave [4] have been thereby investigated.

6.2 PLASMA-WAVE ECHOES

We now wish to study the plasma-wave echo phenomena. As we remarked earlier, the reversible nature of the Vlasov equation is reflected in the microscopic phase evolution of the particle distribution function after the application of a disturbance to the plasma; a macroscopic quantity such as the electric-field potential, on the other hand, would decay through Landau damping. It has been recognized that a macroscopic quantity might then reappear in the plasma if we could reverse the direction of phase evolution of the microscopic elements. Such a reversal is in fact possible through the application of a second disturbance; a macroscopic field (i.e., the echo) will reappear in the plasma many Landau-damping periods after the application of the second disturbance [5]. The plasma echo is related to other known echo phenomena, such as spin echo [6], cyclotron resonance echo [7], and photon echo [8].

As the examples in the previous section have shown, we can reduce a problem involving plasma-wave propagation to a one-dimensional one through an appropriate choice of external perturbation. The Vlasov equation in these circumstances may be expressed as

$$\frac{\partial f}{\partial t}+v\frac{\partial f}{\partial x}-\frac{q}{m}\frac{\partial \varphi}{\partial x}\frac{\partial f}{\partial v}=0 \tag{6.25}$$

where f is a one-dimensional distribution function defined according to Eq. (6.6). The potential field is the sum of the external and induced potentials determined from the Poisson equation

$$\frac{\partial^2\varphi}{\partial x^2}=\frac{\partial^2\varphi_{\text{ext}}}{\partial x^2}-4\pi qn\left[\int f\,dv-1\right]. \tag{6.26}$$

The last term on the left-hand side of (6.25), therefore, represents the *nonlinear term* of the Vlasov equation. When two separate disturbances are applied to the plasma, this nonlinear term acts to mix the effects of such disturbances. The plasma echo appears as a consequence of such a mixing.

A. Fundamental Processes

Consider an external disturbance

$$\varphi_1(x,t)=\varphi_1\delta(\omega_p t)[\exp(ik_1x)+\text{cc}] \tag{6.27}$$

applied to the plasma at $t=0$. According to calculations in Section 6.1B, the induced potential field decays with a time constant $|\gamma(k_1)|^{-1}$; for $t\gg|\gamma(k_1)|^{-1}$, the induced field vanishes (see Fig. 6.3). The perturbation $\delta f_1(x,v;t)$ induced in the distribution function still remains, however; from the first term on the right-hand side of Eq. (6.19), we find its phase evolution as

$$\delta f_1(x,v;t)\sim\exp[\pm ik_1(x-vt)]. \tag{6.28}$$

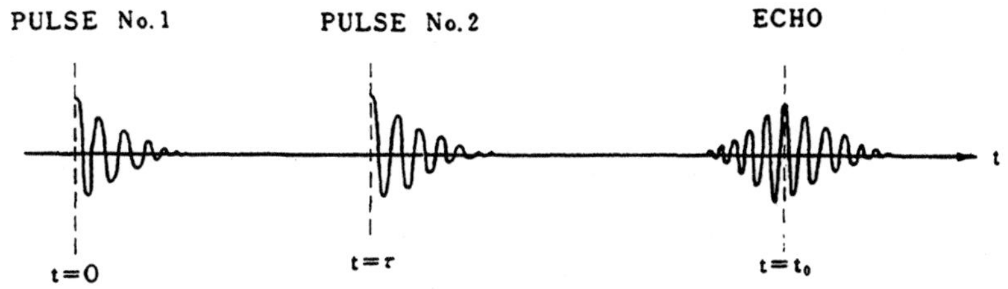

Fig. 6.3 Temporal plasma-wave echo.

We now apply a second pulse

$$\varphi_2(x,t)=\varphi_2\delta[\omega_p(t-\tau)][\exp(ik_2x)+\mathrm{cc}] \tag{6.29}$$

at $t=\tau\gg|\gamma(k_1)|^{-1}$. Through the nonlinear term in Eq. (6.25), this pulse couples with the distribution function $f(x,v;t)$ and creates disturbances in the plasma. The distribution function now consists of two separate contributions: unperturbed part $f(v)$ and a perturbation $\delta f_1(x,v;t)$ due to the application of the first pulse.

That part of the disturbances stemming from coupling between (6.29) and $f(v)$ behaves in the same manner as the disturbance created by the first pulse: The induced potential field decays with a time constant $|\gamma(k_2)|^{-1}$; the long-time behavior of the perturbed part of the distribution function will be

$$\delta f_2(x,v;t)\sim\exp\{\pm ik_2[x-v(t-\tau)]\}. \tag{6.30}$$

Macroscopically, therefore, the pulse will disappear in time $t\gg\tau+|\gamma(k_2)|^{-1}$.

The plasma echo arises as a result of coupling between $\varphi_2(x,t)$ and $\delta f_1(x,v;t)$. The phase evolution of the distribution function perturbed through such a coupling may then be described by a product between (6.28) and (6.30) as

$$\delta f_{12}(x,v;t)\sim\exp\{i[(k_1\pm k_2)x-k_1vt\mp k_2v(t-\tau)]\}+\mathrm{cc} \tag{6.31}$$

Calculation of a macroscopic quantity generally involves an integration of such a distribution function over the velocity space, which would ordinarily result in phase mixing and thereby the disappearance of the macroscopic quantity. The phase mixing can be avoided, however, if the phase is independent of the velocity. Let us assume that such a condition is realized for Eq. (6.31) at $t=t_0$; then $k_1vt_0\pm k_2v(t_0-\tau)=0$, or $t_0=[\pm k_2/(k_1\pm k_2)]\tau$. Since

$t_0 > \tau$, we must choose the lower signs; hence, we have

$$t_0 = [k_2/(k_2 - k_1)]\tau. \tag{6.32}$$

Consequently, when $k_2 > k_1$, a macroscopic field, varying sinusoidally in space with wave number

$$k_3 = k_2 - k_1, \tag{6.33}$$

will reappear around the time $t = t_0$. The phenomenon described above is called a *temporal plasma-wave echo*, since it involves temporal propagations of the plasma waves.

We can consider similar echo phenomena arising from spatial propagations of the plasma waves as well. Thus, we apply a first signal

$$\varphi_1(x,t) = \varphi_1 \delta(k_D x)[\exp(-i\omega_1 t) + \mathrm{cc}] \tag{6.34}$$

at $x = 0$ in the plasma; the field damps away for $|x| \gg |k_i(\omega_1)|^{-1}$. A second signal

$$\varphi_2(x,t) = \varphi_2 \delta[k_D(x-l)][\exp(-i\omega_2 t) + \mathrm{cc}] \tag{6.35}$$

is then applied at $x = l \gg |k_i(\omega_1)|^{-1}$; it will also vanish for $|x - l| \gg |k_i(\omega_2)|^{-1}$. When $\omega_2 > \omega_1$, an echo will appear at

$$x_0 = [\omega_2/(\omega_2 - \omega_1)]l \tag{6.36}$$

with a frequency

$$\omega_3 = \omega_2 - \omega_1. \tag{6.37}$$

This is, therefore, a *spatial plasma-wave echo* phenomenon.

The two examples considered above are referred to as second-order plasma echoes; these derive from the second-order perturbation contributions of the nonlinear term with respect to the external disturbances—first order in the disturbance $\varphi_1(x,t)$ and first order in another disturbance $\varphi_2(x,t)$. We can likewise extend the arguments and consider higher-order echoes. The nth-order effect of the disturbance (6.27) on the distribution is thus calculated as

$$\delta f_1^{(n)}(x,v;t) \sim \varphi_1^n \exp[\pm ink_1(x - vt)] \tag{6.38}$$

and the mth-order effect of the disturbance (6.29) is

$$\delta f^{(m)}(x,v;t) \sim \varphi_2^m \exp\{\pm imk_2[x - v(t-\tau)]\}. \tag{6.39}$$

The $(m+n)$th-order temporal echo appears as a result of coupling between (6.38) and (6.39) at time

$$t_0 = [mk_2/(mk_2 - nk_1)]\tau \tag{6.40}$$

with the wave number

$$k_3 = mk_2 - nk_1. \tag{6.41}$$

The amplitude of the resulting echo is proportional to $\varphi_1^n \varphi_2^m$. Similarly, the $(m+n)$th spatial echo appears at the coordinate

$$x_0 = [m\omega_2/(m\omega_2 - n\omega_1)]l \tag{6.42}$$

with the frequency

$$\omega_3 = m\omega_2 - n\omega_1. \tag{6.43}$$

Those plasma-wave echo phenomena have been investigated experimentally both for the electron plasma oscillation [9] and for the ion-acoustic wave [10, 11].

B. Detailed Calculation

Based on the fundamental ideas outlined in the previous section, we can proceed to calculate the detailed features of the plasma-wave echo from the Vlasov–Poisson equations, (6.25) and (6.26). The calculations that we shall describe in this section are mainly due to O'Neil and Gould [5].

We write the distribution function as a summation of an unperturbed part and perturbation; the latter is Fourier analyzed so that

$$f(x,v;t) = f(v) + \sum_k{}' \int \frac{d\omega}{2\pi} \delta f_{k\omega}(v) \exp[i(kx - \omega t) + \eta t]. \tag{6.44}$$

The prime means omission of the term with $k=0$, which has been included in the unperturbed distribution. The potential field $\varphi(x,t)$ may similarly be expanded into Fourier components; the term with $k=0$ does not contribute because of the presence of neutralizing space-charge background.

These expressions are substituted into the Vlasov equation (6.25) and the Poisson equation (6.26). Retaining all the effects of the nonlinear term in Eq. (6.25), we find

$$\delta f_{k\omega}(v) = -\frac{q}{m}\frac{kf'(v)}{\omega - kv + i\eta}\varphi(k,\omega)$$

$$-\frac{q}{m}\frac{1}{\omega - kv + i\eta}\sum_{k'}{}' \int \frac{d\omega'}{2\pi}(k-k')\varphi(k-k',\omega-\omega')\frac{\partial}{\partial v}[\delta f_{k'\omega'}(v)]. \tag{6.45}$$

The essential difference between Eqs. (6.4) and (6.45) is represented by the last term on the right-hand side of the latter equation. With the aid of (6.45), the potential field is obtained from the Poisson equation as

$$\varphi(k,\omega)=\frac{\varphi_{\rm ext}(k,\omega)}{\epsilon(k,\omega)}-\frac{\omega_{\rm p}^2}{k^2\epsilon(k,\omega)}\int dv\frac{1}{\omega-kv+i\eta}$$

$$\times\sum_{k'}{}'\int\frac{d\omega'}{2\pi}(k-k')\varphi(k-k',\omega-\omega')\frac{\partial}{\partial v}[\delta f_{k'\omega'}(v)] \qquad (6.46)$$

where the dielectric function $\epsilon(k,\omega)$ has been defined by Eq. (6.5). Again, Eq. (6.46) differs from the content of Eq. (6.2) by the presence of the nonlinear contributions on its right-hand side. Equations (6.45) and (6.46) are the basic equations for a theoretical investigation of plasma-wave echo phenomena. With the aid of perturbation-theoretical calculations, we can in principle analyze a temporal or spatial echo to any order of external disturbances.

For the sake of definiteness, we shall henceforth consider the case of a temporal plasma echo. The external disturbance is the summation of (6.27) and (6.29):

$$\varphi_{\rm ext}(x,t)=\varphi_1\delta(\omega_{\rm p}t)[\exp(ik_1x)+{\rm cc}]+\varphi_2\delta[\omega_{\rm p}(t-\tau)][\exp(ik_2x)+{\rm cc}].$$

Its Fourier components are

$$\varphi_{\rm ext}(k,\omega)=(\varphi_1/\omega_{\rm p})[\delta(k,k_1)+\delta(k,-k_1)]$$

$$+(\varphi_2/\omega_{\rm p})[\delta(k,k_2)+\delta(k,-k_2)]\exp(i\omega\tau). \qquad (6.47)$$

We substitute this expression into Eqs. (6.45) and (6.46), and carry out perturbation calculations with respect to φ_1 and φ_2.

To the first order, Eqs. (6.45) and (6.46) yield standard results from a linear response analysis,

$$\varphi^{(1)}(k,\omega)=\frac{\varphi_1}{\omega_{\rm p}\epsilon(k,\omega)}[\delta(k,k_1)+\delta(k,-k_1)]$$

$$+\frac{\varphi_2}{\omega_{\rm p}\epsilon(k,\omega)}[\delta(k,k_2)+\delta(k,-k_2)]\exp(i\omega\tau), \qquad (6.48a)$$

$$\delta f_{k\omega}{}^{(1)}(v)=-\frac{q}{m}\frac{kf'(v)}{\omega-kv+i\eta}\varphi^{(1)}(k,\omega). \qquad (6.48b)$$

No echoes are expected to take place in this order.

To the second order in the perturbation, the potential field is given by

$$\varphi^{(\mathrm{II})}(k,\omega) = -\frac{q\omega_p^2}{mk\epsilon(k,\omega)}\sum_{k'}{}'\int\frac{d\omega'}{2\pi}\int dv\frac{k'(k-k')f'(v)}{(\omega-kv+i\eta)^2(\omega'-k'v+i\eta)}$$
$$\times\varphi^{(\mathrm{I})}(k',\omega')\varphi^{(\mathrm{I})}(k-k',\omega-\omega'). \qquad (6.49)$$

We substitute (6.48a) into this equation; a second-order temporal echo will appear from those terms containing cross products between φ_1 and φ_2. Retaining such terms only, we obtain

$$\varphi^{(\mathrm{II})}(k,\omega) = \frac{q\varphi_1\varphi_2k_1k_2}{mk\epsilon(k,\omega)}\int\frac{d\omega'}{2\pi}\int dv\frac{f'(v)}{(\omega-kv+i\eta)^2}$$
$$\times\left\{\frac{\delta(k,k_3)\exp(i\omega'\tau)}{(\omega'-k_2v+i\eta)\epsilon(k_2,\omega')\epsilon(-k_1,\omega-\omega')}+\frac{\delta(k,-k_3)\exp(i\omega'\tau)}{(\omega'+k_2v+i\eta)\epsilon(-k_2,\omega')\epsilon(k_1,\omega-\omega')}\right.$$
$$\left.+\frac{\delta(k,k_3)\exp[i(\omega-\omega')\tau]}{(\omega'+k_1v+i\eta)\epsilon(-k_1,\omega')\epsilon(k_2,\omega-\omega')}+\frac{\delta(k,-k_3)\exp[i(\omega-\omega')\tau]}{(\omega'-k_1v+i\eta)\epsilon(k_1,\omega')\epsilon(-k_2,\omega-\omega')}\right\}. \qquad (6.50)$$

where k_3 has been given by Eq. (6.33).

The ω' integration for the first two terms on the right-hand side may be carried out by closing the contour in the upper half-plane. The only poles that can contribute to such integrations are the zeros of $\epsilon(\pm k_1,\omega-\omega')$; they give rise to exponential decay $\sim\exp[-|\gamma(k_1)|\tau]$. Since $\tau\gg|\gamma(k_1)|^{-1}$, those terms may be neglected. The ω' integration for the last two terms may be calculated by closing the contour in the lower half-plane. Two different kinds of poles are now involved: the poles from the zeros of $\epsilon(\pm k_1,\omega')$ and those at $\omega'=\pm k_1v-i\eta$. The contributions from the former poles behave in the same way as the first two terms of (6.50); these may be neglected. The only significant contributions, therefore, come from the latter poles at $\omega'=\pm k_1v-i\eta$.

Taking account of the simplifications described above, we now carry out inverse Fourier transformation of Eq. (6.50). Expressing the result as

$$\varphi^{(\mathrm{II})}(x,t)=\varphi^{(\mathrm{II})}(t)\exp(ik_3x)+\mathrm{cc},$$

we find

$$\varphi^{(\mathrm{II})}(t) = -i\frac{q\varphi_1\varphi_2k_1k_2}{mk_3}\int\frac{d\omega}{2\pi}\int dv\frac{f'(v)}{(\omega-k_3v+i\eta)^2}$$
$$\times\frac{\exp\{i[k_1v\tau-\omega(t-\tau)]\}}{\epsilon(-k_1,-k_1v)\epsilon(k_2,\omega+k_1v)\epsilon(k_3,\omega)}.$$

The ω integration can be calculated by closing the contour in the lower half-plane. Three kinds of poles are involved: the double pole at $\omega = k_3 v - i\eta$, the zeros of $\epsilon(k_2, \omega + k_1 v)$, and the zeros of $\epsilon(k_3, \omega)$. The contributions from the last two kinds of poles produce exponentially decaying factors $\exp[-|\gamma(k_2)|(t-\tau)]$ and $\exp[-|\gamma(k_3)|(t-\tau)]$. Assuming $|\gamma(k_2)|(t-\tau) \gg 1$ and $|\gamma(k_3)|(t-\tau) \gg 1$, we neglect those contributions. The equation above thus becomes†

$$\varphi^{(\mathrm{II})}(t) = -\frac{q\varphi_1\varphi_2 k_1 k_2}{mk_3} \int dv \frac{f'(v)}{\epsilon(-k_1, -k_1 v)} \left\{ \frac{\partial}{\partial \omega} \left[\frac{\exp\{i[k_1 v\tau - \omega(t-\tau)]\}}{\epsilon(k_2, \omega + k_1 v)\epsilon(k_3, \omega)} \right] \right\}_{\omega = k_3 v}. \tag{6.51}$$

Equation (6.51) will be split into three terms when the differentiation with respect to ω is actually carried out. Among them, the terms involving the derivatives of the dielectric functions simply oscillate in time with constant amplitudes. The term arising from the derivative of the exponential function, on the other hand, exhibits a secular growth in time proportional to $t-\tau$. In the vicinity of the echo peak

$$t - \tau = t_0 - \tau = (k_1/k_3)\tau, \tag{6.52}$$

this secular term represents the major contribution. Keeping only this term and using (6.52) approximately for $t-\tau$ appearing in its coefficient, we find that Eq. (6.51) reduces to

$$\varphi^{(\mathrm{II})}(t_0 + t') \cong -\frac{q\varphi_1\varphi_2 k_1^2 k_2 \tau}{imk_3^2} \int_{-\infty}^{\infty} dv \frac{f'(v)\exp(-ik_3 v t')}{\epsilon(-k_1, -k_1 v)\epsilon(k_2, k_2 v)\epsilon(k_3, k_3 v)}, \tag{6.53}$$

where $t' = t - t_0$ is the time measured from the peak position t_0. We thus find the shape of the echo profile by carrying out the velocity-space integration in Eq. (6.53). The integration contour may be closed in the upper half-plane for $t' < 0$ and in the lower half-plane for $t' > 0$; the singularities of $f'(v)$ at $v = \pm i\infty$ will be canceled by similar singularities associated with the dielectric functions (see footnote, p. 113). The zeros of the dielectric functions involved in Eq. (6.53) therefore determine the shape of the echo buildup and its decay.

For a Maxwellian plasma, we may use the result of Eq. (6.11). The echo builds up as $\exp[-(k_3/k_1)\gamma(k_1)(t-t_0)]$; it decays as $\exp[(k_3/k_2)\gamma(k_2)(t-t_0)]$ and $\exp[\gamma(k_3)(t-t_0)]$. The echo shape is not symmetric with respect to the buildup and decay.

†Note that $f'(Z) = (2\pi i)^{-1} \oint [f(t)/(t-Z)^2] dt$.

C. Effects of Collisions

As the arguments in the preceding sections may have amply demonstrated, the existence of the echo depends critically on the preservation of delicate phase memory during the evolution of microscopic elements. We would, therefore, anticipate that deflections of particle trajectories caused by collisional processes, however small they may be, will have sensitive effects on the echo phenomena by destroying the delicate phase relationship among the microscopic elements. In this section, we briefly investigate such collisional effects on plasma-wave echoes; more detailed investigations are available in the literature [12–15].

To take the collisions into account, we must add to the right-hand side of Eq. (6.25) the collision term obtained in Section 2.5. The solution of this kinetic equation for the plasma-wave propagation is ordinarily a rather complicated problem. The situation is simplified, however, by the fact that we are here interested in those problems associated with a change in the distribution function arising from relatively small momentum transfers; relaxation processes may be described by a collision term of Fokker–Planck type, having a diffusion behavior in the velocity space. (The Fokker–Planck equation will be discussed in Chapter 10.) This observation enables us to replace the exact collision term by the model collision term proposed by Lenard and Bernstein [16]

$$\left.\frac{\partial f}{\partial t}\right]_{\mathrm{c}}=\nu\frac{\partial}{\partial v}\left[vf+\frac{T}{m}\frac{\partial f}{\partial v}\right], \tag{6.54}$$

where v is the velocity component in the direction of wave propagation and ν is an effective collision frequency. Equation (6.54) retains two of the most important properties of the Coulomb collision term; it has a diffusion behavior and vanishes if we substitute in it a one-dimensional Maxwellian distribution reduced from Eq. (1.24).

We therefore adopt Eq. (6.54) as our collision term. Acting upon the increment (6.28) created by the first pulse (6.27), it produces

$$\left.\frac{\partial}{\partial t}\delta f_1(x,v;t)\right]_{\mathrm{c}}=\nu\left[1-ik_1vt-\frac{T}{m}(k_1t)^2\right]\delta f_1(x,v;t)$$

$$\rightarrow-\nu\left[ik_1vt+\frac{T}{m}(k_1t)^2\right]\delta f_1(x,v;t),$$

where we have been choosing the upper sign of (6.28); a long-time behavior has been singled out in the last step. Hence, in the presence of collisions, we have

$$\delta f_1(x,v;t)_c = \delta f_1(x,v;t)\exp\left[-\nu\left(i\frac{k_1 v t^2}{2} + \frac{T}{m}\frac{k_1^2 t^3}{3}\right)\right]. \tag{6.55}$$

The amount of phase advance during the time interval τ is

$$-k_1 v\tau(1+\nu\tau/2); \tag{6.56}$$

the decay factor due to collisions is

$$\exp[-\nu(T/m)(k_1^2\tau^3/3)]. \tag{6.57}$$

At $t=\tau$, the second pulse (6.29) is applied to the plasma. Without collisions, the increment of distribution would propagate after this time with phase evolution

$$\delta f_3(x,v;t) \sim \exp\{-ik_3[x-v(t-\tau)]\}.$$

With collisions, however, this will be modified as

$$\delta f_3(x,v;t)_c = \delta f_3(x,v;t)\exp\left\{\nu\left[i\frac{k_3 v(t-\tau)^2}{2} - \frac{T}{m}\frac{k_3^2(t-\tau)^3}{3}\right]\right\}. \tag{6.58}$$

The phase advances during the time interval $t-\tau$ by the amount

$$k_3 v(t-\tau)[1+\nu(t-\tau)/2]; \tag{6.59}$$

the amplitude decays by a factor

$$\exp\{-\nu(T/m)[k_3^2(t-\tau)^3/3]\}. \tag{6.60}$$

At the echo peak $t=t_0$, the total phase advance, (6.56) plus (6.59), must vanish. Assuming $\nu\tau \ll 1$ and $\nu(t_0-\tau)\ll 1$, we calculate the peak position from this condition as

$$t_0 = \frac{k_2}{k_3}\tau\left[1-\frac{\nu\tau k_1}{2k_2}\left(1-\frac{k_1}{k_3}\right)\right]. \tag{6.61}$$

Collisions therefore act to shift the position of the echo peak slightly. From Eqs. (6.57) and (6.60), we also find that the echo amplitude is reduced by a factor

$$\exp\{-\nu(T/m)(k_1^2\tau^2\tau_0/3)\} \tag{6.62}$$

because of the particle collisions.

6.3 LARGE-AMPLITUDE PLASMA WAVES

The small-amplitude calculation of the plasma wave in Section 6.1 assumed that the main body of the distribution function remains constant during the process of the wave evolution. The wave thereby decays through Landau damping; the decay constant may be obtained from a comparison between (4.16) and (6.5) as

$$\gamma_L = \frac{\pi\omega_p^2}{(\partial\epsilon_1/\partial\omega)k^2} f'(\omega/k). \tag{6.63}$$

In this derivation, Eq. (B.4) has been used. The imaginary part of Eq. (6.11) is a special case of this general formula with a Maxwellian distribution.

Alternatively, we could consider the other limiting cases [17, 18]: A large-amplitude plasma wave is applied to the plasma; its amplitude remains almost constant. The distribution function, on the other hand, is modulated by the potential field associated with the wave; it now becomes time dependent.

The distinction between those two cases is related to the ways in which the nonlinear term of the Vlasov equation (6.25) may be linearized. Generally, the linearization may be performed according to

$$\frac{\partial\varphi}{\partial x}\frac{\partial f}{\partial v} = \frac{\partial\varphi}{\partial x}\frac{\partial f_0}{\partial v} + \frac{\partial\varphi_0}{\partial x}\frac{\partial f}{\partial v} \tag{6.64}$$

where the quantities with the subscript 0 are to be regarded as time independent. Landau's treatment of a small-amplitude plasma wave corresponds to retaining the first term only through the linearization. For a large-amplitude wave, however, the second term describes the major effects.

The criterion for deciding which of the two treatments is more applicable to a given situation, therefore, depends on the wave amplitude. To obtain such a criterion, we must first look into the behavior of those particles trapped in the potential troughs of the large-amplitude plasma wave.

A. Trapped Particles

Consider a monochromatic wave in a plasma whose potential field is given by

$$\varphi(x,t)=\varphi_0\cos(kx-\omega t). \tag{6.65}$$

Upon carrying out a Galilean transformation

$$X=x-(\omega/k)t \tag{6.66}$$

from the laboratory frame (x,t) to a frame (X,t) comoving with the wave, we find that the potential (6.65) takes a stationary form

$$\varphi(X,t)=\varphi_0\cos(kX). \tag{6.67}$$

Since the particle velocity v is also transformed to

$$V=v-(\omega/k), \tag{6.68}$$

the total energy of a particle in this moving frame of reference is

$$W=\tfrac{1}{2}mV^2+q\varphi_0\cos(kX). \tag{6.69}$$

In the phase space (X,V), a particle thus moves along a trajectory W = constant. (See Fig. 6.4.) Those particles with $W<0$ are thereby trapped by the wave potential, bouncing back and forth in the trough; they perform a periodic motion in the phase space.

The period of such a bouncing motion may be estimated from consideration of those particles near the bottom of the trough. Since the electric field associated with (6.67) is $\varphi_0 k\sin(kX)$, the equation of motion for those particles with $|kX|\ll 1$ is

$$m\ddot{X}=q\varphi_0 k^2 X.$$

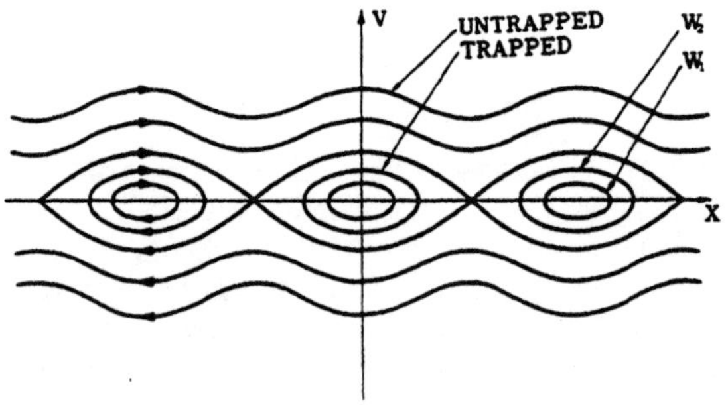

Fig. 6.4 The phase trajectories of trapped and untrapped particles.

Hence, the bounce frequency ω_B is determined as

$$\omega_B = |q\varphi_0 k^2/m|^{1/2}. \tag{6.70}$$

As the wave amplitude increases, so does the bounce frequency.

The criterion for the validity of Landau's treatment of small-amplitude plasma waves may be found from a comparison between this bounce frequency and Landau's decay constant (6.63). Thus, when the wave amplitude is so small that

$$\omega_B \ll |\gamma_L|, \tag{6.71}$$

the wave will have Landau-damped substantially by the time a trapped particle has completed one cycle of the bounce motion. The trapping is not an important effect here; a small-amplitude treatment neglecting the last term of Eq. (6.64) is applicable in these circumstances.

On the other hand, if the wave amplitude is so large that

$$\omega_B \gg |\gamma_L|, \tag{6.72}$$

then the trapped particles will have completed many cycles of their bounce motion before the wave starts to decay. The second term on the right-hand side of Eq. (6.64) now represents the major effects. In the next section, we consider qualitative consequences arising from such trapping effects.

B. Amplitude Oscillation

Suppose that the distribution function $f(v)$ initially has the form depicted in Fig. 6.5; the wave field (6.65) or (6.67) is also present in the plasma. The particles in the shaded area are then being trapped in the wave potential. Since $f'(\omega/k)<0$, the γ_L calculated according to Eq. (6.63) takes on a negative value at $t=0$.

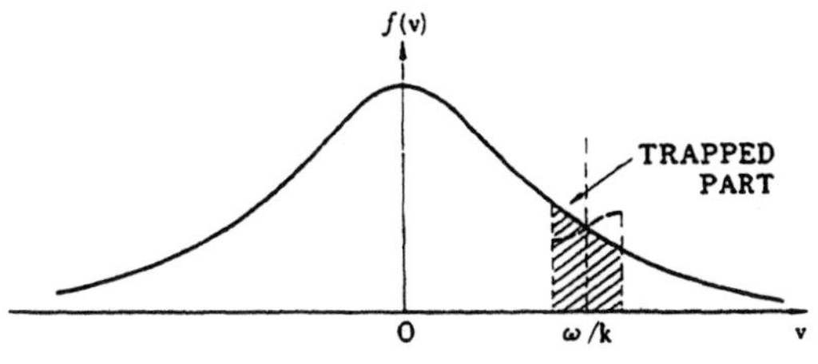

Fig. 6.5 Initial distribution of the particles and the trapped part.

Let us now follow the motion of the trapped particles for a half period, π/ω_B. The particles will have completed a half cycle of their bounce motion; their distribution may assume a shape inverted with respect to the axis $v=\omega/k$ in the trapped domain. This is not exactly true, however, for those particles close to the outer boundary of the trapped domain; the bounce frequency (6.70) has been calculated only for those particles near the bottom of the trough. The equation of motion for those particles close to the periphery of the trapped domain deviates significantly from that of harmonic motion. As a consequence, the shape of the distribution for the trapped particles at $t=\pi/\omega_B$ will be slightly dulled from the exact mirror image of the distribution at $t=0$, as the broken line in Fig. 6.5 illustrates. Nevertheless, we note that $f'(\omega/k)$ is now positive; hence, $\gamma_L>0$ at $t=\pi/\omega_B$.

If we wait another half period, most of the trapped particles will return to their original phase relationship at $t=0$; γ_L will become negative again. Thus, the value of γ_L is expected to oscillate around $\gamma_L=0$ with the frequency ω_B, as the distribution in the trapped domain is modulated by the wave with the same frequency.

It is apparent, however, that this sort of oscillation cannot continue indefinitely. For one thing, we have noted the effects of anharmonicity which act to smooth out the distribution in the trapped domain. In addition, we must note that ample time is available for those particles trapped in the same potential trough to interact with each other. The particle interactions, however weak, will eventually act to phase-mix between those trapped particles on the same trajectory ($W=$constant) as well as those on the slightly different trajectories W_1 and W_2 in Fig. 6.4. All these effects working together will contribute to the flattening of the distribution or the formation of a plateau in the trapped domain. Consequently, the time-dependent behavior of γ_L should resemble that of a damped oscillation, asymptotically approaching $\gamma_L=0$ (see Fig. 6.6).

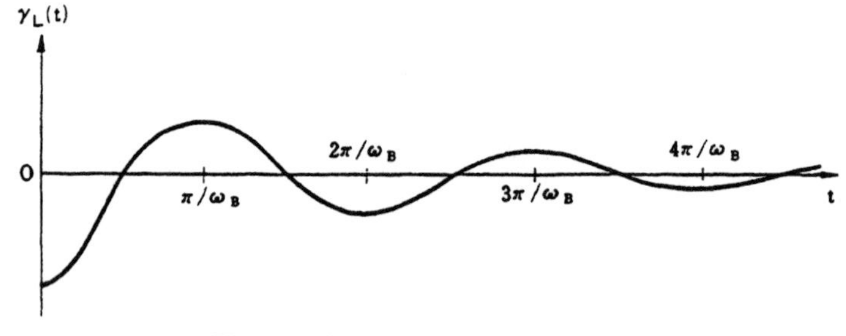

Fig. 6.6 Schematic behavior of γ_L.

Thus far we have not been concerned with the change in the wave amplitude; it has been assumed constant, large enough to satisfy (6.72). In the limit as $|\gamma_L|/\omega_B \to 0$, the wave amplitude indeed remains constant. If, however, we take into account the first-order effects of $|\gamma_L|/\omega_B$, then the wave amplitude φ_0 in Eq. (6.65) no longer remains constant. Instead, it should vary in accord with the equation

$$\frac{d|\varphi_0|^2}{dt} = 2\gamma_L(t)|\varphi_0|^2 \tag{6.73}$$

which may be solved as

$$|\varphi_0(t)|^2 = |\varphi_0(0)|^2 \exp\left[2\int_0^t \gamma_L(t')\,dt'\right]. \tag{6.74}$$

In the light of Fig. 6.6, we may thus expect that $|\varphi_0(t)|$ also contains an oscillatory component with frequency $\sim\omega_B$.

This phenomenon is called the amplitude oscillation; it reflects the importance of the particle-trapping effects in large-amplitude plasma waves. The period of the amplitude oscillation as well as the criterion, (6.71) versus (6.72), has been investigated experimentally for both the electron plasma oscillation [19] and the ion-acoustic wave [20].

C. The BGK Solutions

In the previous section, we argued that a plateau formation is expected as a time asymptotic solution for the distribution function in the trapped domain of a large-amplitude plasma wave. We may also note that when such a plateau is formed, $f'(\omega/k)=0$; the wave amplitude in turn remains stationary. Based on such an observation, we may attempt to construct an exact stationary solution to the Vlasov–Poisson equations containing an arbitrary electrostatic wave with finite amplitude.

We work in the phase space (X, V) defined according to Eqs. (6.66) and (6.68), the frame comoving with the electrostatic wave (6.65). In that part of the phase space in which the trapping takes place (i.e., $W<0$), we set the distribution function to be constant

$$f(X,V) = f_p \qquad (W<0). \tag{6.75}$$

A plateau with the height f_p has thus been formed; f_p is related to the number of the particles trapped in the potential troughs, which should be determined later in a self-consistent way. In the untrapped domain, we assume that $f(X, V)$ is a function of W and the sign of V only:

$$f(X,V) = \begin{cases} F_1(W) & (W\geqslant 0, V>0) \\ F_2(W) & (W\geqslant 0, V<0) \end{cases} \tag{6.76}$$

The distribution function must be continuous, so that

$$F_1(0) = F_2(0) = f_p. \tag{6.77}$$

The time-independent version of the Vlasov equation in the phase space (X, V) is

$$V\frac{\partial f}{\partial X} - \frac{q}{m}\frac{\partial \varphi}{\partial X}\frac{\partial f}{\partial V} = 0. \tag{6.78}$$

In the light of Eqs. (6.67) and (6.69), it is clear that the distribution function (6.75) and (6.76) indeed satisfies the Vlasov equation (6.78). The existence of the plateau (6.75) also guarantees the stationarity of the wave.

The remaining task, therefore, is to examine the solution in terms of the Poisson equation, which reads

$$\frac{\partial^2 \varphi}{\partial X^2} = -4\pi qn\left[\int_{-\infty}^{\infty} f(X, V)\, dV - 1\right]. \tag{6.79}$$

Substitution of (6.75) and (6.76) into Eq. (6.79) yields an integral equation which determines the shapes of $F_1(W)$ and $F_2(W)$ and the level of f_p. From a solution of this equation we can thus determine an exact solution of the Vlasov–Poisson equations for arbitrary stationary waves. It is in fact possible to extend this approach and construct waves of quite arbitrary shape, for instance, isolated pulses and sinusoidal waves. The existence and the properties of the solutions to such nonlinear problems have been investigated by Bernstein, Greene, and Kruskal [21]; they are called the *BGK solutions*.

References

1. G. Van Hoven, *Phys. Rev. Letters* **17**, 169 (1966).
2. H. Derfler and T. C. Simonen, *Phys. Rev. Letters* **17**, 172 (1966).
3. J. H. Malmberg and C. B. Wharton, *Phys. Rev. Letters* **17**, 175 (1966).
4. A. Y. Wong, N. D'Angelo, and R. W. Motley, *Phys. Rev. Letters* **9**, 415 (1962); A. Y. Wong, R. W. Motley, and N. D'Angelo, *Phys. Rev.* **133**, A436 (1964).
5. R. W. Gould, T. M. O'Neil, and J. H. Malmberg, *Phys. Rev. Letters* **19**, 219 (1967); T. M. O'Neil and R. W. Gould, *Phys. Fluids* **11**, 134 (1968).
6. E. L. Hahn, *Phys. Rev.* **80**, 580 (1950).
7. R. M. Hill and D. E. Kaplan, *Phys. Rev. Letters* **14**, 1062 (1965).
8. I. D. Abella, N. A. Kurnit, and S. R. Hartmann, *Phys. Rev.* **141**, 391 (1966).
9. J. H. Malmberg, C. B. Wharton, R. W. Gould, and T. M. O'Neil, *Phys. Rev. Letters* **20**, 95 (1968); *Phys. Fluids* **11**, 1147 (1968).
10. H. Ikezi and N. Takahashi, *Phys. Rev. Letters* **20**, 140 (1968); H. Ikezi, N. Takahashi, and K. Nishikawa, *Phys. Fluids* **12**, 853 (1969).
11. D. R. Baker, N. R. Ahern, and A. Y. Wong, *Phys. Rev. Letters* **20**, 318 (1968).
12. V. I. Karpman, *Zhur. Eksptl. Teoret. Fiz.* **51**, 907 (1966) [*Soviet Phys. JETP* **24**, 603 (1967)].

13. C. H. Su and C. Oberman, *Phys. Rev. Letters* **20**, 427 (1968); F. L. Hinton and C. Oberman, *Phys. Fluids* **11**, 1982 (1968).
14. T. M. O'Neil, *Phys. Fluids* **11**, 2420 (1968).
15. T. H. Jensen, J. H. Malmberg, and T. M. O'Neil, *Phys. Fluids* **12**, 1728 (1969).
16. A. Lenard and I. B. Bernstein, *Phys. Rev.* **112**, 1456 (1958).
17. L. M. Al'tshul' and V. I. Karpman, *Zhur. Eksptl. Teoret. Fiz.* **49**, 515 (1965) [*Soviet Phys. JETP* **22**, 361 (1966)].
18. T. M. O'Neil, *Phys. Fluids* **8**, 2255 (1965).
19. J. H. Malmberg and C. B. Wharton, *Phys. Rev. Letters* **19**, 775 (1967); C. B. Wharton, J. H. Malmberg, and T. M. O'Neil, *Phys. Fluids* **11**, 1761 (1968).
20. N. Sato, H. Ikezi, Y. Yamashita, and N. Takahashi, *Phys. Rev. Letters* **20**, 837 (1968); *Phys. Rev.* **183**, 278 (1969).
21. I. B. Bernstein, J. M. Greene, and M. D. Kruskal, *Phys. Rev.* **108**, 546 (1957).

Problems

6.1. Derive Eqs. (6.2) and (6.4).

6.2. Consider the behavior of the dielectric response function (6.5) along the imaginary axis on the ω plane. Assuming $k>0$, show that

$$\epsilon(k,iy)\to 1+(\omega_p^2/y^2) \qquad (y\to+\infty),$$

$$\epsilon(k,iy)\to 1+(\omega_p^2/y^2)-2\pi i(\omega_p^2/k^2)f'(iy/k) \qquad (y\to-\infty).$$

Examine similar properties for the W function defined by Eq. (4.3).

6.3. Consider an electron plasma without a magnetic field whose one-dimensional velocity distribution (in the x direction) is given by a Lorentzian

$$f(v)=\frac{1}{\pi}\frac{\zeta}{v^2+\zeta^2}.$$

At $t=0$, we apply a pulse field $\varphi_{\text{ext}}(x,t)=\varphi_1\cos(kx)\delta(\omega_p t)$ to the plasma, and follow its linear response; the pulse shape satisfies $(k\zeta)^2\ll\omega_p^2$.

(a) Investigate the time-dependent behavior of the induced field $\varphi_{\text{ind}}(x,t)$.

(b) Calculate the perturbed part $\delta f(x,v;t)$ of the distribution function for $t\gg 1/k\zeta$.

6.4. Derive Eqs. (6.45) and (6.46).

6.5. Following the treatment described in Section 6.2C, investigate the effects of collisions on the temporal plasma-wave echo when the collision term takes the form

$$\left.\frac{\partial f}{\partial t}\right]_c=-\nu(f-f_0).$$

Here, ν is an effective collision frequency and f_0 is an equilibrium distribution function.

CHAPTER 7

INSTABILITIES IN HOMOGENEOUS PLASMAS

As we studied in the previous chapters, the collective modes in plasmas near thermodynamic equilibrium usually represent stable elementary excitations. These collective modes suffer a certain amount of damping through resonant interactions with individual particles; their lifetimes thereby remain finite.

When the plasma is away from thermodynamic equilibrium, a collective mode may in some cases become unstable; the amplitude of such a collective mode tends to grow exponentially. Most of the plasmas found in nature are in fact significantly far from thermodynamic equilibrium. For example, a spatial variation of the physical quantities, such as the density, the temperature, and the intensity of the magnetic field, may be involved in laboratory plasmas as well as in astrophysical plasmas. Plasmas may also be subjected to the influence of external force fields, which would create flows of particles, momentum, and energy in plasmas. Possibilities of plasma instabilities exist in any of these nonequilibrium circumstances.

When such instability sets in, the properties of the plasma change drastically. As against the relatively quiescent state of the plasma near thermodynamic equilibrium, the plasma now is characterized by the presence of significantly increased fluctuations; such a state is referred to as plasma turbulence. Among other things, the level of the fluctuations affects the rates of transport processes in the plasma. We may thus be speaking of the anomalous transport processes associated with the onset of a plasma instability. In this and the following chapters, we shall mainly be concerned with investigation of the conditions under which the onset of various plasma instabilities may be expected. In the subsequent chapters, we shall consider some topics closely related to the consequences of such instabilities, namely, fluctuations, transport processes, and turbulence.

The source of energy responsible for the onset of a plasma instability may be sought in the excess of the free energy contained in a plasma away from thermodynamic equilibrium [1]. In this regard, we find it meaningful to classify various plasma instabilities into two groups: instabilities in homogeneous plasmas, and those in inhomogeneous plasmas. In the former cases, possible deviations from equilibrium occur in the velocity space, but not in the configuration space. The kinetic energies associated with particle streaming or anisotropic velocity distribution, for example, are the major sources of the extra free energy in these circumstances. The instabilities in this category are treated in this chapter. The instabilities to be considered in the following chapter will involve additional effects of spatial variations; energy released by plasma expansion, energy associated with the diamagnetic current in inhomogeneous plasma, the magnetic energy, and potential energy liberated from a plasma displacement are examples of the free energy available in such cases.

7.1 THE PENROSE CRITERION FOR PLASMA WAVE INSTABILITY

The stability of a plasma wave may be investigated through examination of the imaginary part γ_k of its characteristic frequency. When $\gamma_k<0$, the plasma wave decays; the plasma is stable against an excitation of such a collective mode. When $\gamma_k>0$, on the other hand, the plasma wave grows exponentially; the plasma is said to be unstable against such an excitation.

For a given one-dimensional velocity distribution $f(v)$ of the electrons, we can investigate the properties of the plasma oscillations through the dielectric response function (6.5). With the aid of Eq. (B.4), the dispersion relation in the vicinity of the real axis (i.e., $|\gamma_k/\omega_k|\ll 1$) becomes

$$1-\frac{\omega_\mathrm{p}^2}{k^2}P\int_{-\infty}^{\infty}\frac{f'(v)}{v-(\omega/k)}\,dv-i\pi\frac{\omega_\mathrm{p}^2}{k^2}f'\left(\frac{\omega}{k}\right)=0. \tag{7.1}$$

Assuming that the phase velocity $|\omega/k|$ of the wave is much greater than the average velocity of the electrons, we may expand $[v-(\omega/k)]^{-1}$ in Eq. (7.1) with respect to vk/ω, and then carry out partial integrations with respect to v. The result is

$$1-\frac{\omega_\mathrm{p}^2}{\omega^2}\left[1+2k\frac{\langle v\rangle}{\omega}+3k^2\frac{\langle v^2\rangle}{\omega^2}+\cdots\right]-i\pi\frac{\omega_\mathrm{p}^2}{k^2}f'\left(\frac{\omega}{k}\right)=0 \tag{7.2}$$

where

$$\langle v^l\rangle\equiv\int_{-\infty}^{\infty}dv\,v^l f(v).$$

The frequency ω_k may be determined from the real part of Eq. (7.2); from its imaginary part, we thereby find

$$\gamma_k=\frac{\pi}{2}\frac{\omega_\mathrm{p}^2\omega_k}{k^2}f'\left(\frac{\omega_k}{k}\right). \tag{7.3}$$

This equation is essentially the same as Eq. (6.63). It then follows that the plasma wave is stable when

$$(\omega_k/k)f'(\omega_k/k)<0. \tag{7.4}$$

The foregoing criterion has the merit of simplicity and transparency; it can be understood physically in terms of resonant coupling between the wave and particles. As stated in (7.4), however, the criterion suffers a drawback in that it is not invariant under Galilean transformation. Consider as an example the situation as depicted in Fig. 7.1; a given velocity distribution is observed from two different frames of reference, 0 and 0′. In the frame 0, $vf'(v)<0$ everywhere. In the frame 0′, however, $vf'(v)$ takes on positive values in the domain between 0 and 0′ in Fig. 7.1. The magnitude and the sign of the function $vf'(v)$, therefore, depend on the frame of reference.

A stability criterion free of such a shortcoming has been obtained by Penrose [2]. Consider again the dispersion relation (7.1) on the real axis in the ω plane; it represents the condition for marginal stability, which is expressed as

$$f'(\omega/k)=0, \tag{7.5a}$$

$$\int_{-\infty}^{\infty} \frac{f'(v)}{v-(\omega/k)}\,dv=\frac{k^2}{\omega_p^2}. \tag{7.5b}$$

The principal part integral is not needed in Eq. (7.5b) because of Eq. (7.5a). The stability criterion may thereby be stated as follows: Exponentially growing modes exist if, and only if, there is a minimum of $f(v)$ at $v=v_m$ such that

$$\int_{-\infty}^{\infty} \frac{f'(v)}{v-v_m}\,dv>0. \tag{7.6}$$

It follows from this criterion that a single-humped velocity distribution, as shown in Fig. 7.1, cannot sustain a growing plasma oscillation.

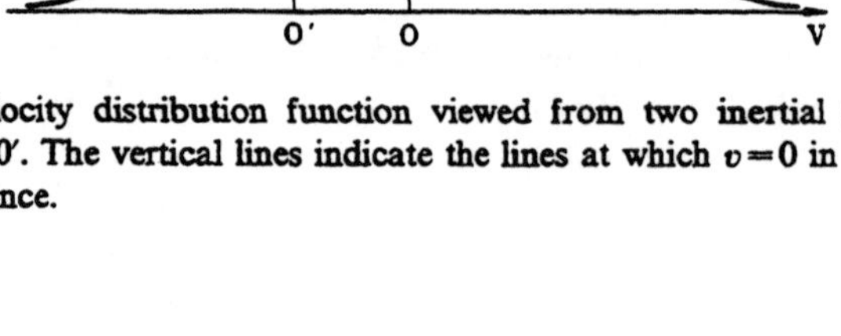

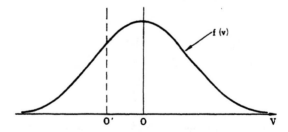

Fig. 7.1 A velocity distribution function viewed from two inertial frames of reference, 0 and 0′. The vertical lines indicate the lines at which $v=0$ in the respective frames of reference.

The criterion just stated may be proved most clearly with the aid of the Nyquist criterion in the servomechanism theory. Consider the mapping between two complex planes ω and Z, specified by the functional relationship

$$Z(\omega)=\int_{\bar{C}}\frac{f'(v)}{v-(\omega/k)}\,dv, \tag{7.7}$$

where $\bar{C}$ is the contour defined in Fig. 4.1. The dispersion relation $\epsilon(k,\omega)=0$ is then expressed as

$$k^2/\omega_p^2=Z(\omega). \tag{7.8}$$

Hence, we may conclude that, if the shape of $f(v)$ is such that a portion of the positive real axis in the Z plane is mapped through Eq. (7.7) into the upper half of the ω plane, then the excitations with positive γ_k can exist for those values of k corresponding to that portion of the real Z axis. If, on the other hand, no portions of the positive real Z axis fall into the upper half of the ω plane through the mapping, then no solutions of ω from Eq. (7.8) have positive imaginary parts; plasma waves would be stable.

When $f(v)$ is a Maxwellian,

$$Z(\omega)=-(m/T)W\left[\omega/k(T/m)^{1/2}\right].$$

The upper half of the ω plane is mapped through this function into the Z plane as shown in Fig. 7.2. The plasma is stable, since no portions of the positive real Z axis are enclosed by the contour. If we deform the shape of $f(v)$ from the Maxwellian so that the mapping now goes as shown in Fig. 7.3(a) or in Fig. 7.3(b), for example, then a portion of the positive real Z axis is found inside the mapped contour. Exponentially growing plasma oscillations are possible in these circumstances.

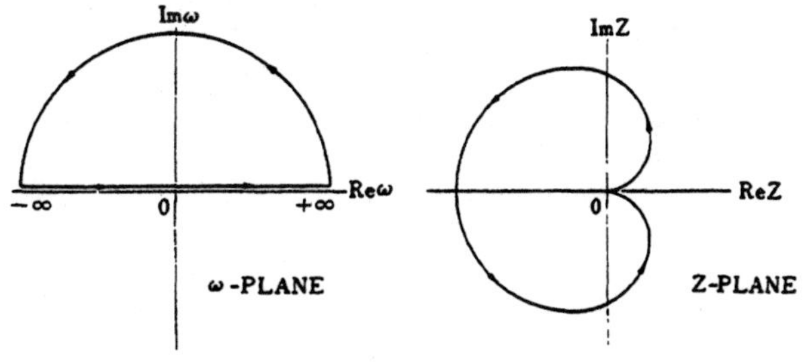

Fig. 7.2 Mapping between ω and Z according to Eq. (7.7) for a Maxwellian plasma.

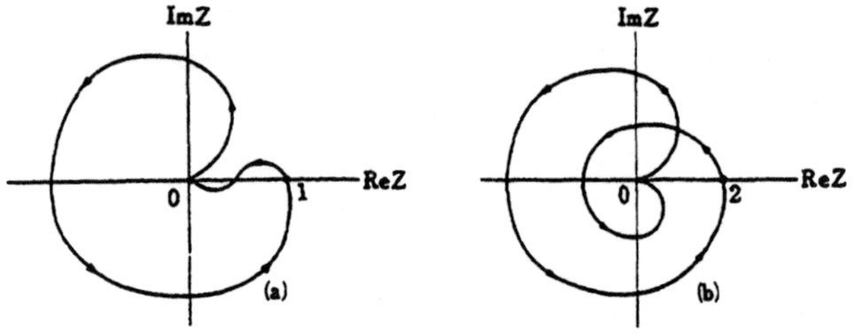

Fig. 7.3 Examples of mapping indicating instabilities.

The important point to be noted in Fig. 7.3 is the intersection between the contour and the positive real Z axis at 1 or 2. The features common to all those intersections are (1) $\mathrm{Im}\,Z(\omega)=0$; (2) the contour crosses such an intersection from $\mathrm{Im}\,Z(\omega)<0$ to $\mathrm{Im}\,Z(\omega)>0$; and (3) $\mathrm{Re}\,Z(\omega)>0$. The conditions (1) and (2) imply that $f'(\omega/k)=0$ and $f''(\omega/k)>0$ at the intersection; hence it corresponds to a minimum point of $f(v)$. The condition (3) is equivalent to (7.6). Consequently, the Penrose criterion amounts to the condition for the existence of an intersection like 1 and 2 when the real ω axis is mapped into the Z plane through Eq. (7.7).

When such an intersection exists, a portion of the positive real Z axis should certainly be enclosed by the contour. (The upper half of the ω plane is mapped onto the left-hand side of the contour in the Z plane.) Conversely, when some portion of the positive real Z axis is mapped into the upper half of the ω plane, the right-side boundary of that portion must correspond to a point which satisfies the three conditions listed earlier. The proof of the Penrose criterion is thus completed.

7.2 BEAM–PLASMA INSTABILITY

A typical case of plasma instability, perhaps the simplest from a mathematical viewpoint, occurs in a beam–plasma system [3]. Consider a monochromatic beam of charged particles injected into a single-component plasma consisting of cold electrons; for simplicity, we assume that the beam also consists of electrons, although cases involving other kinds of charged particles can be treated similarly with slight and obvious modifications. The distribution function is

$$f(\mathbf{v})=(n_0/n)\delta(\mathbf{v})+(n_b/n)\delta(\mathbf{v}-\mathbf{v}_d), \qquad n\equiv n_0+n_b \tag{7.9}$$

where $\mathbf{v}_d$ is the beam velocity, n_0 and n_b are the densities of the cold electrons and the beam electrons.

The dielectric response function, Eq. (3.50), may then be calculated as

$$\epsilon(\mathbf{k},\omega)=1-\frac{\omega_0^2}{\omega^2}-\frac{\omega_b^2}{(\omega-kv_d)^2} \tag{7.10}$$

where

$$\omega_0^2=\frac{4\pi n_0e^2}{m},\qquad \omega_b^2=\frac{4\pi n_be^2}{m}.$$

and $v_d=\mathbf{v}_d\cdot\mathbf{k}/k$ is the component of the drift velocity in the direction of the wave vector. Introducing the dimensionless variables

$$x\equiv\frac{\omega}{\omega_0},\qquad z\equiv\frac{kv_d}{\omega_0},\qquad \alpha\equiv\frac{\omega_b^2}{\omega_0^2}=\frac{n_b}{n_0}, \tag{7.11}$$

we find that the dispersion relation takes the form

$$1=1/x^2+[\alpha/(x-z)^2]. \tag{7.12}$$

Since this is a quadruple equation with respect to x, we can solve it exactly by an algebraic means. Stability of the system can be analyzed simply by a graphical means, however.

We denote the right-hand side of Eq. (7.12) as $y(x;z)$, and plot it in Fig. 7.4, where z is varied as parameter. The solutions of Eq. (7.12) should then be

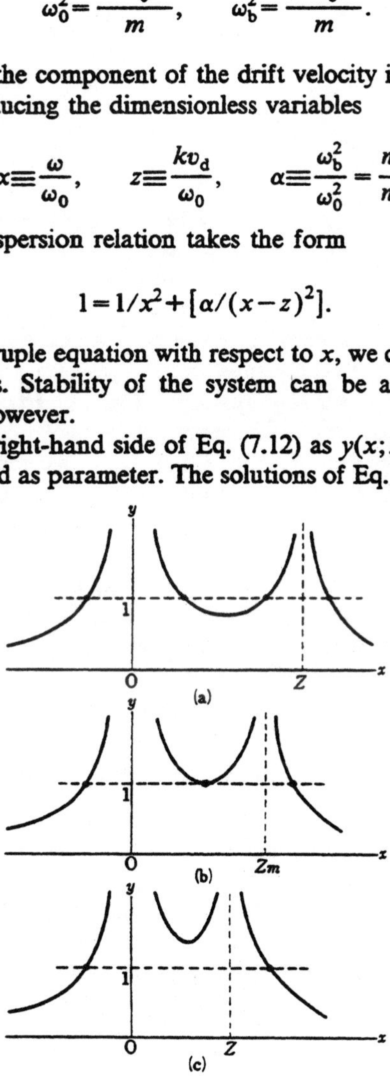

Fig. 7.4 Beam–plasma instability.

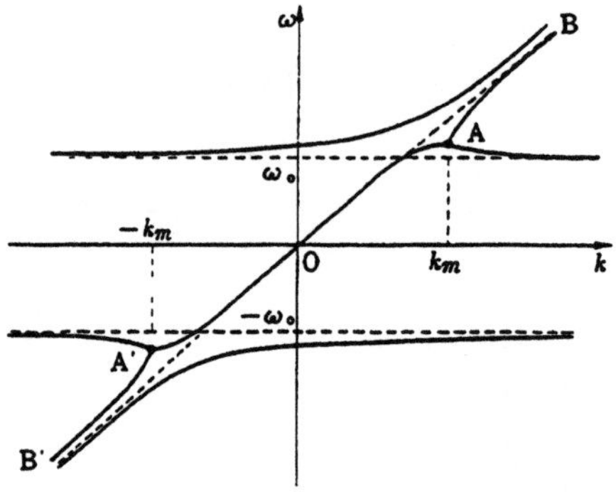

Fig. 7.5 Dispersion relations of a beam–plasma system (schematic) with $\omega_b^2 \ll \omega_0^2$. The broken line BOB' represents the relation $\omega = kv_d$.

given by the intersections between those curves and the line $y = 1$. When z (i.e., the wave number) is sufficiently large, $y(x; z)$ may behave as shown in Fig. 7.4(a), so that four real solutions for x (i.e., the frequency) exist; hence, the system is stable. If, on the contrary, z is very small, we may have a situation like that described by Fig. 7.4(c); only two real solutions are now available. The rest of two solutions must therefore be complex conjugate to each other, representing a growing mode and a damped one. The system is unstable in these circumstances.

The maximum value of z for such an instability can be calculated from the condition that the curve $y(x; z_m)$ be tangential to the line $y = 1$, as shown in Fig. 7.4(b). From a simple calculation, we find

$$z_m = (1 + \alpha^{1/3})^{3/2},$$

or recovering the original variables through (7.11), we obtain

$$k_m = (\omega_0 / v_d)\left[1 + (n_b/n_0)^{1/3}\right]^{3/2}. \tag{7.13}$$

In most practical cases, we can assume that $\omega_b^2 \ll \omega_0^2$. Figure 7.5 schematically describes the dispersion relations of the beam-plasma system under these circumstances. The branch AOA' represents the real part of the complex solutions. The beam–plasma instabilities have been extensively investigated experimentally.

7.3 ION-ACOUSTIC WAVE INSTABILITY

In Section 4.4D, we saw that in a two-component plasma with $T_e \gg T_i$ a pair of well-defined acoustic modes with the frequencies $\omega = \pm sk$ exist in the long-wavelength domain; each of these modes suffers a small and equal amount of damping. When we impart a drift motion of the electrons as a whole relative to the ions, one of the acoustic modes acquires a longer lifetime; the other mode, on the contrary, suffers stronger damping. When the drift velocity is increased further, the former mode becomes unstable. The existence of such an ion-acoustic wave instability under the action of an applied electric field has been considered by many investigators both for gaseous plasmas [4–8] and for solid-state plasmas [9, 10].

We fix our frame of reference on the rest frame of the ions. The one-dimensional distributions in the direction of the wave vector **k** may then be expressed as

$$f_e(v) = \frac{1}{(2\pi T_e/m_e)^{1/2}} \exp\left[-\frac{(v-v_d)^2}{2(T_e/m_e)} \right], \tag{7.14a}$$

$$f_i(v) = \frac{1}{(2\pi T_i/m_i)^{1/2}} \exp\left[-\frac{v^2}{2(T_i/m_i)} \right], \tag{7.14b}$$

where v_d is the component of the drift velocity also in the **k** direction. The dielectric response function of this system is

$$\epsilon(k,\omega) = 1 + \frac{k_e^2}{k^2} W\left(\frac{\omega - kv_d}{k(T_e/m_e)^{1/2}} \right) + \frac{k_i^2}{k^2} W\left(\frac{\omega}{k(T_i/m_i)^{1/2}} \right). \tag{7.15}$$

For the ion-acoustic wave, we consider the frequency domain

$$(T_i/m_i)^{1/2} \ll |\omega/k| \ll (T_e/m_e)^{1/2} + v_d, \tag{7.16}$$

so that the dielectric response function (7.15) may be approximated as

$$\epsilon(k,\omega) = 1 + \frac{k_e^2}{k^2} - \frac{\omega_i^2}{\omega^2} + i\left(\frac{\pi}{2}\right)^{1/2} \frac{k_e^2(\omega - kv_d)}{k^3(T_e/m_e)^{1/2}}$$
$$+ i\left(\frac{\pi}{2}\right)^{1/2} \frac{k_i^2\omega}{k^3(T_i/m_i)^{1/2}} \exp\left[-\frac{\omega^2}{2k^2(T_i/m_i)} \right]. \tag{7.17}$$

The equation $\epsilon(k,\omega)=0$ can then be solved in a standard way; we find

$$\omega=\pm\omega_k=\pm\frac{\omega_i k}{(k^2+k_e^2)^{1/2}}, \tag{7.18a}$$

$$\gamma_k=-\left(\frac{\pi}{8}\right)^{1/2}\frac{\omega^3}{k^3\omega_i^2}\left\{\frac{k_e^2(\omega-kv_d)}{(T_e/m_e)^{1/2}}+\frac{k_i^2\omega}{(T_i/m_i)^{1/2}}\exp\left[-\frac{\omega^2}{2k^2(T_i/m_i)}\right]\right\}\Bigg|_{\omega=\pm\omega_k}. \tag{7.18b}$$

Equation (7.18a) is the same as Eq. (4.65); Eq. (7.18b) differs from the imaginary part of Eq. (4.60a) essentially by the presence of the drift velocity. Of the two solutions described by Eqs. (7.18), the branch $\omega=-\omega_k$ is always damped; its rate of damping is greater than the one in the absence of the drift motion. For this reason, we shall no longer pay attention to this particular branch.

To investigate a possible instability, we thus substitute $\omega=\omega_i k/(k^2+k_e^2)^{1/2}$ in Eq. (7.18b). In the limit of long wavelengths, we obtain

$$\gamma_k=-\left(\frac{\pi}{8}\right)^{1/2}\left(\frac{m_e}{m_i}\right)^{1/2}k\left\{(s-v_d)+s\left(\frac{m_iT_e^3}{m_eT_i^3}\right)^{1/2}\exp\left(-\frac{T_e}{2T_i}\right)\right\}. \tag{7.19}$$

When $T_e \gg T_i$, the last term in Eq. (7.19), due to the ionic Landau damping, may be negligible. In these circumstances, we find that the ion-acoustic wave becomes unstable when the drift velocity exceeds the sound velocity (i.e., $v_d > s$).

The condition for the marginal stability, $\gamma_k=0$, determines the boundary between the growing and damped waves; it may be described by a curve $v_d(k)$ in the v_d–k plane. To calculate such a boundary curve, we must go back to Eq. (7.15) and solve the equation $\epsilon(k,\omega)=0$ for real values of ω.

A graphical method of obtaining the boundary curve is explained in Fig. 7.6. We first rewrite Eq. (7.15) so that

$$-\frac{k^2}{k_e^2}-W\left(\frac{\omega-kv_d}{k(T_e/m_e)^{1/2}}\right)=\frac{k_i^2}{k_e^2}W\left(\frac{\omega}{k(T_i/m_i)^{1/2}}\right). \tag{7.20}$$

For a fixed value of k^2/k_e^2, the left- and right-hand sides of Eq. (7.20) are plotted in Fig. 7.6(a) as curve I and curve II, respectively; two intersections, A and B, may thereby be found. Each intersection determines two numbers according to its reading along curve I and curve II. For the point A, let us assume that the reading is a_1 on I and a_2 on II. Note that $a_1<0$ in the case of

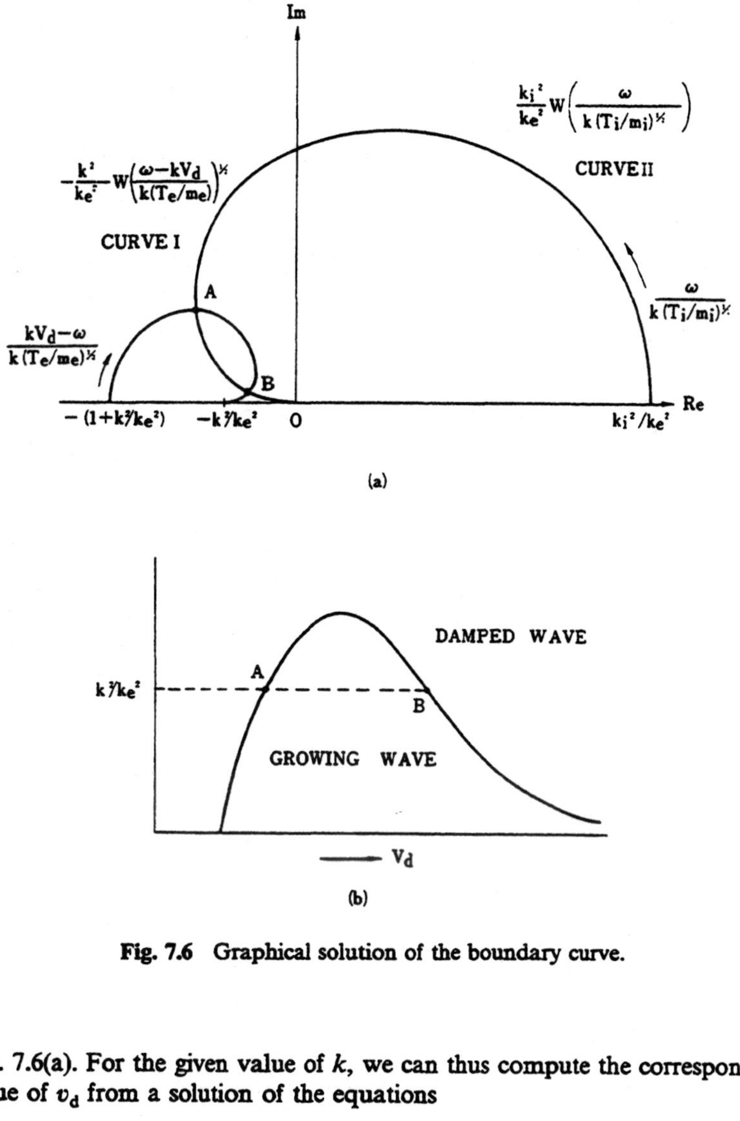

Fig. 7.6 Graphical solution of the boundary curve.

Fig. 7.6(a). For the given value of k, we can thus compute the corresponding value of v_d from a solution of the equations

$$(\omega - kv_d)/k(T_e/m_e)^{1/2} = a_1, \qquad \omega/k(T_i/m_i)^{1/2} = a_2.$$

Similarly, another value of v_d may be computed from intersection B. Consequently, we can determine two points, A and B, on the boundary curve, as in Fig. 7.6(b). Repetition of these procedures for different values of k will then produce the entire boundary curve in the v_d–k plane. Some examples of such boundary curves calculated for an electron–hole plasma in InSb are shown in Fig. 7.7.

When $T_e \gg T_i$, it is possible to use Eqs. (7.18) approximately to calculate the front edge of the boundary curve, that is, the side determined from the intersections A of Fig. 7.6. From the condition that $\gamma_k = 0$, we thus obtain

$$v_d(k) = \frac{\omega_k}{k}\left\{1 + \left(\frac{m_i}{m_e}\right)^{1/2}\left(\frac{T_e}{T_i}\right)^{3/2} \exp\left[-\frac{1}{2}\left(\frac{T_e/T_i}{1+k^2/k_e^2} + 3\right)\right]\right\} \tag{7.21}$$

where ω_k has been given by Eq. (7.18a).

When such a boundary curve is given, it is straightforward to determine the condition for the onset of an ion-acoustic instability. A *critical point* may

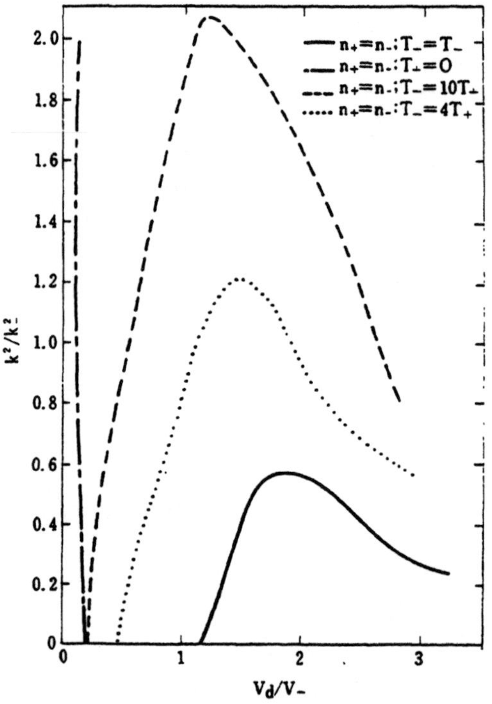

Fig. 7.7 The boundary between growing waves and damped waves for an electron–hole plasma. $n_\pm$ and $T_\pm$ are the densities and temperatures, respectively, of the holes and the electrons: $k_-^2 = 4\pi n_- e^2/\epsilon_0 T_-$, $V_- = (2T_-/m_-)^{1/2}$, and $m_+ = 14m_-$ is assumed ($\epsilon_0 \cong 16$ for InSb). After Pines and Schrieffer [9].

be defined as that at which, with increasing v_d, the acoustic wave first becomes unstable. Associated with it, the critical wave number k_c and the minimum drift velocity $v_c = v_d(k_c)$ may also be defined. In the vicinity of such a critical point, we can approximately express the dielectric response function (7.17) as

$$\epsilon(k,\omega) \cong \left(1+\frac{k_e^2}{k^2}\right)\left(1-\frac{\omega_k^2}{\omega^2}\right)+i\left(\frac{\pi}{2}\right)^{1/2}\frac{k_e^2}{k^2}\frac{v_d(k)-v_d}{(T_e/m_e)^{1/2}}. \tag{7.22}$$

As v_d approaches the critical velocity v_c, the imaginary part of $\epsilon(k,\omega)$ tends to vanish at $k=k_c$; this would imply an infinite lifetime for the ion-acoustic wave, or the onset of an instability.

7.4 INSTABILITIES NEAR HARMONICS OF THE CYCLOTRON FREQUENCIES

In the presence of a strong magnetic field, the nature of the longitudinal collective modes becomes remarkably different from that in the absence of the magnetic field. As we noted in Section 4.6, the electron plasma oscillation and the ion-acoustic wave propagate preferentially in the direction of the magnetic field. The energy source of such oscillations is, therefore, the kinetic energy of the particles associated with their parallel motion along the magnetic field. We have also seen the emergence of a new collective mode, the Bernstein mode, which propagates in a direction almost perpendicular to the magnetic field. This is a collective mode associated with the cyclotron motion of the particles; its energy source may be sought in the kinetic energy of the particle motion perpendicular to the magnetic field.

When the plasma is in thermodynamic equilibrium, the energy content in each of those collective modes is balanced; the system is manifestly stable. When an imbalance of energy contents between two collective modes exists, there arises a possibility of instability through coupling of those two modes. We begin this section with a short discussion of the instability from the viewpoint of such a coupling.

A. Coupling between Collective Modes and Instability

When the frequencies and the wave vectors of two collective modes match each other, these collective modes couple strongly. If there is an imbalance in the energy contents between the two modes involved, the excessive amount of energy in one mode may thereby be fed into the excitation of the other mode through this coupling. To establish a growing tendency of the wave, this energy input must exceed the ordinary decay rate of the wave inherent in the plasma (e.g., Landau damping). Nevertheless, these arguments indicate a possibility of instability associated with imbalance in the energy contents between mutually coupling collective modes.

In fact, we can interpret the ion-acoustic wave instability of the previous section in terms of such a coupling between two collective modes. Generally, when a group of particles moves as a whole with a velocity $\mathbf{v}_d$, it is possible to define another collective mode

$$\omega = \mathbf{k} \cdot \mathbf{v}_d \tag{7.23}$$

associated with such a drift motion. When the frequency and the wave vector of this collective mode match those of the ion-acoustic wave, an instability takes place. (For the moment, we ignore the ion Landau damping.) As the drift velocity further increases, energy is fed from the drift mode (7.23) to the ion-acoustic wave, resulting in the growth of the latter wave.

A situation quite similar to the foregoing may take place in a plasma with a magnetic field when the Bernstein modes, on the one hand, and the electron plasma oscillation or the ion-acoustic wave, on the other, couple each other. The comparison to be made here is between the energy contained in the cyclotron motion of the particles and that associated with the parallel motion. When the former exceeds the latter, for example, a possibility arises that the energy of cyclotron motion is fed into the electron plasma oscillation or the ion-acoustic wave in the vicinity of the harmonics of the cyclotron frequency. Such instabilities have been extensively investigated by Harris and his coworkers [11]. Let us therefore consider some of the instabilities in this category.

B. Anisotropic Velocity Distributions

A situation involving energy imbalance between the parallel and perpendicular degrees of freedom (with respect to the magnetic field) may be described by assigning two different temperatures, $T_\parallel$ and $T_\perp$, to the respective degrees of freedom. Hence, in place of the isotropic distribution (4.66), we may now have

$$f(v_\perp, v_\parallel) = \left(\frac{m}{2\pi T_\perp}\right)\left(\frac{m}{2\pi T_\parallel}\right)^{1/2} \exp\left[-\frac{v_\perp^2}{2(T_\perp/m)} - \frac{v_\parallel^2}{2(T_\parallel/m)}\right]. \tag{7.24}$$

In the light of the relation (4.44), such an anisotropic distribution is more likely to occur in the ions rather than in the electrons. For this reason, we shall assume Eq. (7.24) only for the ions; the electron distribution is still assumed to be isotropic, described by Eq. (4.66).†

†When the electron distribution also becomes anisotropic and takes the form of Eq. (7.24), the results obtained in this section are to be modified by replacing T_e by $T_{e\parallel}$, the electron temperature parallel to the magnetic field.

The instability under consideration takes place in the vicinity of the ion cyclotron harmonics; the dynamics of the ions plays a more important part than that of the electrons. For the sake of notational simplicity, we shall henceforth omit the subscript i except for ω_i and $k_i \equiv (4\pi n_i e^2/T_\parallel)^{1/2}$; the quantities without subscript refer to those associated with the ions.

The dielectric response function of the system is then calculated as

$$\epsilon(\mathbf{k},\omega) = 1 + \frac{k_e^2}{k^2}\left\{1 + \sum_n \frac{\omega}{\omega - n\Omega_e}\left[W\left(\frac{\omega - n\Omega_e}{|k_\parallel|(T_e/m_e)^{1/2}}\right) - 1\right]\Lambda_n(\beta_e)\right\}$$

$$+ \frac{k_i^2}{k^2}\left\{1 + \sum_n \left[1 + \frac{T_\parallel}{T_\perp}\frac{n\Omega}{\omega - n\Omega}\right]\left[W\left(\frac{\omega - n\Omega}{|k_\parallel|(T_\parallel/m)^{1/2}}\right) - 1\right]\Lambda_n(\beta)\right\} \tag{7.25}$$

where $\beta = k_\perp^2 T_\perp/\Omega^2 m$. We consider the situation in which a strong magnetic field is involved; an average Larmor radius of the electrons may be assumed to be much smaller than the wavelength of the phenomenon under consideration. Hence, $\beta_e \ll 1$. We take this inequality further to its limit and assume $\beta_e \to 0$; we shall thereby adopt the zero Larmor-radius approximation for the electrons. Equation (7.25) then reduces to

$$\epsilon(\mathbf{k},\omega) = 1 + \frac{k_e^2}{k^2} W\left(\frac{\omega}{|k_\parallel|(T_e/m_e)^{1/2}}\right)$$

$$+ \frac{k_i^2}{k^2}\left\{1 + \sum_n \left[1 + \frac{T_\parallel}{T_\perp}\frac{n\Omega}{\omega - n\Omega}\right]\left[W\left(\frac{\omega - n\Omega}{|k_\parallel|(T_\parallel/m)^{1/2}}\right) - 1\right]\Lambda_n(\beta)\right\}. \tag{7.26}$$

In the following, we investigate those instabilities described by this dielectric response function in the two limiting cases of the temperature ratio between the electrons and the ions.

C. Instabilities of Electron Plasma Oscillations

In certain cases of plasma experiments, the ion temperatures may be maintained at a substantially higher level than those of the electrons. Such a situation may be expected when the major source of the energy supply in the plasma comes through the kinetic energies of the ions. In an ion-injection machine, for example, high-energy ions are injected into a plasma-confinement system. Ion energies are Maxwellized in the system; a high ionic temperature may be attained. The transfer of the ion energies to the electrons is a process much slower than the establishment of an ion temperature. In a stationary situation, the electron temperature may thus remain relatively low.

Let us therefore consider a cold-electron case such that

$$T_e/m_e \ll T_\parallel/m. \tag{7.27}$$

Since $k_e^2 \gg k_i^2$, the real part $\epsilon_1(\mathbf{k},\omega)$ of the dielectric function (7.26) can be expressed as

$$\epsilon_1(\mathbf{k},\omega) = 1 - (\omega_e^2/\omega^2)(k_\parallel^2/k^2).$$

We thereby find the frequencies of the collective mode as

$$\omega_k = \omega_e(k_\parallel/k) = \omega_e \cos\theta \tag{7.28}$$

where θ is the angle between the wave vector and the magnetic field. This collective mode is the electron plasma oscillation discussed in Section 4.6A.

The imaginary part of the dielectric function can be similarly calculated from Eq. (7.26). The growth rate γ_k of the collective mode is thereby obtained as

$$\gamma_k = -\left(\frac{\pi}{8}\right)^{1/2} \frac{\omega_k^2 k_e^2}{|k_\parallel| k^2 (T_e/m_e)^{1/2}} \exp\left[-\frac{\omega_k^2}{2k_\parallel^2 (T_e/m_e)}\right]$$

$$-\left(\frac{\pi}{8}\right)^{1/2} \frac{k_i^2 \omega_k}{k^2} \sum_n \left[\frac{\omega_k - n\Omega}{|k_\parallel|(T_\parallel/m)^{1/2}} + \frac{T_\parallel}{T_\perp} \frac{n\Omega}{|k_\parallel|(T_\parallel/m)^{1/2}}\right]$$

$$\times \exp\left[-\frac{(\omega_k - n\Omega)^2}{2k_\parallel^2 (T_\parallel/m)}\right] \Lambda_n(\beta). \tag{7.29}$$

The first term on the right-hand side represents a negative definite term arising from the electronic Landau damping. Since we are interested in the collective mode (7.28) in the vicinity of a harmonic frequency $n\Omega$ of the ion cyclotron motion, this Landau damping term becomes negligibly small when

$$(n\Omega)^2 \gg k_\parallel^2 (T_e/m_e). \tag{7.30}$$

We shall assume this condition, which is in agreement with the spirit of the cold electron approximation (7.27).

We now investigate the possibility of instability in the vicinity of $\omega_k = n\Omega$, a condition for the coupling between the electron plasma oscillation and the ion Bernstein mode. In view of Eq. (7.28), we find that such a coupling can take place only when

$$\omega_e > n\Omega. \tag{7.31a}$$

This requirement determines a critical density or a minimum electron density n_e for the instability as

$$n_e > n^2 B^2 m_e / 4\pi m^2 c^2. \tag{7.31b}$$

Having thus established the minimum density, we next examine the sign of the growth rate in the vicinity of $\omega_k = n\Omega$. In the light of (7.30), we may neglect the first term of Eq. (7.29) and write it as

$$\gamma_k \cong -\left(\frac{\pi}{8}\right)^{1/2} \frac{k_i^2 \omega_k}{k^2} \left[\frac{\omega_k - n\Omega}{|k_\parallel|(T_\parallel/m)^{1/2}} + \frac{T_\parallel}{T_\perp} \frac{n\Omega}{|k_\parallel|(T_\parallel/m)^{1/2}} \right]$$

$$\times \exp\left[-\frac{(\omega_k - n\Omega)^2}{2k_\parallel^2 (T_\parallel/m)} \right] \Lambda_n(\beta). \tag{7.32}$$

The second term on the right-hand side of Eq. (7.32) is always negative; it therefore contributes to the damping of the wave. The first term changes its sign as ω_k passes $n\Omega$. In the domain $\omega_k > n\Omega$, this term is negative and hence contributes to additional damping. In the domain $\omega_k < n\Omega$, on the other hand, it becomes positive and thus indicates the possibility of an instability. When the growth rate arising from the first term of Eq. (7.32) exceeds the damping rate of the second term, an instability will take place. The first term gives rise to a maximum growth rate at $\omega_k = n\Omega - |k_\parallel|(T_\parallel/m)^{1/2}$. This frequency must first of all be greater than $(n - \frac{1}{2})\Omega$; otherwise, the growth rate would be completely masked by the strong damping term stemming from the contribution of the $(n-1)$th Bernstein mode. Hence, we have a condition

$$|k_\parallel| < (m/T_\parallel)^{1/2} (\Omega/2). \tag{7.33}$$

Note that a combination of (7.27) and (7.33) essentially yields (7.30) unless n is an extremely large number. The maximum value of Eq. (7.32) in these circumstances is therefore given by

$$\gamma_{\max} \cong \left(\frac{\pi}{8}\right)^{1/2} \frac{k_i^2 (n\Omega)}{k^2} \left[1 - \frac{T_\parallel}{T_\perp} \frac{n\Omega}{|k_\parallel|(T_\parallel/m)^{1/2}} \right] \exp(-\tfrac{1}{2}) \Lambda_n(\beta). \tag{7.34}$$

For an instability to take place, this quantity must be positive; combining this requirement with (7.33), we obtain

$$\left(\frac{T_{\parallel}}{T_{\perp}}\right) < \frac{|k_{\parallel}|(T_{\parallel}/m)^{1/2}}{n\Omega} < \frac{1}{2n} \tag{7.35}$$

for the conditions of plasma-wave instability in the vicinity of a harmonic frequency $n\Omega$ of the ion cyclotron motion.

To summarize, we have found the following: When the electron temperature is relatively low, so that (7.27) is satisfied, the electron plasma oscillation (7.28) is a well-defined elementary excitation of the system. Since its frequency may take on a value between 0 and ω_e, depending on its direction of propagation, the electron plasma oscillation can couple with the nth ion Bernstein mode ($\omega \cong n\Omega$) if the electron density is above a threshold value as described by (7.31b). When the cyclotron motion of the ions contains excessive kinetic energy, so that $T_{\perp} > T_{\parallel}$, a possibility exists that this portion of the energy may be fed into the electron plasma oscillation through the foregoing coupling mechanism at a frequency ω_k slightly less than the nth harmonic frequency of Bernstein mode. The conditions for instability resulting from such excitations of the electron plasma oscillation are given by (7.35).

D. Instabilities of Ion-Acoustic Waves

Instead of the cold-electron limit of (7.27), we may consider the other limiting cases of hot electrons and assume

$$T_e/m_e \gg T_{\parallel}/m. \tag{7.36}$$

This sort of situation may take place when the major supply of the energy to the plasma is achieved through the electrons first. Contrary to the cases discussed in the previous section, it is not a straightforward task to specify relevant physical processes which could create an anisotropic ion distribution in these circumstances. Let us, nevertheless, investigate briefly the consequences arising from the condition (7.36).

The real part of the dielectric function (7.26) now takes the form [in the frequency domain of (4.85a)]

$$\epsilon_1(\mathbf{k},\omega) = 1 + \frac{k_e^2}{k^2} - \frac{\omega_i^2}{\omega^2}\frac{k_{\parallel}^2}{k^2}$$

so that the ion-acoustic wave

$$\omega_k = \frac{\omega_i k_{\parallel}}{(k^2 + k_e^2)^{1/2}} \tag{7.37}$$

emerges as an elementary excitation of the system. This wave will couple with the nth ion Bernstein mode; the condition for such a coupling determines the threshold density of the ions:

$$n_i > n^2 B^2 / 4\pi m c^2. \tag{7.38}$$

Note that (7.38) is more restrictive than (7.31b) by a factor of the mass ratio (m_e/m). Except for this, the conditions for instability in the present case are found to be basically the same as (7.35). A slight modification to be noted here is that the electron Landau damping, the first term of Eq. (7.29), remains finite and takes on a value $-(\pi/8)^{1/2}(m_e/m)^{1/2}\omega_k$ [see Eq. (4.63)]. Within the domain (7.35), the growth rate arising from the second term of (7.28) must therefore overcome this additional damping to create an unstable situation.

Theoretically, we can likewise extend the argument and consider the instabilities in the vicinity of the harmonics of the electron cyclotron frequency associated with an anisotropic distribution of the electrons. The basic physics and mathematical procedures involved here are quite the same as those described above. The conditions for physical realization of such instabilities, however, become much restrictive and in some cases prohibitive.

E. Instabilities Associated with the Drift Motion of the Electrons

In Section 7.3, we saw the possibility of the ion-acoustic wave instability arising as a result of coupling between the drift mode (7.23) and the ion-acoustic wave. Similar instabilities may occur in the vicinity of the harmonics of the ion cyclotron frequency when such a drift mode couples with the ion Bernstein mode in the plasma with magnetic field.

Consider a situation in which the electrons as a whole drift with a velocity v_d relative to the ions in the direction parallel to the magnetic field. Adopting once again the zero Larmor-radius approximation for the electrons (i.e., $\beta_e \to 0$) and assuming for simplicity an isotropic Maxwellian, (4.66), for the ions as well as for the electrons, we find that the dielectric response function becomes

$$\epsilon(\mathbf{k},\omega) = 1 + \frac{k_e^2}{k^2} W\left(\frac{\omega - k_\parallel v_d}{|k_\parallel|(T_e/m_e)^{1/2}} \right)$$
$$+ \frac{k_i^2}{k^2} \left\{ 1 + \sum_n \frac{\omega}{\omega - n\Omega} \left[W\left(\frac{\omega - n\Omega}{|k_\parallel|(T/m)^{1/2}} \right) - 1 \right] \Lambda_n(\beta) \right\}. \tag{7.39}$$

In the frequency domain such that

$$|\omega - k_\parallel v_d| \ll |k_\parallel|(T_e/m_e)^{1/2}, \qquad |\omega - n\Omega| \gg |k_\parallel|(T/m)^{1/2}, \tag{7.40}$$

this dielectric function may be approximated as

$$\epsilon(\mathbf{k},\omega)=1+\frac{k_e^2}{k^2}+\frac{k_i^2}{k^2}\left[1-\sum_n \frac{\omega}{\omega-n\Omega}\Lambda_n(\beta)\right]+i\left(\frac{\pi}{2}\right)^{1/2}\frac{k_e^2}{k^2}\frac{\omega-k_\parallel v_d}{|k_\parallel|(T_e/m_e)^{1/2}}. \tag{7.41}$$

From the real part of Eq. (7.41), we recover the frequencies of the ion Bernstein mode as given by Eq. (4.91). When the drift velocity exceeds its phase velocity along the magnetic field, that is,

$$v_d > \omega/k_\parallel, \tag{7.42}$$

the energy of the drifting electrons is channeled into the Bernstein mode; a growing Bernstein mode may result. This growth rate must then overcome the Landau damping of the ions to produce an unstable situation. Since a Bernstein mode propagates predominantly in the directions almost perpendicular to the magnetic field, a substantially large value of drift velocity may be required to satisfy the condition (7.42).

7.5 LOSS-CONE INSTABILITY

In the previous section, we were concerned mostly with the anisotropic situations as described by the two-temperature Maxwellian (7.24). In many cases of plasmas occurring in the laboratory or in nature, however, it is also quite likely that the velocity distributions deviate substantially from the Maxwellian. Such a situation may arise, for example, when only those particles located in a certain domain of the phase space can be confined, and other particles are allowed to escape away from the system. The particle distribution function resulting from such a special constraint in the phase space will naturally have a shape quite different from a Maxwellian.

The most typical example of such a non-Maxwellian situation is the loss-cone distribution of the plasma confined in the so-called magnetic mirror configuration. The system tends to keep those particles which have more kinetic energy in the perpendicular direction than in the parallel direction with respect to the magnetic field. The resulting imbalance in the energy contents will then create a plasma-wave instability, the loss-cone instability of Rosenbluth and Post [12]. To study these instabilities, we begin with a brief consideration of magnetic-mirror properties.

A. Magnetic Mirror

A charged particle in the magnetic field performs a helical motion around a magnetic line of force. Associated with such a gyration motion of the charged particle, a magnetic moment μ_B may be defined; its magnitude is given by

$$\mu_B = \frac{1}{c}(\text{current})\cdot(\text{area})$$

$$= \frac{1}{c}\left(\frac{q\Omega}{2\pi}\right)\pi\left(\frac{mv_\perp c}{qB}\right)^2 = \frac{w_\perp}{B} \tag{7.43}$$

where $w_\perp = \frac{1}{2}mv_\perp^2$ is the kinetic energy in the perpendicular directions. The direction of the magnetic moment is opposite that of the local magnetic field.

The magnetic moment (7.43) is proportional to the action integral associated with the periodic cyclotron motion of the charged particle. Hence, it is an adiabatic invariant. For an infinitesimally small rate of change in the magnetic field in space and time, the magnetic moment is kept invariant.†

One of the important consequences of the adiabatic invariance of the magnetic moment is the notion of the magnetic-mirror field constructed by converging magnetic lines of force, as illustrated in Fig. 7.8. As the lines of force converge, the field strength increases. This increasing magnetic field then acts to reflect some of the charged particles impinging on it; hence, it is called the magnetic mirror. Let us see how such a reflection mechanism is related to the adiabatic invariance of the magnetic moment.

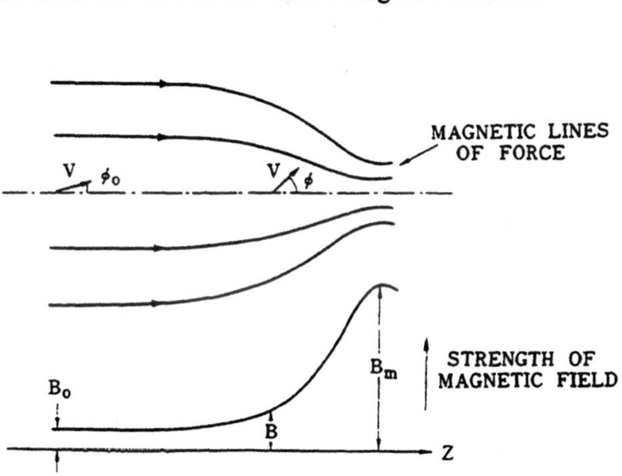

Fig. 7.8 Magnetic mirror.

†For a detailed discussion of the adiabatic invariants in the motion of charged particles, see, e.g., S. Chandrasekhar, *Plasma Physics*, compiled by S. K. Trehan (Univ. of Chicago Press, Chicago, 1960).

Consider a particle moving with velocity $\mathbf{v}$ at an angle ϕ relative to the direction of the magnetic field $\mathbf{B}$; we then have

$$v_\perp / v_\parallel = \tan\phi \tag{7.44}$$

and the magnetic moment is written as

$$\mu_B = (w/B)\sin^2\phi. \tag{7.45}$$

First, we note that the total kinetic energy w of the particle is constant in a static magnetic field; we also evoke the adiabatic invariance of μ_B. Equation (7.45) thus determines a relationship between B and ϕ for a given particle. When the initial values B_0 and ϕ_0 of the magnetic field and of the angle are known, Eq. (7.45) may also be written as

$$\sin^2\phi = (B/B_0)\sin^2\phi_0. \tag{7.46}$$

If the field configuration is such that the strength of the magnetic field increases gradually from B_0 to a maximum value B_m, then the angle ϕ should also increase according to Eq. (7.46); as the particle travels toward the maximum point, the parallel energy $w_\parallel$ is thereby converted into $w_\perp$. When

$$(B_m/B_0)\sin^2\phi_0 > 1, \tag{7.47}$$

the particle can reach only as far as the point where the magnetic field satisfies

$$(B/B_0)\sin^2\phi_0 = 1. \tag{7.48}$$

There, $\phi = \pi/2$, i.e., $v_\parallel = 0$; the particle is reflected at this point and starts to move away from the magnetic mirror.†

If, on the other hand, the mirror ratio B_m/B_0 or the initial angle ϕ_0 is not large enough, so that (7.47) is not satisfied, the particle is no longer reflected by such a field configuration; it escapes away from the system. For a given mirror ratio B_m/B_0, the maximum angle ϕ_L for such an escape to take place is therefore given by

$$\phi_L = \sin^{-1}(B_0/B_m)^{1/2}. \tag{7.49}$$

This angle determines the *loss-cone* of the system; those particles whose

†Fermi in [13] has proposed a theory of cosmic-ray accelerations by collisions against moving magnetic-field irregularities in the interstellar space; in this theory, he considers such a mirror mechanism (called a "Type A" reflection) and a reflection following a curve of the line of force (a "Type B" reflection).

velocity vectors lie in this cone are immediately lost from the field configuration.

When two magnetic mirrors are separated at a distance to produce a magnetic-field configuration of the shape in Fig. 7.9, then those charged particles with sufficiently large ϕ are reflected back and forth between the mirrors and are thus trapped. Such a scheme of plasma confinement is called a magnetic bottle; intensive research has been carried out on the possible uses of such a mirror-confinement scheme for controlled thermonuclear fusion devices.

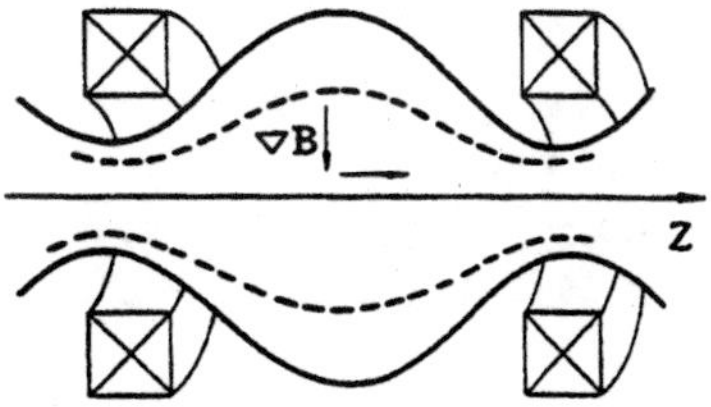

Fig. 7.9 A magnetic bottle.

B. Instability

For the plasma confined in a magnetic-mirror device, the velocity distribution, therefore, deviates significantly from a Maxwellian. It may now be described by a loss-cone distribution, in which for a given value of $v_{\|}$ those particles with $v_{\perp}<\kappa v_{\|}$ are absent from the system (see Fig. 7.10). The parameters κ and v_m characterizing the loss-cone distribution should be functions of the magnetic mirror ratio.

The plasma-wave instability associated with such a loss-cone distribution may be understood simply in terms of those instabilities from a doubly

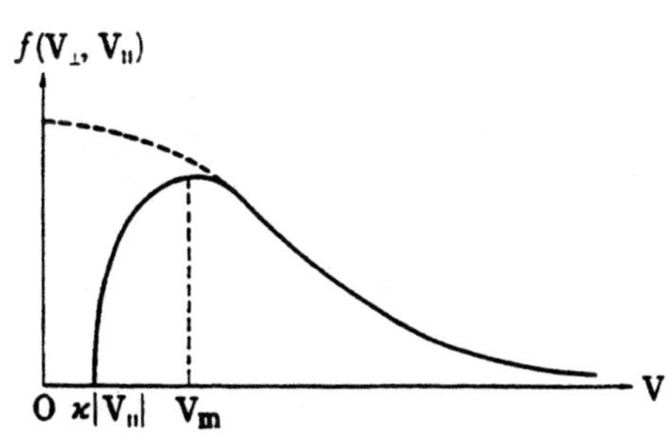

Fig. 7.10 Loss-cone distribution.

humped velocity distribution. Consider a one-dimensional distribution defined by

$$f(v_x) \equiv \int_{-\infty}^{\infty} dv_y \int_{-\infty}^{\infty} dv_{\parallel} f(v_{\perp}, v_{\parallel}). \tag{7.50}$$

Its gradient at $v_x = v_0$ is then calculated as

$$f'(v_x)|_{v_x = v_0} = \int_{v_0}^{\infty} dv_{\perp} \int_{-\infty}^{\infty} dv_{\parallel} \frac{v_0}{(v_{\perp}^2 - v_0^2)^{1/2}} \frac{\partial f}{\partial v_{\perp}}. \tag{7.51}$$

In view of Fig. 7.10 and the character of the weighting function $v_0/(v_{\perp}^2 - v_0^2)^{1/2}$ in Eq. (7.51), we find that $f'(v_0)$ can be positive if $v_0 < v_m$. In fact, the one-dimensional distribution (7.50) may now take a doubly humped shape, as depicted in Fig. 7.11. The system is therefore unstable against excitations of

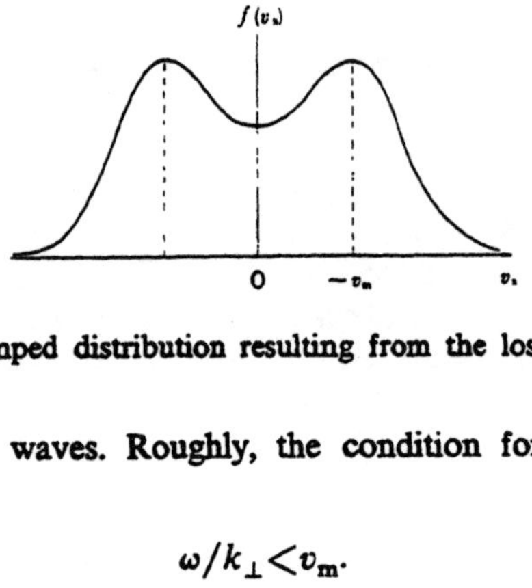

Fig. 7.11 Doubly humped distribution resulting from the loss-cone distribution of Figure 10.

electrostatic plasma waves. Roughly, the condition for instability is then obtained as

$$\omega / k_{\perp} < v_{\mathrm{m}}. \tag{7.52}$$

To carry out a more accurate analysis of the instability, we start from the dielectric response function as calculated in Eq. (3.74). We adopt the cold-electron approximation, so that $|k_{\perp} v_{\perp} / \Omega_e| \ll 1$ and $|k_{\parallel} v_{\parallel} / \omega| \ll 1$ for the electrons. Because of the former condition, only the $n = 0$ term remains for the electrons in the dielectric function; the latter condition enables us to neglect the electron Landau damping. The dispersion relation thus becomes

$$1 - \frac{\omega_e^2}{\omega^2} \frac{k_{\parallel}^2}{k^2} - \frac{\omega_i^2}{k^2} 2\pi \int_0^{\infty} v_{\perp} dv_{\perp} \int_{-\infty}^{\infty} dv_{\parallel} \sum_n \left[\frac{n\Omega}{v_{\perp}} \frac{\partial f}{\partial v_{\perp}} + k_{\parallel} \frac{\partial f}{\partial v_{\parallel}} \right] \frac{J_n^2(k_{\perp} v_{\perp} / \Omega)}{n\Omega + k_{\parallel} v_{\parallel} - \omega - i\eta} = 0 \tag{7.53}$$

where we are still using the notation of Section 7.4 so that the quantities without subscript refer to those associated with the ions.

The frequency of the collective mode is determined from the real part of Eq. (7.53). Since $m \gg m_e$, we have

$$\omega_k = \omega_e (k_\parallel / k) = \omega_e \cos\theta, \tag{7.54}$$

which is just the electron plasma oscillation, Eq. (4.78a) or (7.28). To be able to pick up instabilities in the vicinity of $\omega_k = n\Omega$, we must have $\omega_e > n\Omega$; this condition sets a limiting density for the electrons as (7.31).

We may calculate the imaginary part of Eq. (7.53) by setting $\omega_k = n\Omega$; the major contribution arises from the term with this particular harmonic number. The imaginary part γ_k of the frequency is thereby determined as

$$\gamma_k = \pi^2 \frac{(n\Omega)^2 \omega_i^2}{k^2 |k_\parallel|} \int_0^\infty dv_\perp \left[\frac{\partial f}{\partial v_\perp} \right]_{v_\parallel = 0} J_n^2\left(\frac{k_\perp v_\perp}{\Omega} \right) \tag{7.55}$$

where we have assumed a certain symmetry in the loss-cone distribution so that $(\partial f / \partial v_\parallel)_{v_\parallel = 0} = 0$. Each function in the integrand of Eq. (7.55) behaves as shown in Fig. 7.12. As may be clear from these behaviors, the major contribution to the integral of Eq. (7.55) arises from the vicinity of the first peak of J_n^2. When the position α of this peak falls into the positive domain of $(\partial f / \partial v_\perp)_{v_\parallel = 0}$, that is, when $\beta > \alpha$, the γ_k given by Eq. (7.55) will take on a positive value; an exponentially growing wave or an instability is indicated. Since $\alpha \cong n$ and $\beta \cong k_\perp v_m / \Omega$, the criterion for the instability is approximately expressed as

$$k_\perp v_m > n\Omega. \tag{7.56}$$

This is essentially identical to (7.52).

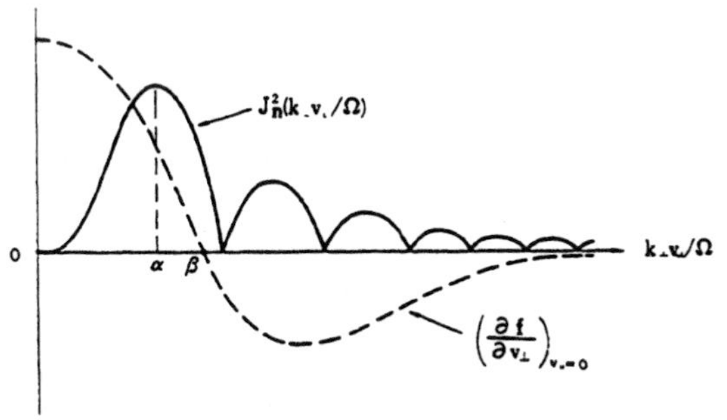

Fig. 7.12 Behavior of J_n^2 and $(\partial f / \partial v_\perp)_{v_\parallel = 0}$.

From a detailed investigation of Eq. (7.55), we find that for fixed values of ω_e^2 (i.e., density) and v_m (i.e., magnetic-mirror ratio) a maximum growth rate may be achieved when the harmonic number n is small (i.e., $n=1$) and $k_\perp$ is large (i.e., $k_\perp \cong k$). In these circumstances, we estimate

$$\gamma_k \cong \omega_e \omega_i^2 / \Omega^2. \tag{7.57}$$

A remarkably large growth rate is thus indicated.

7.6 ELECTROMAGNETIC INSTABILITIES

Thus far we have been concerned with instabilities of longitudinal electrostatic oscillations. We can extend the physical ideas and mathematical techniques employed therein to cases involving the instabilities of transverse electromagnetic waves. The possibilities of instabilities are again quite diversified. In this section we confine ourselves to the simplest and perhaps most important cases of electromagnetic instabilities propagating along the magnetic field.

A. Dielectric Response Function for Circularly Polarized Waves

Since $k_\perp = 0$ and $|k_\parallel| = k$ for propagation parallel to the magnetic field, the tensor $\Pi_\sigma(v_\perp, v_\parallel; n)$ involved in the general expression of the dielectric tensor (3.71) may be simplified as

$$\Pi_\sigma(v_\perp, v_\parallel; n) = \begin{bmatrix} \frac{v_\perp^2}{4}[\delta(n,1)+\delta(n,-1)] & i\frac{v_\perp^2}{4}[\delta(n,1)-\delta(n,-1)] & 0 \\ -i\frac{v_\perp^2}{4}[\delta(n,1)-\delta(n,-1)] & \frac{v_\perp^2}{4}[\delta(n,1)+\delta(n,-1)] & 0 \\ 0 & 0 & v_\parallel^2 \delta(n,0) \end{bmatrix}. \tag{7.58}$$

Following the argument in Section 5.3A, we now carry out a transformation of the dielectric tensor through the unitary matrix (5.26). The dielectric response functions appropriate to the right or left circularly polarized waves may thereby be calculated as

$$\epsilon_{r(l)}(k,\omega) = 1 + \sum_\sigma \frac{\omega_\sigma^2}{\omega^2} \int d\mathbf{v} \frac{v_\perp}{2} \left[(\omega - k v_\parallel) \frac{\partial f_\sigma}{\partial v_\perp} + k v_\perp \frac{\partial f_\sigma}{\partial v_\parallel} \right] \frac{1}{\omega - k v_\parallel \pm \Omega_\sigma + i\eta} \tag{7.59}$$

where the velocity integration has been defined by (3.73).

B. Fire-Hose Instability

Let us investigate the dielectric functions (7.59) for an electron-ion plasma in the limit of low frequencies such that $|\omega - kv_{\|}| \ll \Omega_i \ll |\Omega_e|$. We may then expand the denominator of (7.59) and carry out partial integrations to obtain

$$\epsilon_{r(l)}(k,\omega) \to 1 \pm \sum_\sigma \frac{\omega_\sigma^2}{\omega^2 \Omega_\sigma} \int d\mathbf{v}(kv_{\|} - \omega) f_\sigma + \sum_\sigma \frac{\omega_\sigma^2}{\omega^2 \Omega_\sigma^2} \int d\mathbf{v}(kv_{\|} - \omega)^2 f_\sigma + \cdots$$

$$- \sum_\sigma \frac{\omega_\sigma^2}{\omega^2 \Omega_\sigma^2} \int d\mathbf{v} \frac{k^2 v_\perp^2}{2} f_\sigma - \cdots$$

Assuming no net flows of the particles, $\int d\mathbf{v}\, v_{\|} f_\sigma = 0$, as well as compensation of the net space charges, $\sum_\sigma \omega_\sigma^2 / \Omega_\sigma = 0$, we find that the leading terms of the foregoing expansion may be expressed

$$\epsilon_{r(l)}(k,\omega) \to 1 + \frac{4\pi \rho_{\mathrm{m}} c^2}{B^2} + \frac{4\pi k^2 c^2}{\omega^2 B^2}(p_{\|} - p_\perp). \tag{7.60}$$

Here, $\rho_{\mathrm{m}} \equiv \sum_\sigma n_\sigma m_\sigma$ is the mass density of the plasma, and

$$p_{\|} = \sum_\sigma n_\sigma m_\sigma \int v_{\|}^2 f_\sigma d\mathbf{v}, \tag{7.61a}$$

$$p_\perp = \sum_\sigma n_\sigma m_\sigma \int \frac{v_\perp^2}{2} f_\sigma d\mathbf{v} \tag{7.61b}$$

are the kinetic pressures of the plasma parallel and perpendicular to the magnetic field.

The dispersion relation for the electromagnetic wave has been given by Eq. (5.29). Substitution of (7.60) into this equation then yields

$$\omega = \pm \frac{kcc_{\mathrm{A}}}{(c^2 + c_{\mathrm{A}}^2)^{1/2}} \left[1 - \frac{p_{\|} - p_\perp}{B^2/4\pi} \right]^{1/2}$$

$$\cong \pm kc_{\mathrm{A}} \left[1 - \frac{p_{\|} - p_\perp}{B^2/4\pi} \right]^{1/2}. \tag{7.62}$$

In the last expression of Eq. (7.62), we have assumed that $c \gg c_{\mathrm{A}}$, the Alfvén velocity defined by Eq. (5.48). When we are dealing with a tenuous plasma such that $B^2/4\pi \gg p_{\|}, p_{\perp}$, or a dense isotropic plasma such that $p_{\|} = p_{\perp} \gtrsim B^2/4\pi$, Eq. (7.62) describes a propagation of a stable Alfvén wave. If, however, we consider a dense anisotropic plasma such that $p_{\|} > p_{\perp} + B^2/4\pi$, Eq. (7.62) becomes

$$\omega = \pm i \frac{k}{\sqrt{\rho_m}} \left[p_{\|} - p_{\perp} - \frac{B^2}{4\pi} \right]^{1/2} . \tag{7.63}$$

Onset of an instability is thus predicted. This instability is called the *fire-hose instability*.

The physical mechanism of such an instability and hence the reason for this nomenclature are easily understood: Suppose that the magnetic lines of force are bent slightly due to electromagnetic perturbations. Since the charged particles are tied to the magnetic lines of force, their velocity components parallel to the magnetic field exert centrifugal force upon the lines of force and thereby act to enhance the original perturbation. We here observe a similarity between this instability and the well-known unstable motion of a fire hose when the flow velocity of water inside it becomes very high. Opposing the action of the parallel pressure to excite the instability, the perpendicular pressure and the tension of the magnetic lines of force act to stabilize the disturbances and bring the system back to the state without perturbations. The stabilizing action of the perpendicular pressure arises in the following way. When a curvature of the lines of force is created, the intensity of the magnetic field also deviates from uniformity. Since the magnetic moment (7.43) is an adiabatic invariant, the density of the perpendicular energy becomes greater where the magnetic field is stronger. Increased perturbations will be resisted by the local concentration of the perpendicular energy density.

C. Two-Temperature Distribution with Drift

Similar to the cases of the longitudinal instabilities, it is possible to consider electromagnetic instabilities associated with anisotropic distribution or drift motion of the particles. To take account of these situations, we substitute a two-temperature distribution with drift

$$f(v_{\perp}, v_{\|}) = \left[(2\pi T_{\perp}/m)(2\pi T_{\|}/m)^{1/2} \right]^{-1} \exp\left\{ -\left[\frac{v_{\perp}^2}{2(T_{\perp}/m)} + \frac{(v_{\|} - v_d)^2}{2(T_{\|}/m)} \right] \right\} \tag{7.64}$$

into Eq. (7.59); the result is

$$\epsilon_{\mathrm{r(l)}}(k,\omega)=1-\frac{\omega_{\mathrm{p}}^2}{\omega^2}\left\{\frac{\tilde{\omega}}{\tilde{\omega}\pm\Omega}\left[1-W\left(\frac{\tilde{\omega}\pm\Omega}{k(T_{\|}/m)^{1/2}}\right)\right]+\left(1-\frac{T_{\perp}}{T_{\|}}\right)W\left(\frac{\tilde{\omega}\pm\Omega}{k(T_{\|}/m)^{1/2}}\right)\right\}$$

(7.65)

where $\tilde{\omega}$ is a Doppler-shifted frequency

$$\tilde{\omega}=\omega-kv_{\mathrm{d}}. \tag{7.66}$$

We extend Eq. (7.65) to the cases of a multicomponent plasma in a standard way.

Generally, when the imaginary part of a dielectric function changes its sign, the possibility of a growing wave or instability is expected. Let us, therefore, examine the imaginary part of Eq. (7.65) in two separate cases, paying attention to the change of its sign.

First, we consider the case in which no drift motion is involved; $v_{\mathrm{d}}=0$ and hence $\tilde{\omega}=\omega$. The only possible source of instability here is the temperature anisotropy. The imaginary part arising from the first term in the curly bracket of Eq. (7.65) does not change its sign under these conditions; it gives rise to the damping due to the Doppler-shifted cyclotron resonances, as discussed in Section 5.3. The imaginary part associated with the second term does change its sign, however, according to the sign of the frequency $\omega\pm\Omega$. A possibility of electromagnetic instability associated with the temperature anisotropy is therefore indicated. To confirm the existence of instability, we must make sure that the growth rate arising from the second term indeed exceeds the damping rate due to the first term.

We next consider the case in which the temperatures are isotropic; the second term in the curly bracket of Eq. (7.65) vanishes. This case thereby singles out the effects of the drift velocity. The imaginary part of the dielectric function is now found to change its sign as $\tilde{\omega}$ does. When the drift velocity of the particles exceeds the phase velocity of the electromagnetic wave under consideration, the particles feed their energy to the wave; a growing-wave situation may result. To determine the actual growth or damping rate of the electromagnetic wave, we must take account of all the contributions from various particle species in the plasma.

References

1. M. N. Rosenbluth, in *Plasma Physics* (IAEA, Vienna, 1965), p. 485.
2. O. Penrose, *Phys. Fluids* 3, 258 (1960).
3. R. J. Briggs, *Electron-Stream Interaction with Plasmas* (MIT Press, Cambridge, Mass., 1964).
4. O. Buneman, *Phys. Rev. Letters* 1, 8 (1958); *Phys. Rev.* 115, 503 (1959).

5. J. D. Jackson, *J. Nuclear Energy* **C1**, 171 (1960).
6. I. B. Bernstein, E. A. Frieman, R. M. Kulsrud, and M. N. Rosenbluth, *Phys. Fluids* **3**, 136 (1960).
7. E. A. Jackson, *Phys. Fluids* **3**, 786 (1960).
8. B. D. Fried and R. W. Gould, *Phys. Fluids* **4**, 139 (1961).
9. D. Pines and J. R. Schrieffer, *Phys. Rev.* **124**, 1387 (1961).
10. M. J. Harrison, *J. Phys. Chem. Solids* **23**, 1079 (1962).
11. E. G. Harris, *Phys. Rev. Letters* **2**, 34 (1959); E. G. Harris, *J. Nuclear Energy* **C2**, 138 (1961); G. K. Soper and E. G. Harris, *Phys. Fluids* **8**, 984 (1965).
12. M. N. Rosenbluth and R. F. Post, *Phys. Fluids* **8**, 547 (1965); R. F. Post and M. N. Rosenbluth, *Phys. Fluids* **9**, 730 (1966).
13. E. Fermi, *Phys. Rev.* **75**, 1169 (1949).

Problems

7.1. When the velocity distribution functions are written as functions of the particle energy alone, as in the case of Problem 3.4, show that the plasma is stable against excitations of the plasma waves.

7.2. Derive Eq. (7.13).

7.3. Consider an electron plasma without a magnetic field whose one-dimensional velocity distribution is given by a summation of two Lorentzians

$$f(v)=\frac{0.75}{\pi}\frac{\zeta}{v^2+\zeta^2}+\frac{0.25}{\pi}\frac{\zeta}{(v-v_{\mathrm{d}})^2+\zeta^2}.$$

Determine the boundary curve of the growing and damped plasma oscillations on the plane of v_{d}/ζ versus $k^2\zeta^2/\omega_{\mathrm{p}}^2$.

7.4. As Fig. 7.7 illustrates, the boundary curve exhibits a characteristic folding back toward the k axis when the temperature ratio $(T_{\mathrm{e}}/T_{\mathrm{i}})$ is sufficiently large; the critical wave number k_{c} departs from zero and takes on a finite value in these circumstances. Using Eq. (7.21), find the relation between the mass ratio $(m_{\mathrm{e}}/m_{\mathrm{i}})$ and the critical temperature ratio $(T_{\mathrm{e}}/T_{\mathrm{i}})_{\mathrm{c}}$ above which such a folding back takes place.

7.5. Find the threshold density of the electrons above which a coupling between the electron plasma oscillation and the electron Bernstein mode can take place.

7.6. Derive Eq. (7.51).

7.7. Carry out the calculations leading to Eq. (7.59).

7.8. For the dispersion relation of Eq. (5.3a), we spoke about the cutoff of wave propagation as $(kc/\omega)^2$ changes its sign at $\omega=\omega_{\mathrm{p}}$; no mention was made about an onset of instability. In the case of the dispersion relation (7.62), however, we are speaking of instability rather than cutoff or resonance when $p_{\parallel}$ exceeds $p_{\perp}+B^2/4\pi$. Consider the reason for this apparent difference.

CHAPTER 8

INSTABILITIES IN INHOMOGENEOUS PLASMAS

We now turn to consideration of those instabilities characteristic of inhomogeneous plasmas. As we remarked earlier, additional sources of free energy associated with the spatial variations of physical quantities are available in these circumstances. A nonuniform system, therefore, has a natural tendency to release this extra amount of free energy and thereby to approach a uniform state of thermodynamic equilibrium. The magnetic field applied to the plasma usually places certain constraints on the motion of charged particles; ordinary relaxation processes through collisions may not provide a sufficiently effective mechanism by which such an equilibrium may be approached. The onset of an instability may then have to be looked upon as an alternative avenue through which the plasma finds it preferable to release the extra free energy; so-called anomalous relaxations would thereby result in the plasma.

The nature of the instabilities therefore depends critically on the interplay between the magnetic field and the spatial inhomogeneities. In many cases, the effects of force fields upon the motion of charged particles may also have to be taken into consideration. Examples of such forces include the gravity, the electric field, and the centrifugal force due to the curvature of the magnetic lines of force. These force fields or inhomogeneities acting together with the magnetic field then produce various drift motions of charged particles [1]. As in the cases of instabilities in homogeneous plasmas, such particle drifts play a very essential part in the interpretation and understanding of various instabilities in inhomogeneous plasmas.

We begin this chapter, therefore, with a brief survey of particle drifts in a magnetic field. We shall then calculate the dielectric response function for inhomogeneous plasmas by taking account of various possible drifts. The result of this calculation will be applied to investigations of such instabilities as the interchange instability, the drift wave instability, and the drift cyclotron instability.

8.1 DRIFTS OF PARTICLES IN A MAGNETIC FIELD

In this section, we briefly describe the two kinds of drift motion that are most closley related to the instabilities in inhomogeneous plasmas.

A. Diamagnetic Drift

A diamagnetic drift arises as a result of interplay between a spatial inhomogeneity and the finiteness of the Larmor radius. When there is a density gradient, the current contributions from the cyclotron orbits in the high-density side exceed those coming from the low-density side; a net current flow appears as illustrated in Fig. 8.1a. Similarly, when a temperature gradient is involved, the difference in the average velocities in the high- and low-temperature domains accounts for the appearance of current flow, as shown in Fig. 8.1b. As we shall describe in Section 8.2 and in Problem 8.1, when a Maxwellian velocity distribution is assumed locally, the average velocity $\mathbf{v}_D$ of the charged particles associated with such diamagnetic currents is calculated to be [2]

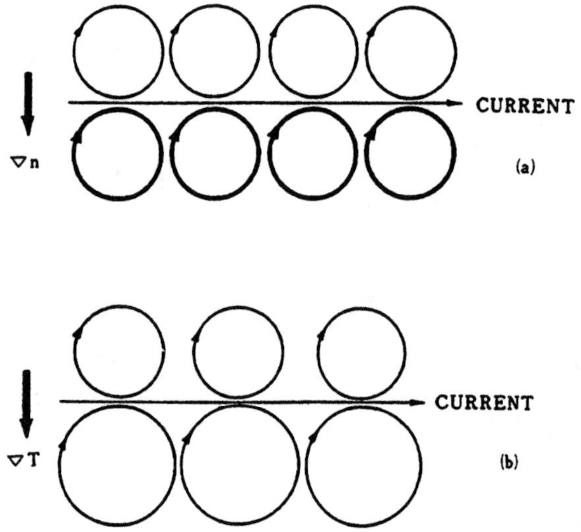

Fig. 8.1 Diamagnetic currents. The line thickness indicates the number of particles contained in each orbit.

$$\mathbf{v}_D = -\frac{cT}{qB^2}\left(\frac{\nabla n}{n} + \frac{\nabla T}{T}\right)\times\mathbf{B}. \tag{8.1}$$

In the light of estimation (1.15), we may express the kinetic pressure p of the plasma according to the ideal-gas law

$$p = nT. \tag{8.2}$$

Equation (8.1) may then be simplified as

$$\mathbf{v}_D = -\frac{c}{nqB^2}(\nabla p)\times\mathbf{B}. \tag{8.3}$$

This is therefore a drift velocity arising as a result of the pressure gradient.

B. Effects of Constant External Forces

When an external force field **F** is applied to a plasma in a magnetic field **B**, there appears a drift motion of the charged particles in the direction perpendicular to both **F** and **B**. The equation of motion for a charged particle is

$$m\frac{d\mathbf{v}}{dt} = \mathbf{F} + \frac{q}{c}\mathbf{v}\times\mathbf{B}. \tag{8.4}$$

For the components parallel to the magnetic field, Eq. (8.4) describes a trivial motion of constant acceleration.

For the components perpendicular to the magnetic field, we have

$$m\frac{d\mathbf{v}_\perp}{dt} = \mathbf{F}_\perp + \frac{q}{c}\mathbf{v}_\perp\times\mathbf{B}. \tag{8.5}$$

Defining

$$\mathbf{v}_F \equiv \frac{c}{qB^2}\mathbf{F}_\perp\times\mathbf{B}, \tag{8.6}$$

we substitute $\mathbf{v}_\perp = \mathbf{v}_F + \mathbf{v}_1$ into Eq. (8.5); it then reduces to

$$m\frac{d\mathbf{v}_1}{dt} = \frac{q}{c}\mathbf{v}_1\times\mathbf{B},$$

an equation of motion for a simple cyclotron motion. Equation (8.5), therefore, describes a superposition of a uniform drift motion $\mathbf{v}_F$ and a simple cyclotron motion $\mathbf{v}_1$.

When the force in particular arises from the electric field **E**, we substitute $\mathbf{F} = q\mathbf{E}$ in Eq. (8.6) to obtain

$$\mathbf{v}_E = c\frac{\mathbf{E}\times\mathbf{B}}{B^2}, \tag{8.7}$$

the well-known *$E\times B$ drift*. When the gravitational acceleration **g** is considered, we have

$$\mathbf{v}_g = \frac{mc}{qB^2}\mathbf{g}\times\mathbf{B}, \tag{8.8}$$

the *gravitational drift*.

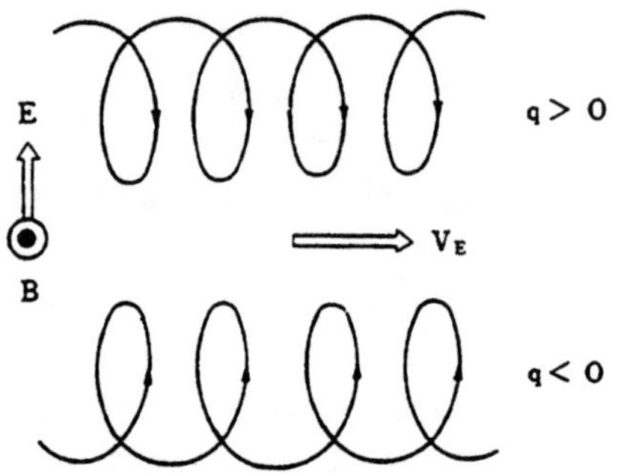

Fig. 8.2 The $E\times B$ drift.

The physical origin of such a drift is easily understood. Consider, for example, the case of $E\times B$ drift as illustrated in Fig. 8.2. A particle gyrates in the direction depending upon the sign of its charge with the radius proportional to its perpendicular velocity $v_\perp$. It is also accelerated by the electric field. On the side where the kinetic energy of the particle is raised by the electric field, the particle orbit assumes a larger radius of curvature; on the other side, where the particle energy is lowered, the particle moves with a smaller radius of curvature. As a result, the drift motion shown in Fig. 8.2 appears. If the sign of the charge is changed, both the direction of acceleration and the direction of gyration are to be inverted; hence, there is no change in the direction of the drift. The gravitational drift can be similarly understood. In this case, a change in the sign of the charge results in an inversion in the drift direction.

We might add in passing that the diamagnetic drift (8.3) can be obtained from Eq. (8.6) if we take the effective force due to the pressure gradient as

$$\mathbf{F} = -\frac{1}{n}\nabla p, \tag{8.9}$$

a natural choice from a hydrodynamic consideration.

8.2 LONGITUDINAL DIELECTRIC FUNCTION FOR AN INHOMOGENEOUS PLASMA

Various instabilities may be investigated through calculation of the relevant dielectric response functions. Since we are here interested in inhomogeneous plasmas, however, the results obtained in Section 3.4 are no longer applicable; we have to modify the calculations so that the salient features of inhomogeneous plasmas may be properly taken into account [3–5].

A. Unperturbed Distribution Functions

We assume the geometrical situation of the plasma under consideration to be as follows. A uniform magnetic field $\mathbf{B}=B\hat{z}$ is applied in the z direction. Inhomogeneities exist in the x direction; a uniform force field $\mathbf{F}=F\hat{x}$ is applied also in the x direction. In the light of the survey in the previous section, drift motion of the particles then appears in the y direction.

In the absence of collisions, the total energy of a particle

$$w=\tfrac{1}{2}mv^2-Fx \tag{8.10a}$$

is a constant of motion. Since the system is translationally invariant in the y and z directions, the canonical momenta

$$p_y=mv_y+(qB/c)x, \tag{8.10b}$$

$$p_z=mv_{\parallel}, \tag{8.10c}$$

should also be conserved. Unperturbed distributions must then be expressible as functions of those three conserved quantities. The simplest choice of such a function, which would reduce to the Maxwellian (1.24) in the absence of the inhomogeneities and the force field, may be given by

$$f(x,\mathbf{v})=\left(\frac{m}{2\pi T}\right)^{3/2}\left[1+\alpha\left(x+\frac{v_y}{\Omega}\right)\right]\exp\left[-\left(\frac{mv^2}{2T}-\frac{Fx}{T}\right)\right], \tag{8.11}$$

where α is a parameter characterizing the spatial variation of the density. We are interested in this distribution function in the vicinity of $x=0$. In expressing the distribution functions in this form, we are disregarding the possibilities of taking into account the spatial variations of temperatures or their anisotropy. This shortcoming may be compensated by the relative simplicity in mathematics which ensues from our adoption of Eq. (8.11). Furthermore, most of the essential features of the inhomogeneous plasmas can be studied through the analyses based on Eq. (8.11).

The density gradient at $x=0$ may be calculated from Eq. (8.11) as

$$\frac{1}{n}\frac{dn}{dx}=\alpha+\frac{F}{T}. \tag{8.12}$$

We first assume that the spatial variation of the density is small over a distance of the average Larmor radius, so that

$$\left|\frac{1}{n}\frac{dn}{dx}\right|\frac{\langle v_\perp\rangle}{|\Omega|}\ll 1. \tag{8.13}$$

We further assume that the space-charge neutrality is maintained not only at $x=0$ but also in the vicinity of $x=0$, to the first order in the inhomogeneity and the force field; for an electron-ion plasma, this condition is expressed as

$$\alpha_e + (F_e/T_e) = \alpha_i + (F_i/T_i). \tag{8.14}$$

We distinguish between F_e and F_i because a force field may act differently on an electron and on an ion.

The drift velocity may be calculated from Eq. (8.11) as

$$\mathbf{v}_d = \int \mathbf{v} f(x, \mathbf{v})\, d\mathbf{v}. \tag{8.15}$$

To the lowest order in α and F, we obtain

$$\mathbf{v}_d = \hat{y}\alpha \frac{T}{m\Omega} = \hat{y}\left(\frac{T}{m\Omega}\frac{1}{n}\frac{dn}{dx} - \frac{F}{m\Omega}\right). \tag{8.16}$$

The first term on the right-hand side of Eq. (8.16) corresponds to a special case of the diamagnetic drift (8.1) in which $\nabla T=0$; the second term represents the $\mathbf{F}\times\mathbf{B}$ drift, (8.6).

The Vlasov equation for the plasma under consideration may be written as

$$\frac{\partial f}{\partial t} + \mathbf{v}\cdot\frac{\partial f}{\partial \mathbf{r}} + \left[\frac{q}{m}\left(\mathbf{E} + \frac{\mathbf{v}}{c}\times\mathbf{B}\right) + \frac{1}{m}\mathbf{F}\right]\cdot\frac{\partial f}{\partial \mathbf{v}} = 0 \tag{8.17}$$

where E contains the electric field produced by the space-charge distribution in the plasma. By virtue of (8.14), this field vanishes to the first order in the inhomogeneities when the unperturbed state described by the distribution functions (8.11) is considered. It is then straightforward to show that Eq. (8.11) is a stationary solution of the Vlasov equation (8.17) in the absence of any additional perturbations. We may use Eq. (8.11) as the unperturbed distribution functions from which the dielectric response function of the system may be calculated.

B. Perturbation

Let us now consider application of an external perturbation to the plasma, which is characterized by the distribution function (8.11). In so doing we particularly adopt the electrostatic approximation, so that only the longitudinal (linear) response of the plasma to an electrostatic disturbance will be considered. The dielectric response function, introduced in Section 3.2, is the relevant response function in these circumstances. Such an approximation is known to be valid [4] in those cases of plasmas for which the magnetic pressure $B^2/8\pi$ is much greater than the kinetic pressure, i.e., in the so-called low β plasmas.†

Since the system is inhomogeneous in the x direction, a Fourier decomposition of physical variables in that direction is no longer a useful concept. Instead, we write the electrostatic potential in the plasma as

$$\varphi(\mathbf{r},t)=\varphi(x)\exp[i(k_y y+k_z z-\omega t)+\eta t]. \tag{8.18}$$

An appropriate functional form of $\varphi(x)$ should then be determined from the propagation characteristics of the field potential in the x direction. The distance over which such a propagation can have a significant effect may be measured by the average Larmor radius of the particles. If one passes to the limit of very strong magnetic field, then such a distance would vanish. Perturbations may not effectively propagate in the x direction; we can approximate

$$\varphi(x)\cong\varphi(0) \tag{8.19}$$

in these circumstances. The situation is quite different in the y direction, however. Here, the drift modes arising from the drift motion Eq. (8.16) are quite capable of maintaining communication between two points separated in the y direction. In fact, the presence of these drift modes is one of the most essential features in the inhomogeneous plasmas. The effects of spatial dispersion in the y direction must therefore be fully taken into account.

On the strength of the foregoing arguments, we shall henceforth adopt the approximation (8.19) for simplicity. General mathematical procedures going beyond this approximation are also available in literature [4,5]. Writing thus the distribution function in the vicinity of $x=0$ as a summation of the unperturbed part (8.11) and a perturbation,

$$f(\mathbf{r},\mathbf{v};t)=f(x,\mathbf{v})+\delta f(\mathbf{v})\exp[i(k_y y+k_z z-\omega t)+\eta t]$$

and linearizing the Vlasov equation (8.17) with respect to the perturbations, we find

$$\left\{\frac{\partial}{\partial t}+\mathbf{v}\cdot\frac{\partial}{\partial \mathbf{r}}+\left[\Omega(\mathbf{v}\times\hat{z})+\frac{F}{m}\hat{x}\right]\cdot\frac{\partial}{\partial \mathbf{v}}\right\}\delta f(\mathbf{v})\exp[i(k_y y+k_z z-\omega t)+\eta t]$$

$$=i\varphi(0)\frac{q}{m}(k_y\hat{y}+k_z\hat{z})\cdot\frac{\partial f}{\partial \mathbf{v}}\exp[i(k_y y+k_z z-\omega t)+\eta t]. \tag{8.20}$$

We wish to solve this equation for $\delta f(\mathbf{v})$, so that the relationship between the total field (8.18) and the induced field in the plasma may be established.

†In this terminology, β refers to the ratio of the plasma kinetic pressure to the magnetic pressure of the confinement field.

C. Integration along the Unperturbed Trajectory

To solve Eq. (8.20), we employ the method of integration along the unperturbed trajectory in the phase space as described in Section 3.4A. The trajectory should be determined from a solution of the equations

$$\frac{d\mathbf{r}}{dt}=\mathbf{v}, \qquad \frac{d\mathbf{v}}{dt}=\Omega(\mathbf{v}\times\hat{z})+\frac{F}{m}\hat{x}.$$

Following the notation of Section 3.4A, we express the solution to these equations as

$$\mathbf{v}'=\mathbf{B}(t'-t)\cdot(\mathbf{v}-\mathbf{v}_F)+\mathbf{v}_F, \tag{8.21}$$

$$\mathbf{r}'=\mathbf{r}+\Omega^{-1}\mathbf{H}(t'-t)\cdot(\mathbf{v}-\mathbf{v}_F)+\mathbf{v}_F(t'-t) \tag{8.22}$$

where $\mathbf{v}_F$ is the drift velocity defined by Eq. (8.6). In the present case, the force field is in the x direction; we have

$$\mathbf{v}_F=-\frac{F}{m\Omega}\hat{y}\equiv v_F\hat{y}. \tag{8.23}$$

This velocity appeared in the last term of Eq. (8.16).

We now integrate Eq. (8.20) along the phase-space trajectory, Eqs. (8.21) and (8.22), to obtain

$$\delta f(\mathbf{v})=i\varphi(0)\frac{q}{m}\int_0^\infty d\tau\left(k_y\hat{y}+k_z\hat{z}\right)\cdot\frac{\partial f(x',\mathbf{v}')}{\partial\mathbf{v}'}\exp[-i\phi(\tau)-\eta\tau] \tag{8.24}$$

where

$$\phi(\tau)=k_y\left[\frac{v_x}{\Omega}(1-\cos\Omega\tau)+\frac{v_y-v_F}{\Omega}\sin\Omega\tau+v_F\tau\right]+k_z v_\parallel\tau-\omega\tau, \tag{8.25}$$

and $\tau=t-t'$. Since

$$\frac{\partial f(x',\mathbf{v}')}{\partial\mathbf{v}'}=\left(\frac{\alpha}{\Omega}\hat{y}-\frac{m}{T}\mathbf{v}'\right)f(x',\mathbf{v}')$$

$$\frac{d}{d\tau}\exp[-i\phi(\tau)-\eta\tau]=\left[-i\left(k_y\hat{y}+k_z\hat{z}\right)\cdot\mathbf{v}'+i\omega\right]\exp[-i\phi(\tau)-\eta\tau],$$

we calculate Eq. (8.24) through a partial integration with respect to τ. With the aid of the fact that the unperturbed distribution is invariant along the unperturbed trajectory, i.e., $f(x',\mathbf{v}')=f(x,\mathbf{v})$, we thus find

$$\delta f(\mathbf{v}) = -\varphi(0)\frac{q}{T}f(x,\mathbf{v})\left\{1+i(\omega-\omega^*)\int_0^\infty \exp[-i\phi(\tau)-\eta\tau]\,d\tau\right\} \quad (8.26)$$

where

$$\omega^* = \frac{\alpha T}{m\Omega}k_y = \mathbf{k}\cdot\mathbf{v}_d \quad (8.27)$$

is the characteristic frequency of the drift mode. The exponential function in the integrand of Eq. (8.26) can be treated in a way identical to the previous calculation of a similar function, Eq. (3.70). In the present case, we have

$$\exp[-i\phi(\tau)-\eta\tau] = \sum_{n=-\infty}^{\infty}\sum_{n'=-\infty}^{\infty} J_n(\mathfrak{z})J_{n'}(\mathfrak{z})$$

$$\times\exp\left\{-i\left[n(\Omega\tau+\theta)-n'\theta+k_z v_{\parallel}\tau-\tilde{\omega}\tau\right]-\eta\tau\right\}. \quad (8.28)$$

Here, $\mathfrak{z}=k_y v_\perp/\Omega$,

$$\tilde{\omega}=\omega-k_y v_F=\omega-\mathbf{k}\cdot\mathbf{v}_F \quad (8.29)$$

is a Doppler-shifted frequency due to the $\mathbf{F}\times\mathbf{B}$ drift, and θ is the azimuthal angle of $\mathbf{v}-\mathbf{v}_F$ with respect to the y axis.

D. Dielectric Response Function

The induced potential field at $x=0$ can be calculated from (8.26) as

$$\varphi_{\text{ind}}(0) = \frac{4\pi q n}{k^2}\left[\int \delta f(\mathbf{v})\,d\mathbf{v}\right]_{x=0}$$

$$= -\varphi(0)\frac{k_D^2}{k^2}\left\{1+(\omega-\omega^*)\sum_n(\tilde{\omega}-n\Omega)^{-1}\right.$$

$$\left.\times\left[W\left(\frac{\tilde{\omega}-n\Omega}{|k_z|(T/m)^{1/2}}\right)-1\right]\Lambda_n(\beta)\right\} \quad (8.30)$$

where $k^2=k_y^2+k_z^2$, and $\beta=k_y^2T/m\Omega^2$. The dielectric response function is then obtained from the relation [compare Eq. (6.2)]

$$\epsilon(k_y,k_z;\omega)=1-(\varphi_{\text{ind}}(0)/\varphi(0)).$$

For a multicomponent plasma, we thus find

$$\epsilon(k_y,k_z;\omega)=1+\sum_\sigma \frac{k_\sigma^2}{k^2}\left\{1+(\omega-\omega_\sigma^*)\sum_n(\tilde{\omega}_\sigma-n\Omega_\sigma)^{-1}\right.$$

$$\left.\times\left[W\left(\frac{\tilde{\omega}_\sigma-n\Omega_\sigma}{|k_z|(T_\sigma/m_\sigma)^{1/2}}\right)-1\right]\Lambda_n(\beta_\sigma)\right\}. \quad (8.31)$$

Specific choices of the force field and the inhomogeneity will determine the situation under consideration, and hence, the dielectric function (8.31). Combinations of those choices together with the frequency domain of the collective mode will then offer various possibilities of investigating instabilities in inhomogeneous plasmas.

8.3 ORBIT-THEORETICAL TREATMENT OF THE FLUTE INSTABILITY

The instability of a heavy liquid supported by a light liquid against the gravitational field is a classic problem in hydrodynamics. A similar problem can be considered for a plasma supported by a magnetic field against gravity or some other force field. Such instabilities are generally categorized as the Rayleigh–Taylor instability; in particular, those involving gravity are called the gravitational instability. In many instances, the Rayleigh–Taylor instability is also called the flute instability or interchange instability. All of these designations denote typical magnetohydrodynamic instabilities involving macroscopic mass motion of plasmas.

The dielectric response function calculated microscopically in the previous section is in fact capable of describing the macroscopic instabilities in this category. As we shall see in the next section, the analysis based on the dielectric response function not only reproduces the results of macroscopic calculations but also reveals additional features of importance arising from microscopic effects. Before entering into such a microscopic analysis, however, we consider in this section an orbit-theoretical treatment of the flute instabilities. We thereby intend to clarify some of the basic physical processes involved in such macroscopic instabilities.

A. Gravitational Instability [1, 6]

Consider a simple configuration of the plasma and the magnetic field as shown in Fig. 8.3a. The plasma, occupying the region above the plane $x=0$, is

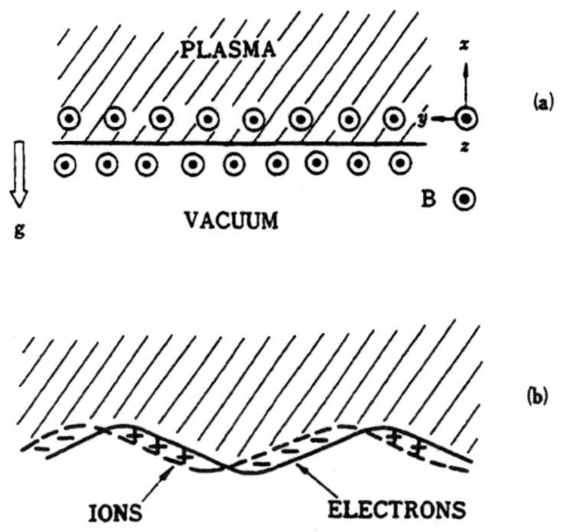

Fig. 8.3 Gravitational instability.

sustained by the magnetic field in the z direction against the gravitational field pointed downward in the x direction. If the magnetic field is strong enough, then the only effect of the gravitational field is to induce drift motion of the charged particles in the y direction. The plasma could conceivably float above the vacuum supported by the magnetic field. The question remains if such a configuration is indeed a stable one.

To examine such a question, we apply a small perturbation

$$\Delta x = \xi \sin ky \tag{8.32}$$

to the boundary plane $x=0$ of the macroscopically compensated plasma consisting of the electrons and ions. The magnetic field $\mathbf{B}=B\hat{z}$ and the gravitation acceleration $\mathbf{g}=-g\hat{x}$ induce drift motion of charged particles in the y direction as given by Eq. (8.8); for a particle of the σ species, the velocity is

$$v_{g\sigma} = c\frac{m_\sigma g}{q_\sigma B}. \tag{8.33}$$

Since the direction of this drift depends on the sign of the charge, a surface-charge accumulation takes place on the perturbed boundary, as illustrated in Fig. 8.3b. The density of the surface charge $\delta\sigma_e(y)$ accumulated during a time interval δt is then calculated as

$$\delta\sigma_e(y)=\sum_\sigma n_\sigma q_\sigma \delta(\Delta x)_\sigma=\sum_\sigma n_\sigma q_\sigma \frac{\partial(\Delta x)}{\partial y}\delta y_\sigma$$

$$=\sum_\sigma n_\sigma q_\sigma \frac{\partial(\Delta x)}{\partial y} v_{g\sigma}\delta t.$$

Hence,

$$\frac{\delta\sigma_e(y)}{\delta t}=\frac{\rho_m c}{B} gk\xi \cos ky$$

or

$$\sigma_e(y)=\frac{\rho_m c}{B} gk\left[\int^t \xi\, dt\right]\cos ky \tag{8.34}$$

where $\rho_m=\sum_\sigma n_\sigma m_\sigma$ is the mass density.

Associated with the surface charge (8.34), the electric field is produced; it is determined from the Poisson equation

$$\nabla\cdot(\epsilon \mathbf{E})=4\pi\sigma_e(y)\delta(x). \tag{8.35}$$

For slowly varying perturbations in directions perpendicular to the magnetic field, we may use Eq. (5.49) for the dielectric constant. With the aid of Eq. (8.34), we find the solution to Eq. (8.35) in the domain $x>0$ to be

$$E_x=\frac{4\pi\rho_m c}{\epsilon_A B} gk\left[\int^t \xi\, dt\right]\cos ky \exp(-kx),$$

$$E_y=\frac{4\pi\rho_m c}{\epsilon_A B} gk\left[\int^t \xi\, dt\right]\sin ky \exp(-kx).$$

Those electric fields produced by the surface charge in turn induce drift motion of the charged particles according to Eq. (8.7). Let us recall that the velocity of this drift is independent of the particle species; it therefore creates the mass motion of the plasma as a whole. The flow velocity is accordingly calculated as

$$v_x=\frac{4\pi\rho_m c^2}{\epsilon_A B^2} gk\left[\int^t \xi\, dt\right]\sin ky \exp(-kx), \tag{8.36a}$$

$$v_y=-\frac{4\pi\rho_m c^2}{\epsilon_A B^2} gk\left[\int^t \xi\, dt\right]\cos ky \exp(-kx). \tag{8.36b}$$

Note that Eqs. (8.36) satisfy $\nabla\cdot\mathbf{v}=0$; this flow pattern does not induce any density irregularities inside the plasma. It does act to shift the boundary plane of the plasma, however; the speed $(\dot{\Delta x})$ of this motion is then determined from Eq. (8.36a) as

$$\dot{\Delta x}=\frac{4\pi\rho_m c^2}{\epsilon_A B^2}gk\left[\int^t \xi\,dt\right]\sin ky.$$

Substituting Eqs. (5.49) and (8.32) in this equation and differentiating once again with respect to time, we find

$$\ddot{\xi}=gk\xi. \tag{8.37}$$

Equation (8.37) has the solutions $\exp(\pm\sqrt{gk}\ t)$; an instability with the growth rate of $\sqrt{gk}$ is thus indicated. The plasma cannot be supported by the magnetic field alone in a stable manner against the gravitational field.

It may then be instructive to consider what would happen if the plasma occupied the region $x<0$ in Fig. 8.3a rather than the region $x>0$. Such a situation can be treated by simply inverting the direction of the gravity in the analysis above, or by replacing g by $-g$. The displacement ξ now exhibits an oscillatory behavior; the system is therefore stable, as we might naturally have expected.

B. Instability Associated with the Curvature of the Magnetic Field

The driving force of the gravitational instability discussed above has been provided essentially by the electric field produced by the charge separation at the boundary. This charge separation in turn was brought about by the gravitational field, whose direction of acceleration is insensitive to the sign of the electric charge of a particle. A charge-insensitive force is not limited to the gravity. In fact, we can develop a similar line of arguments to derive a stability criterion, whenever a charge-insensitive force field is involved in a direction perpendicular to the magnetic field. Hence, for a plasma confined in a magnetic-field configuration, if there is a force acting on the charged particles, regardless of their signs, in the directions from the plasma to the vacuum region, then the system is found to be unstable with respect to the Rayleigh–Taylor mode. If the force is directed in the opposite directions, the system may be stable as far as the Rayleigh–Taylor mode is concerned.

A most important example of such a charge-insensitive field arises when the magnetic lines of force that confine the plasma have finite radii of curvature. Since the radius vector $\mathbf{R}$ of the curvature is given by

$$\frac{\mathbf{R}}{R^2}=-\left(\frac{\mathbf{B}}{B}\cdot\nabla\right)\frac{\mathbf{B}}{B}, \tag{8.38}$$

a charged particle moving along the magnetic field experiences a centrifugal force

$$\mathbf{F}_R = mv_\parallel^2 \frac{\mathbf{R}}{R^2} = 2w_\parallel \frac{\mathbf{R}}{R^2}. \tag{8.39}$$

The curvature also involves the field gradient in the directions perpendicular to the magnetic lines of force. Within the range of validity for the adiabatic invariance of the magnetic moment (7.43), such a field gradient then produces an effective force field in the perpendicular directions as given by

$$\mathbf{F}_G = -\mu_B \nabla_\perp B = -\frac{w_\perp}{B} \nabla_\perp B \tag{8.40}$$

where $\nabla_\perp$ is the differential operator in the directions perpendicular to the magnetic field. From a vacuum field relationship $\nabla \times \mathbf{B} = 0$, which we can reasonably assume for the plasma-confining system, we then find

$$\frac{\nabla_\perp B}{B} = -\frac{\mathbf{R}}{R^2}. \tag{8.41}$$

Hence, collecting the two forces, Eqs. (8.39) and (8.40), we may apply the results of the gravitational-instability analysis to the present situation through a replacement

$$\mathbf{g} \to \frac{w_\perp + 2w_\parallel}{m} \frac{\mathbf{R}}{R^2} = -\left(v_\parallel^2 + \frac{v_\perp^2}{2}\right) \frac{\nabla_\perp B}{B}. \tag{8.42}$$

From the criterion of the Rayleigh–Taylor instability, we then find that a system in which the magnetic field confines the plasma in a convex shape (as in Fig. 8.4a) is unstable because **R** in this case is directed from the plasma to the vacuum; on the contrary a system like Fig. 8.4b is stable.

C. Magnetic Well

Equation (8.42) describes the effective force acting upon a particle in directions perpendicular to the magnetic field when the invariance of the magnetic moment is assured. Looking at this equation, we note that the field strength B plays the role of an effective potential field for the particle. Based on such observations, we may be led to regard the following two features as the essence of designing a magnetohydrodynamically stable field configuration for plasma confinement: (1) The field is nowhere zero; (2) the field strength

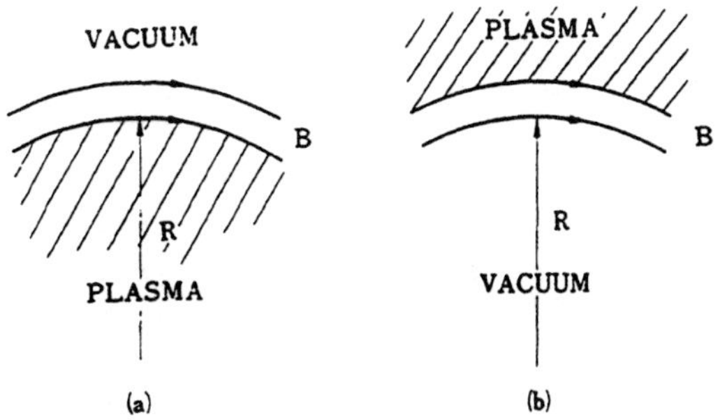

Fig. 8.4 Stability of magnetic confinement.

increases outward from the center toward the periphery of the system. Field configurations satisfying those properties are known as the *minimum-B geometries*.

Property 1 is necessary to maintain the meaning, and thus the adiabatic invariance, of the magnetic moment throughout the system. Property 2 means that B increases outward from a point, or a closed curve, so that surfaces on which B is constant, the magnetic isobars, are closed and nested about this point or curve. A contour plot of B therefore gives the shape of a three-dimensional magnetic well. It should be noted that the isobars are not the same as the surfaces constructed by the magnetic lines of force, although they may coincide in a few very simple geometries.

The best-known example of the minimum-B geometry is the hybrid configuration of the mirror and cusp fields first constructed by Ioffe and his co-workers [7]. The problem of equilibrium and stability of plasma in such a system has been theoretically investigated by Taylor [8]. A simple mirror field like Fig. 7.9 increases in strength outward along the z axis from the center plane but decreases radially. To obtain Property 2 of a minimum-B configuration we may superpose a multipole field produced by an even number of straight rods laid parallel to the z axis (see Fig. 8.5). Odd and even rods carry current in alternate directions to produce a series of cusped fields (see Fig. 8.5b); these fields obviously increase radially. A minimum-B geometry is thus realized when the rod current exceeds a certain critical value.

Increasing the current through this critical value, therefore, offers a simple and unambiguous experimental test for the stability properties of the hybrid configuration. Such experiments by Ioffe *et al.* have indeed shown an abrupt increase of the plasma lifetime by more than an order of magnitude when the strength of the cusp fields reaches a certain value, a clear demonstration of the stability achieved by the minimum-B principle.

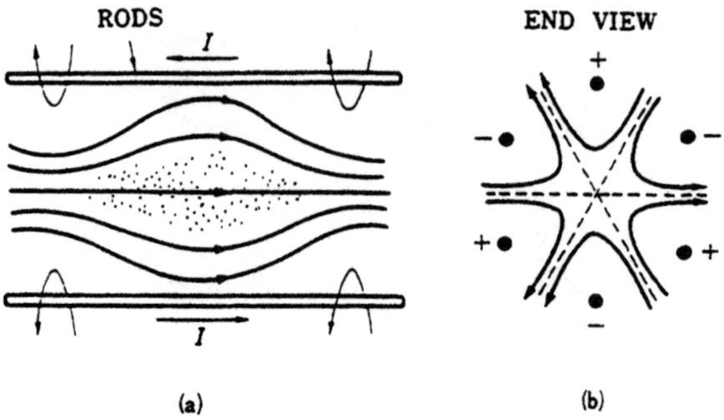

Fig. 8.5 Mirror–cusp field.

8.4 KINETIC-THEORETICAL TREATMENT OF THE FLUTE INSTABILITY AND FINITE LARMOR-RADIUS STABILIZATION

Let us now return to the dielectric response function (8.31) and apply it to the cases of the flute instabilities which we have just studied. In so doing we shall discover additional effects arising from the finiteness of the Larmor radii of the ions which act to stabilize the flute instabilities [3, 9].

The situation appropriate to the flute mode may be analyzed through the following specifications: The plasma is macroscopically neutral; hence, we write $n_e = n_i \equiv N$, and $(1/n_e)dn_e/dx = (1/n_i)dn_i/dx \equiv (1/N)dN/dx$.† The latter condition has been expressed by Eq. (8.14). The magnetic field is so strong that $\beta_e = k_y^2 T_e / m_e \Omega_e^2 \to 0$; we are thereby neglecting the finite Larmor-radius effects of the electrons. The perturbation modes are of flute type, constant along the magnetic lines of force; $k_z = 0$, and hence, the W functions in the dielectric function vanish. Finally, the low-frequency magnetohydrodynamic modes are to be considered; $\tilde{\omega} \ll \Omega_i$, and hence, only the $n=0$ terms remain in Eq. (8.31).

Taking account of the foregoing specifications, we find that the dielectric function appropriate to the flute mode is

$$\epsilon(k,0;\omega) = 1 + \frac{k_e^2}{k^2}\left[1 - \frac{\omega - (\alpha_e T_e/m_e\Omega_e)k}{\omega + (F_e/m_e\Omega_e)k}\right] + \frac{k_i^2}{k^2}\left[1 - \frac{\omega - (\alpha_i T_i/m_i\Omega_i)k}{\omega + (F_i/m_i\Omega_i)k}\Lambda_0(\beta_i)\right]. \tag{8.43}$$

† In order to avoid a possible confusion with the harmonic number n, we employ the capital letter N in this chapter to denote the number density of the electrons or the ions when they are mutually compensated.

The dispersion relation, $\epsilon(\mathbf{k},0;\omega)=0$, then reduces to a quadratic equation for ω,

$$\omega^2-\left[\frac{1}{N}\frac{dN}{dx}\frac{T_i}{m_i\Omega_i}\frac{k(1-\Lambda_0)}{k^2\lambda_D^2+(1-\Lambda_0)}-\left(\frac{F_e}{m_e\Omega_e}+\frac{F_i}{m_i\Omega_i}\right)k\right]\omega$$

$$+\frac{F_eF_i}{m_em_i\Omega_e\Omega_i}k^2-\frac{1}{N}\frac{dN}{dx}\frac{T_i}{m_i\Omega_i}\left(\frac{F_i}{m_i\Omega_i}-\frac{F_e}{m_e\Omega_e}\Lambda_0\right)\frac{k^2}{k^2\lambda_D^2+(1-\Lambda_0)}=0$$

where $\lambda_D\equiv 1/k_i$, and we have suppressed the argument β_i of the function $\Lambda_0(\beta_i)$. This equation is solved as

$$\omega=-\frac{F_ik}{m_i\Omega_i}\left(X+\frac{F_em_i\Omega_i}{F_im_e\Omega_e}\right) \tag{8.44}$$

where

$$X=\tfrac{1}{2}\left(\mu+\nu(1-\Lambda_0)\pm\left\{[\mu+\nu(1-\Lambda_0)]^2-4\mu\nu\right\}^{1/2}\right), \tag{8.45}$$

$$\mu=1-\frac{F_em_i\Omega_i}{F_im_e\Omega_e}, \tag{8.46}$$

$$\nu=-\frac{1}{N}\frac{dN}{dx}\frac{T_i}{F_i}[k^2\lambda_D^2+(1-\Lambda_0)]^{-1}. \tag{8.47}$$

When the frequency ω takes on complex values, that is, when the inside of the square root in Eq. (8.45) becomes negative, the system develops an instability. We now apply these general formulas to two specific examples of the flute instabilities.

A. Gravitational Instability

As the external force field, let us choose the gravitational acceleration $\mathbf{g}=-g\hat{x}$; when $g>0$, the gravity is directed toward the negative x direction. The forces on the electron and the ion are given by

$$F_e=-m_eg, \qquad F_i=-m_ig. \tag{8.48}$$

Since $|\Omega_e|\gg\Omega_i$, we have $\mu=1$. As we shall find shortly, the most important cases of a magnetohydrodynamic instability take place in the long-wavelength domain such that $\beta_i \ll 1$. Hence, we may express

$$1-\Lambda_0(\beta_i)\cong k^2R_L^2 \tag{8.49}$$

where

$$R_L=(T_i/m_i\Omega_i^2)^{1/2} \tag{8.50}$$

is an average Larmor radius of the ions. In terms of the dielectric constant ϵ_A defined by Eq. (5.49), R_L is related to λ_D as

$$R_L^2/\lambda_D^2=\epsilon_A. \tag{8.51}$$

In the light of numerical estimation through Eq. (5.50), we can ordinarily assume that $R_L^2 \gg \lambda_D^2$. We may then write approximately

$$\nu\cong\frac{1}{N}\frac{dN}{dx}\frac{\Omega_i^2}{k^2g}, \qquad \nu(1-\Lambda_0)\cong\frac{1}{N}\frac{dN}{dx}\frac{T_i}{m_ig}.$$

Hence, the condition for instability is obtained from Eq. (8.45) as

$$k^2<4\frac{1}{N}\frac{dN}{dx}\frac{\Omega_i^2}{g}\Bigg/\left(1+\frac{1}{N}\frac{dN}{dx}\frac{T_i}{m_ig}\right)^2. \tag{8.52}$$

The instability, when it takes place, is thus confined in the long-wavelength domain. To establish such a domain, the right-hand side of (8.52) should be positive, i.e.,

$$g\frac{dN}{dx}>0. \tag{8.53}$$

For the instability, the gravitational field must be directed toward the direction in which the plasma density decreases; otherwise, the system will be gravitationally stable. We also note that criterion (8.52) already demonstrates the stabilizing action of the finite ion Larmor radius. As the thermal velocity $(T_i/m_i)^{1/2}$, and hence the Larmor radius, of the ions increases, the upper bound of the unstable k domain determined from the right-hand side of (8.52) decreases. If the values of the wave numbers available to the system are also limited from below by some other reasons (geometrical ones, for example), then the unstable domain of the wave numbers will diminish and may eventually vanish as the ion Larmor radius increases. We shall later come back to a further discussion of these effects in connection with the stability of a magnetic-mirror confinement system.

In the limit of long wavelengths and low temperatures, Eq. (8.44) reduces to

$$\omega = \frac{g}{2\Omega_i}k \pm i\left(\frac{1}{N}\frac{dN}{dx}g\right)^{1/2}. \tag{8.54}$$

The growth rate in this equation differs from that obtained in Eq. (8.37). The reason is simply that the case treated in Section 8.3A involves a discontinuous change of the plasma density at $x=0$, while in the present case we are concerned with a case of a continuous density variation. The growth rate of Eq. (8.54) can be recovered if we apply the method of Section 8.3A to the present situation (see Problem 8.4).

B. Instability Associated with the Curvature of the Magnetic Field

In the light of Eq. (8.42), we may now use

$$F_e = -2T_e/R, \qquad F_i = -2T_i/R. \tag{8.55}$$

The direction of the radius-of-curvature vector has been chosen in such a way as to make the system unstable when $dN/dx > 0$. We find $\mu = 1 + (T_e/T_i)$; the growth rate γ of the instability in the limit of the long wavelengths is

$$\gamma = \left[\frac{1}{N}\frac{dN}{dx}\frac{2(T_e+T_i)}{m_i R}\right]^{1/2}. \tag{8.56}$$

C. Finite Larmor-Radius Stabilization

In order to study the stabilizing effects of the finite Larmor radius, let us consider the plasma confined in a magnetic bottle of Fig. 7.9. In such an axially symmetric system, the wave numbers k of the flute mode describe variations in the azimuthal direction. Geometrically, therefore, the available values of k are bounded from below by approximately an inverse of the radius of the confined plasma. We also recall that the most unstable flute mode occurs in the small k domain. Hence, we may estimate

$$k \cong \frac{1}{N}\frac{dN}{dx} \equiv \frac{1}{r}. \tag{8.57}$$

It then follows that

$$\nu(1-\Lambda_0) \cong \frac{R}{2r}\frac{R_L^2}{R_L^2+\lambda_D^2} = \frac{R}{2r}\frac{\epsilon_A}{\epsilon_A+1}$$

$$\nu \cong \frac{Rr}{2R_L^2}\cdot\frac{\epsilon_A}{\epsilon_A+1}.$$

From Eq. (8.45), the condition for stability may thus be obtained as

$$\mu+\frac{R}{2r}\frac{\epsilon_A}{\epsilon_A+1}>(2\mu)^{1/2}\frac{(Rr)^{1/2}}{R_L}\left(\frac{\epsilon_A}{\epsilon_A+1}\right)^{1/2}. \tag{8.58}$$

For a high-density plasma such that $\epsilon_A \gg 1$, this criterion reduces to

$$\frac{R_L}{r}\left(1+2\mu\frac{r}{R}\right)>2(2\mu)^{1/2}\left(\frac{r}{R}\right)^{1/2}. \tag{8.59}$$

When the ratio r/R is substantially smaller than unity, this condition may be satisfied by a reasonably large value of the Larmor radius R_L; this is the finite Larmor-radius stabilization effect. For a low-density plasma such that $\epsilon_A \ll 1$, the stability criterion becomes

$$\lambda_D\left(1+\frac{R}{2\mu r}\frac{R_L^2}{\lambda_D^2}\right)>\left(\frac{2}{\mu}\right)^{1/2}(Rr)^{1/2}. \tag{8.60}$$

Again, a stabilizing action of the finite Larmor radius is indicated.

Physically, those stabilization effects may be understood in the following way: As we have argued in Section 8.3B, the driving force of a flute instability is the $E\times B$ drift arising from the space-charge perturbation created in the plasma. When the Larmor radius is finite, the electric-field perturbation which an ion feels is different, both in magnitude and in phase, from that observed at its guiding center and therefore by an electron. (For the electrons, the Larmor-radius effects have been neglected.) In addition, the actual space-charge perturbation also differs from that obtained solely from a consideration of the guiding-center drifts such as Eq. (8.34). The combined effects of these differences are such as to cause slight out-of-phase motion between the ions and the electrons which may lead to stabilization.

8.5 DRIFT WAVE AND INSTABILITY

When a group of particles move with a velocity $\mathbf{v}_d$, we define a collective mode associated with such a drift motion according to Eq. (7.23). The frequency (8.27) may, therefore, be looked upon as the collective mode associated with a diamagnetic current in an inhomogeneous plasma; the source of energy in such a collective mode may be traced to the extra free energy contained in the inhomogeneities.

The collective properties of inhomogeneous plasmas are thus substantially modified from those of homogeneous plasmas. An especially remarkable feature is the appearance of the drift wave and the associated instability [9–12]. We consider these topics in this section.

A. Elementary Survey

In order to introduce the drift wave in an elementary way, let us consider the space–time perturbations of the potential and the densities in the plasma:

$$\varphi(\mathbf{r},t)=\varphi\exp[i(k_y y+k_z z-\omega t)],$$

$$n_e(\mathbf{r},t)=N+\delta n_e\exp[i(k_y y+k_z z-\omega t)],$$

$$n_i(\mathbf{r},t)=N+\delta n_i\exp[i(k_y y+k_z z-\omega t)].$$

The equilibrium density N changes gradually in the x direction. We assume that the space–time variation under consideration is so slow that the electrons can follow the potential perturbation adiabatically; hence,

$$\delta n_e/N=e\varphi/T_e. \tag{8.61}$$

The behavior of the ions is governed by the continuity equation

$$\frac{\partial n_i}{\partial t}+\nabla\cdot(n_i\mathbf{v}_i)=0 \tag{8.62}$$

and the equation of motion

$$\frac{\partial \mathbf{v}}{\partial t}=-\frac{e}{m_i}\nabla\varphi+\Omega_i(\mathbf{v}\times\hat{z}). \tag{8.63}$$

In the directions perpendicular to $\hat{z}$, Eq. (8.63) may be written as

$$\mathbf{v}_\perp=-ik_y\frac{c\varphi}{B}\hat{x}+k_y\frac{c\varphi}{B}\frac{\omega}{\Omega_i}\hat{y}$$

$$\cong -ik_y\frac{c\varphi}{B}\hat{x}$$

where we have assumed that $|\omega|\ll\Omega_i$. In the z direction, Eq. (8.63) becomes

$$v_z=(k_z/\omega)(e\varphi/m_i).$$

With the aid of these expressions for the ion velocity, Eq. (8.62) may be linearized and solved as

$$\frac{\delta n_i}{N}=\left(\frac{\omega_e^*}{\omega}+\frac{T_e}{m_i}\frac{k_z^2}{\omega^2}\right)\frac{e\varphi}{T_e} \tag{8.64}$$

where ω_e^* is the drift frequency (8.27) for the electrons in which $F=0$ [hence, $\alpha=(1/N)(dN/dx)$].

For long-wavelength excitations, we must have

$$\delta n_e = \delta n_i. \tag{8.65}$$

Otherwise, the induced space-charge field would become extremely large; condition (8.65) would be restored. Combination of Eqs. (8.61), (8.64), and (8.65) yields the dispersion relation

$$\omega^2 - \omega_e^* \omega - s^2 k_z^2 = 0 \tag{8.66}$$

where $s = (T_e/m_i)^{1/2}$ is the propagation velocity of the ion-acoustic wave. The solution as a function of k_z is depicted in Fig. 8.6. When k_z is so large that $s^2 k_z{}^2 \gg (\omega_e^*)^2$, the two branches of the solution turn into the ordinary ion-acoustic wave, Eq. (4.87a). As one moves toward the small k_z domain, the upper branch departs from the acoustic mode and approaches ω_e^* at $k_z = 0$. Here, the dispersion relation describes a collective mode propagating in the y direction carried by the diamagnetic drift motion of the electrons; such a collective mode is called the drift wave.

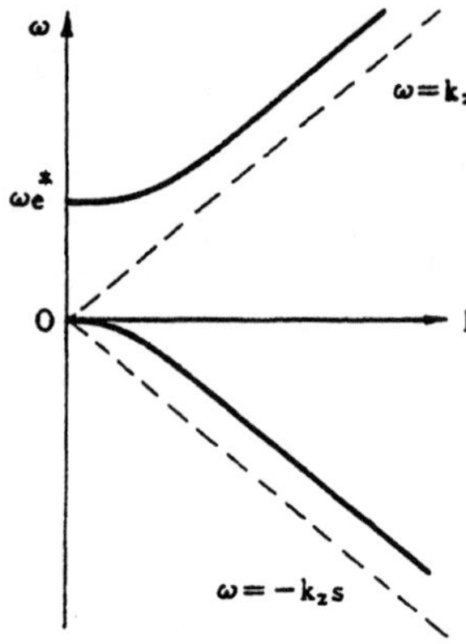

Fig. 8.6 Drift waves.

B. Microscopic Calculations

A microscopic analysis of the drift wave may be carried out with the aid of the dielectric response function (8.31). Here, we assume a situation in which the force field $F = 0$; hence, $\alpha_e = \alpha_i = (1/N)(dN/dx)$. The frequency domain to be considered is the same as the case of the ion-acoustic wave; it is specified by (4.85). The dielectric response function in these circumstances may be simplified as

$$\epsilon(k_y,k_z;\omega)=1+\frac{k_e^2}{k^2}\left[1+i\left(\frac{\pi}{2}\right)^{1/2}\frac{\omega-\omega_e^*}{|k_z|(T_e/m_e)^{1/2}}\exp\left(-\frac{\omega^2}{2k_z^2(T_e/m_e)}\right)\right]$$
$$+\frac{1}{k^2\lambda_D^2}\left\{1-\frac{\omega-\omega^*}{\omega}\left[1+\frac{k_z^2(T_i/m_i)}{\omega^2}-i\left(\frac{\pi}{2}\right)^{1/2}\right.\right. \tag{8.67}$$
$$\left.\left.\times\frac{\omega}{|k_z|(T_i/m_i)^{1/2}}\exp\left(-\frac{\omega^2}{2k_z^2(T_i/m_i)}\right)\right]\Lambda_0(\beta)\right\}$$

where we have omitted the subscript i for some of the ionic quantities so that

$$\omega^*\equiv\omega_i^*=-\omega_e^*(T_i/T_e),$$
$$\lambda_D^2\equiv1/k_i^2=(1/k_e^2)(T_i/T_e), \tag{8.68}$$
$$\beta\equiv k_y^2T_i/m_i\Omega_i^2=k_y^2R_L^2.$$

The real part of the equation, $\epsilon(k_y,k_z;\omega)=0$, becomes

$$k^2\lambda_D^2+\frac{T_i}{T_e}+1-\frac{\omega-\omega^*}{\omega}\left[1+\frac{k_z^2(T_i/m_i)}{\omega^2}\right]\Lambda_0(\beta)=0. \tag{8.69}$$

As we observed in connection with Fig. 8.6, two characteristic frequencies are involved in the present problem: the frequency of a drift mode $\cong|\omega_e^*|$; and the frequency of the acoustic mode $s|k_z|$. The relative magnitude between the two frequencies depends on the ratio $|k_z/k_y|$. In the wave-vector domain such that

$$|k_y|\left(\frac{T_e}{T_i}\right)^{1/2}R_L\frac{1}{N}\left|\frac{dN}{dx}\right|\gg|k_z| \tag{8.70}$$

the drift frequency represents the major effect; the solution of Eq. (8.69) then is

$$\omega=-\frac{\omega^*\Lambda_0(\beta)}{(T_i/T_e)+1-\Lambda_0(\beta)+k_y^2\lambda_D^2}. \tag{8.71}$$

Since $R_L^2 \gg \lambda_D^2$ in many practical cases, the term $k_y^2\lambda_D^2$ in the denominator is usually negligible. When $\beta \ll 1$, Eq. (8.71) reduces to $\omega = \omega_e^*$, a result obtained in the previous section.

From the imaginary part of Eq. (8.67), the growth rate γ of the drift wave may be calculated as

$$\gamma = \left(\frac{\pi}{2}\right)^{1/2} \frac{\omega^2}{\omega^* \Lambda_0(\beta)} \left\{ \frac{T_i}{T_e} \frac{\omega - \omega_e^*}{|k_z|(T_e/m_e)^{1/2}} + \frac{\omega - \omega^*}{|k_z|(T_i/m_i)^{1/2}} \exp\left[-\frac{\omega^2}{2k_z^2(T_i/m_i)} \right] \right\}. \tag{8.72}$$

The second term on the right-hand side of Eq. (8.72) stems from Landau damping of the ions; it is a negative definite quantity and contributes always as damping. This ion Landau-damping term may be negligible when k_z is small so that

$$|\omega/k_z| \gg (T_i/m_i)^{1/2}. \tag{8.73}$$

When $T_e > T_i$, a condition tacitly implied in (4.85), the constraint (8.70) is more restrictive than (8.73). Within the wave-vector domain where the drift mode predominates over the acoustic mode, therefore, we may neglect the ion Landau-damping term.

Thus, by keeping only the first term on the right-hand side of Eq. (8.72) and substituting (8.71) in it, we obtain

$$\gamma = \left(\frac{\pi}{2}\right)^{1/2} \frac{(\omega^*)^2 \Lambda_0(\beta) \{[1 - \Lambda_0(\beta)](1 + T_i/T_e) + k_y^2\lambda_D^2\}}{|k_z|(T_e/m_e)^{1/2}[1 - \Lambda_0(\beta) + (T_i/T_e) + k_y^2\lambda_D^2]^3}. \tag{8.74}$$

Note that this term is a positive definite quantity, indicating a growing drift wave or an instability. As long as there is a density gradient across the magnetic field, the drift wave exists in the wave-vector domain of (8.70). A conclusion from Eq. (8.74), therefore, is that such a drift wave is *always* characterized by a positive growth rate or an instability. It is in this sense that the instability associated with the drift wave is sometimes called the *universal instability*.

As a function of k_y, the growth rate, Eq. (8.74), takes on a maximum value somewhere between

$$1 \ll k_y^2 R_L^2 \ll R_L^2/\lambda_D^2. \tag{8.75}$$

This relation assumes $R_L^2 \gg \lambda_D^2$; if this is not the case, we simply replace (8.75) by $k_y^2\lambda_D^2 \ll 1$.

C. Stabilization of the Drift Wave

In actual plasma-confinement systems, there are a number of factors which act to stabilize the drift waves. The situation, therefore, is not so catastrophic as the name "universal instability" might lead one to suspect.

To see an origin of such a stabilization effect, let us return to (8.73), the condition to avoid the ion Landau damping. We note that the values of $|k_z|$ in actual plasmas are generally limited from below. The total length L of the plasma along the magnetic field, be it in an open-ended system or in a toroidal system, provides a measure of the lower limit for the available values of $|k_z|$. If the geometrical condition of the system is such that, even with this smallest possible value of $|k_z|$, condition (8.73) cannot be satisfied, then we may conclude that the ion Landau-damping term in Eq. (8.72) represents a major effect; the drift wave is thus stabilized.

The critical longitudinal length L_{cr} of the plasma below which the drift waves are stabilized may, therefore, be estimated from the equality $|\omega/k_z| = (T_i/m_i)^{1/2}$ in which $|k_z| \cong \pi/L_{cr}$. The expression for the frequency ω has been obtained as Eq. (8.71) in which we may neglect $k_y^2\lambda_D^2$ by virtue of (8.75). Hence,

$$L_{cr} \cong \frac{\pi}{|k_y|R_L} \frac{1-\Lambda_0(\beta)+(T_i/T_e)}{\Lambda_0(\beta)} r \tag{8.76}$$

where $r \equiv N(dN/dx)^{-1}$ roughly measures a distance of plasma confinement across the magnetic field [compare Eq. (8.57)].

When $|k_y|$ is small so that $k_y^2 R_L^2 \ll 1$, Eq. (8.76) becomes $L_{cr} \cong \pi(T_i/T_e)(r/|k_y|R_L)$. An extremely large critical length is indicated here. Furthermore, the growth-rate contribution stemming from Eq. (8.74), being proportional to $(k_y)^4$, becomes infinitesimally small in this k_y domain. Taking these two effects together, we conclude that drift waves are ordinarily stable in the long-wavelength domain. When $|k_y|$ is large so that (8.75) is satisfied, Eq. (8.76) reduces to

$$L_{cr} \cong \frac{(2\pi)^{3/2}}{2}\left(1+\frac{T_i}{T_e}\right) r \cong 8\left(1+\frac{T_i}{T_e}\right) r. \tag{8.77}$$

Still, a substantially large critical length is indicated. Unless a fairly long thin plasma is constructed, it is rather difficult for the drift waves to avoid the ion Landau damping.

The stabilizing action of the ion Landau damping can also be exploited when the magnetic field has a shear structure. We thus consider a field configuration

$$B_z = B, \qquad B_y = BSx,$$

where S^{-1} is the shearing distance of the magnetic field. A drift wave with a wave vector $[0, k_y, 0]$ at $x=0$ may then be observed as having a slightly different wave vector $[0, k_y(1-S^2d^2)^{1/2}, k_y Sd]$ at $x=d$. The stability criterion $|\omega/k_z|<(T_i/m_i)^{1/2}$ over the distance $d \cong r$ may then be expressed as

$$S > R_L\left(\frac{1}{N}\frac{dN}{dx}\right)^2. \tag{8.78}$$

We thus find that a modest amount of shearing is sufficient to stabilize the drift waves.

8.6 DRIFT CYCLOTRON INSTABILITY

In Section 8.5 we have been concerned with the properties of the drift waves in the low-frequency domain such that $|\omega| \ll \Omega_i$. As the frequency of the drift wave is increased toward the vicinity of Ω_i, we expect a coupling to take place between the drift wave and the fundamental Bernstein mode of the ions. Both the drift wave and the Bernstein mode propagate predominantly in the directions perpendicular to the magnetic field; they may couple quite effectively. The extra free energy in the spatial inhomogeneity of the plasma may thereby be fed into the ion Bernstein mode, resulting in the onset of an instability [13].

To analyze a situation corresponding to such an instability, we assume $F=0$, $k_z=0$ ($|k_y|=k$), and consider the dielectric response function (8.31) in the vicinity of $\omega \cong \Omega_i \ll |\Omega_e|$; we then have

$$\epsilon(k,0;\omega) = 1 + \frac{k_e^2}{k^2}\left[\frac{\omega_e^*}{\omega} + 1 - \Lambda_0(\beta_e)\right] + \frac{1}{k^2\lambda_D^2}\left\{1 - (\omega-\omega^*)\left[\frac{\Lambda_0(\beta)}{\omega} + \frac{\Lambda_1(\beta)}{\omega-\Omega_i}\right]\right\} \tag{8.79}$$

where the notation of (8.68) has been used. Let us further assume that

$$\beta_e = k^2(T_e/T_i)(m_e/m_i)R_L^2 < 1, \tag{8.80a}$$

$$\beta = k^2R_L^2 \gg 1. \tag{8.80b}$$

By virtue of these assumptions, we may expand

$$\Lambda_0(\beta_e) \cong 1 - \beta_e,$$

$$\Lambda_0(\beta) \cong \Lambda_1(\beta) \cong [(2\pi)^{1/2}kR_L]^{-1} \equiv \Delta.$$

We expect that $\Delta \ll 1$. The dispersion relation, $\epsilon(k,0;\omega)=0$, is then calculated from Eq. (8.79) as

$$1+k^2\lambda^2-\frac{\omega^*}{\omega}=\frac{\omega-\omega^*}{\omega-\Omega_i}\Delta \tag{8.81}$$

where

$$\lambda^2=\lambda_D^2+(m_e/m_i)R_L^2. \tag{8.82}$$

Note that $(m_e/m_i)R_L^2/\lambda_D^2=\omega_e^2/\Omega_e^2$ ($\ll 1$).

We now observe the following possibilities of decomposing Eq. (8.81): Unless $\omega \cong \Omega_i$, the right-hand side of Eq. (8.81) is negligibly small; we have a solution corresponding to a drift branch,

$$\omega_1=\omega^*/(1+k^2\lambda^2). \tag{8.83}$$

On the other hand, when $\omega^* \ll \Omega_i$, we may neglect ω^* in (8.81) in the vicinity of $\omega \cong \Omega_i$; hence, we obtain a cyclotron branch

$$\omega_2=\Omega_i\left(1+\frac{\Delta}{1+k^2\lambda^2}\right). \tag{8.84}$$

If these two branches intersect each other, then we expect complex frequency solutions in the vicinity of the intersections, indicating the onset of an instability.

Let us, therefore, seek the condition for such intersections to take place. The frequency ω_1 of the drift branch (8.83) takes on a maximum value

$$\omega_{\max}=\alpha\frac{R_L^2\Omega_i}{2\lambda} \tag{8.85}$$

at $k=1/\lambda$. The values of β_e and β at this wave number are

$$\beta_e=\left(\frac{T_e}{T_i}\right)\frac{(m_e/m_i)R_L^2}{\lambda_D^2+(m_e/m_i)R_L^2}\cong\left(\frac{T_e}{T_i}\right)\frac{\omega_e^2}{\Omega_e^2},$$

$$\beta=\frac{R_L^2}{\lambda_D^2+(m_e/m_i)R_L^2}=\frac{\omega_i^2}{\Omega_i^2},$$

so that the assumptions (8.80) may still be reasonably satisfied there. The condition for the interaction may then be expressed as $\omega_{\max} \geqslant \Omega_i$; we thus find

$$\alpha R_L^2 \geqslant 2[\lambda_D^2+(m_e/m_i)R_L^2]^{1/2}. \tag{8.86}$$

Equation (8.81), being a quadratic equation for ω, can be solved analytically. We first write it as

$$(1+k^2\lambda^2-\Delta)\omega^2-[\Omega_i(1+k^2\lambda^2)+\omega^*(1-\Delta)]\omega+\omega^*\Omega_i=0.$$

The condition for the instability then is

$$4\omega^*\Omega_i(1+k^2\lambda^2-\Delta)>[\Omega_i(1+k^2\lambda^2)+\omega^*(1-\Delta)]^2.$$

In terms of the two characteristic frequencies, Eqs. (8.83) and (8.84), and to the first order in Δ, this condition can be expressed as

$$2k^2\lambda^2\omega_1(\omega_1+\omega_2)\Delta>(\omega_1-\omega_2)^2(1+k^2\lambda^2-2\Delta). \tag{8.87}$$

Because of the smallness of Δ, it is necessary to satisfy the coupling condition $\omega_1 \cong \omega_2$ to observe an instability; only in the vicinity of $\omega_1=\omega_2$ can the drift cyclotron instability take place.

References

1. M. N. Rosenbluth and C. L. Longmire, *Ann. Phys. (N.Y.)* **1**, 120 (1957).
2. E. S. Fradkin, *Zhur. Eksptl. Teoret. Fiz.* **32**, 1176 (1957) [*Soviet Phys. JETP* **5**, 956 (1957)].
3. M. N. Rosenbluth, N. A. Krall, and N. Rostoker, *Nucl. Fusion Suppl.* **1**, 143 (1962).
4. M. N. Rosenbluth, in *Plasma Physics* (IAEA, Vienna, 1965), p. 485.
5. A. B. Mikhailovskii, in *Reviews of Plasma Physics* (M. A. Leontovich, ed.; H. Lashinsky, trans.) Vol. 3, p. 159 (Consultants Bureau, New York, 1967).
6. M. D. Kruskal and M. Schwarzschild, *Proc. Roy. Soc. (London)* **A223**, 348 (1954).
7. Yu. B. Gott, M. S. Ioffe, and V. G. Telkovsky, *Nucl. Fusion Suppl.* **3**, 1045 (1962); M. S. Ioffe, in *Plasma Physics* (IAEA, Vienna, 1965), p. 421.
8. J. B. Taylor, *Phys. Fluids* **6**, 1529 (1963); in *Plasma Physics* (IAEA, Vienna, 1965), p. 449.
9. N. A. Krall and M. N. Rosenbluth, *Phys. Fluids* **6**, 254 (1963).
10. L. I. Rudakov and R. Z. Sagdeev, *Zhur. Eksptl. Teoret. Fiz.* **37**, 1337 (1959) [*Soviet Phys. JETP* **10**, 952 (1960)].
11. A. A. Galeev, V. N. Oraevskii, and R. Z. Sagdeev, *Zhur. Eksptl. Teoret. Fiz.* **44**, 903 (1963) [*Soviet Phys. JETP* **17**, 615 (1963)].
12. A. B. Mikhailovskii and L. I. Rudakov, *Zhur. Eksptl. Teoret. Fiz.* **44**, 912 (1963) [*Soviet Phys. JETP* **17**, 621 (1963)].
13. A. B. Mikhailovskii and A. V. Timofeev, *Zhur. Eksptl. Teoret. Fiz.* **44**, 919 (1963) [*Soviet Phys. JETP* **17**, 626 (1963)].

Problems

8.1. Consider a situation in which the guiding centers of the charged particles in the magnetic field are described by the distribution

$$f_G(\mathbf{r}_\perp,\mathbf{v})=[m/2\pi T(\mathbf{r}_\perp)]^{3/2}\exp[-mv^2/2T(\mathbf{r}_\perp)].$$

Here, the temperature varies in the directions perpendicular to the magnetic field; $\mathbf{r}_\perp$ represents the position vector in these directions. The cyclotron orbit of a charged particle in the phase space may be given by

$$\mathbf{r}=(v_\perp/|\Omega|)(\hat{x}\cos\Omega t-\hat{y}\sin\Omega t)+\hat{z}v_\parallel t,$$

$$\mathbf{v}=-(\Omega/|\Omega|)v_\perp(\hat{x}\sin\Omega t+\hat{y}\cos\Omega t)+\hat{z}v_\parallel.$$

When the density of the guiding centers also varies as $n(\mathbf{r}_\perp)$, the flux of particle flow at the origin is calculated as

$$n\mathbf{v}_\mathrm{D}=\int d\mathbf{v}\, n(-\mathbf{r}_\perp)f_G(-\mathbf{r}_\perp,\mathbf{v})\mathbf{v}$$

where $\mathbf{r}_\perp=(v_\perp/|\Omega|)(\hat{x}\cos\Omega t-\hat{y}\sin\Omega t)$.

We now carry out the Taylor expansion of $n(\mathbf{r}_\perp)$ and $f_G(\mathbf{r}_\perp,\mathbf{v})$ around the origin and retain the lowest-order contributions of the spatial inhomogeneities; we then calculate the average $\langle n\mathbf{v}_\mathrm{D}\rangle$ of the particle flux over a cyclotron period. Carry through all these calculations and show that the particle drift may be written as

$$\mathbf{v}_\mathrm{D}=\frac{\langle n\mathbf{v}_\mathrm{D}\rangle}{n}=-\frac{cT}{qB^2}\left[\frac{\nabla_\perp n}{n}+\frac{\nabla_\perp T}{T}\right]\times\mathbf{B}.$$

8.2. Show that the distribution function (8.11) is a stationary solution of the Vlasov equation (8.17) in the absence of external perturbations.

8.3. Prove relation (8.41).

8.4. When the plasma density changes continuously in the x direction, rather than in the discontinuous way of Fig. 8.3(a), then the space-charge field $\delta\rho_e(y)$ will be created in the plasma as a result of perturbation (8.32). This field can be calculated quite analogously to the surface charge (8.34) and is now given by

$$\rho_e(y)=\frac{1}{N}\frac{dN}{dx}\frac{\rho_m c}{B}gk\left[\int^t \xi\, dt\right]\cos ky.$$

We may neglect the spatial dependence of $(1/N)(dN/dx)$ and determine the induced electric field from the Poisson equation. Thus, following the procedure of Section 8.3A, show that the growth rate of the gravitational instability in these circumstances is given by $[(1/N)(dN/dx)g]^{1/2}$.

8.5. In a fluid-dynamic approach, the behavior of the electrons and the ions in a plasma may be described by the continuity equation

$$\frac{\partial n_\sigma}{\partial t}+\frac{\partial}{\partial \mathbf{r}}\cdot(n_\sigma\mathbf{u}_\sigma)=0 \qquad (\sigma=e,i)$$

and the equation of motion (without collisions)

$$n_\sigma\left(\frac{\partial}{\partial t}+\mathbf{u}_\sigma\cdot\frac{\partial}{\partial \mathbf{r}}\right)\mathbf{u}_\sigma=-\frac{T_\sigma}{m_\sigma}\frac{\partial n_\sigma}{\partial \mathbf{r}}+\frac{q_\sigma}{m_\sigma}n_\sigma\left(\mathbf{E}+\frac{\mathbf{u}_\sigma}{c}\times\mathbf{B}\right),$$

where n is the number density and $\mathbf{u}$ is the flow velocity. Consider a compensated electron–ion plasma with a uniform magnetic field in the z direction; in the unperturbed state, $\mathbf{E}=0$ and the plasma density depends on x coordinate as

$$n_e(x)=n_i(x)=N(1+\alpha x).$$

(a) Assuming an electric-field perturbation of the form

$$\mathbf{E}(\mathbf{r},t)=-\frac{\partial}{\partial \mathbf{r}}\{\varphi(0)\exp[i(k_y y+k_z z-\omega t)]\},$$

calculate the longitudinal dielectric response function $\epsilon(k_y,k_z;\omega)$. [Use the electrostatic approximation.]
(b) Assuming $\alpha=0$, derive the dispersion relation of the collective mode in the domain where $T_e/m_e\gg(\omega/k_z)^2\gg T_i/m_i$ and $\omega^2\ll\Omega_i^2(\ll\Omega_e^2)$.
(c) When $\alpha\neq 0$, the drift frequencies may be defined as

$$\omega_\sigma^*=k_y(cT_\sigma/q_\sigma B)\alpha.$$

Derive the dispersion relation of the collective mode in the vicinity of $\omega=\omega_i{}^*$ under the conditions that $k_z=0$, $\Omega_\sigma^2\gg(\omega-\omega_\sigma{}^*)^2$, and $\Omega_e^2\gg\omega_e^2$.

CHAPTER 9

FLUCTUATIONS

The macroscopic state of a plasma is usually specified in terms of a set of statistically averaged quantities. The number density, kinetic temperature, and current density are examples of such macroscopic quantities. Since the system consists of a large number of randomly moving particles, the instantaneous values of those quantities do deviate from their mean values; the physical quantities are therefore said to fluctuate.

For a system in thermodynamic equilibrium, it is a well-known fact that such fluctuations ordinarily remain small in magnitude as compared with those mean values. When one of the instabilities sets in, however, the fluctuations are expected to grow enormously; the system goes over to a turbulent state. In any case, regardless of the level of fluctuation, the study of fluctuation phenomena represents one of the central topics in statistical physics. The reason may be understood through a consideration of the ways in which fluctuations are fundamentally involved in the description and analysis of the physical properties of a statistical system.

The fluctuations are usually represented by their spectral functions, which are in some cases called form factors. We note, first of all, that the fluctuation spectrum is directly related to the pair correlation function of the relevant physical quantities via a Fourier transformation. This relationship is generally referred to as the Wiener–Khinchine theorem.

The spectral functions of fluctuations are also connected to the linear response functions of the system. When the system is in thermodynamic equilibrium, the fluctuation-dissipation theorem provides the necessary link between the two sets of functions. For a certain class of nonequilibrium stationary plasmas, a corresponding method of calculating the fluctuation spectrum in terms of the response functions is available.

With the knowledge of the fluctuation spectrum, the correlation energy and hence the various thermodynamic quantities of the system can then be calculated. As we saw in Chapter 2, the correlation function enters in the collision term of the plasma kinetic equation. The rates of various relaxation processes are thereby determined from the spectral functions of fluctuations.

Finally, and perhaps most important of all, the fluctuation spectrum is, through a scattering experiment, a directly observable quantity. Generally, within the validity of the Born approximation, the cross section of inelastic scattering, involving energy and momentum transfer between the many-particle system and the incident beam of particles, yields direct information about the form factors of the many-particle system. In the case of plasmas, it is more advantageous to use an electromagnetic wave for the scattering experiment rather than a beam of particles. The incoherent scattering of a monochromatic electromagnetic wave, such as a laser beam, has been developed into a powerful diagnostic technique for the investigation of the basic physical properties of plasmas.

The purpose of the present chapter is, therefore, to elucidate some of the fundamental relationships involving fluctuation phenomena. Topics related to the relaxation processes and turbulence will be discussed in the subsequent chapters. The fluctuation-dissipation theorem [1–4] is treated in Appendix D; a calculation of the scattering cross section for a many-particle system within the Born approximation [5, 6] is described in Appendix E.

9.1 FORM FACTORS AND CORRELATIONS

We begin a mathematical formulation of the fluctuation phenomena by considering a classical system containing N identical particles of charge q and mass m in a box of volume V. The system is assumed to be translationally invariant in space and time, so that $n \equiv N/V$ represents the average number density of the particles. Since we are dealing with point particles, the density field of the system is expressed as [compare Eq. (1.27)]

$$\rho(\mathbf{r},t) = \sum_{i=1}^{N} \delta[\mathbf{r}-\mathbf{r}_i(t)] \tag{9.1}$$

where the $\mathbf{r}_i(t)$ represent the spatial trajectory of the ith particle. The fluctuation of the density from the mean value n is then given by

$$\begin{aligned}\delta\rho(\mathbf{r},t) &= \rho(\mathbf{r},t) - n \\ &= \sum_{i=1}^{N} \delta[\mathbf{r}-\mathbf{r}_i(t)] - n.\end{aligned} \tag{9.2}$$

The spatial Fourier components of the density fluctuations are calculated from Eq. (9.2) as

$$\begin{aligned}\delta\rho_{\mathbf{k}}(t) &= \int_V d\mathbf{r}\, \delta\rho(\mathbf{r},t) \exp(-i\mathbf{k}\cdot\mathbf{r}) \\ &= \sum_{i=1}^{N} \exp[-i\mathbf{k}\cdot\mathbf{r}_i(t)] - N\delta(\mathbf{k},0).\end{aligned} \tag{9.3}$$

where $\delta(\mathbf{k},0)$ is the three-dimensional extension of Kronecker's delta (5.25). Written as a sum of extremely numerous terms with randomly varying phases, these Fourier components represent a set of random variables. Such random variables may be analyzed in terms of their spectral functions defined through statistical averages. We shall thus introduce the form factors of the system in the following.

A. The Static Form Factor and Time-Independent Correlation Function

A simple and nontrivial function describing a spectral distribution of the density fluctuations may be obtained from the mean square values of Eq. (9.3). We thus define the *static form factor* of the many-particle system as

$$S(\mathbf{k})=\frac{1}{N}\langle|\delta\rho_{\mathbf{k}}(t)|^2\rangle=\frac{1}{N}\langle\delta\rho_{\mathbf{k}}(t)\delta\rho_{-\mathbf{k}}(t)\rangle \tag{9.4}$$

where the angular brackets denote a statistical average over an ensemble of the systems. Because the system is stationary, Eq. (9.4) is a function of the wave vector only; we may suppress the time argument there. The static form factor represents a power spectrum of the density fluctuations in the $\mathbf{k}$ space.

For a system consisting of statistically uncorrelated particles, we have

$$S(\mathbf{k})=1-\delta(\mathbf{k},0). \tag{9.5}$$

To prove this, we rewrite Eq. (9.4) with the aid of (9.3) as

$$S(\mathbf{k})=1-N\delta(\mathbf{k},0)+\frac{1}{N}\left\langle\sum_{i\neq j}^{N}\exp[-i\mathbf{k}\cdot(\mathbf{r}_i-\mathbf{r}_j)]\right\rangle. \tag{9.6}$$

When the positions of two different particles are uncorrelated, the last term of Eq. (9.6) yields $(N-1)\delta(\mathbf{k},0)$; hence, Eq. (9.5) has been proved.

The static form factor is closely related to the pair correlation function, which has been defined by Eq. (2.11b). To find the relationship explicitly, it is instructive to consider the Klimontovich distribution (A.2) and investigate its fluctuation components, $\delta N(X;t)$; in view of Eq. (A.13), these may be defined as

$$\delta N(X;t)=N(X;t)-f_1(X;t). \tag{9.7}$$

It then follows that

$$\delta\rho(\mathbf{r},t)=n\int d\mathbf{v}\,\delta N(X;t), \tag{9.8a}$$

$$\delta\rho_{\mathbf{k}}(t)=n\int_V d\mathbf{r}\int d\mathbf{v}\,\delta N(X;t)\exp(-i\mathbf{k}\cdot\mathbf{r}). \tag{9.8b}$$

In terms of the density fluctuations in the phase space, $\delta N(X;t)$, the static form factor (9.4) may thus be calculated as

$$S(\mathbf{k})=\frac{n^2}{N}\int_V d\mathbf{r}\int_V d\mathbf{r}'\int d\mathbf{v}\int d\mathbf{v}'\langle\delta N(X;t)\delta N(X';t)\rangle\exp[-i\mathbf{k}\cdot(\mathbf{r}-\mathbf{r}')].$$

The volume integrations may be transformed into those with respect to $(\mathbf{r}-\mathbf{r}')$ and $(\mathbf{r}+\mathbf{r}')/2$; because of the translational invariance of the system, the latter integration simply produces a volume factor V. Hence, we have

$$S(\mathbf{k})=n\int_V d(\mathbf{r}-\mathbf{r}')\int d\mathbf{v}\int d\mathbf{v}'\langle\delta N(X;t)\delta N(X';t)\rangle\exp[-i\mathbf{k}\cdot(\mathbf{r}-\mathbf{r}')]. \tag{9.9}$$

The statistical average in the integrand of Eq. (9.9) can be expressed in terms of multiparticle distribution functions as we follow the procedure of Appendix A. With the aid of (A.15), (2.11a), and (2.11b), we thus find

$$\langle\delta N(X;t)\delta N(X';t)\rangle=(1/n)\delta(X-X')F(X)+G(X,X'). \tag{9.10}$$

Substitution of this expression into Eq. (9.9) yields

$$S(\mathbf{k})=1+n\int d\mathbf{v}\int d\mathbf{v}'\,G_{\mathbf{k}}(\mathbf{v},\mathbf{v}') \tag{9.11}$$

where $G_{\mathbf{k}}(\mathbf{v},\mathbf{v}')$ represents the spatial Fourier components of $G(X,X')$. The first term, unity, on the right-hand side of Eq. (9.11) comes from the self-correlations of the particles; the second term describes the mutual correlations between two different particles.

To single out the net correlation effects involving two different particles, we find it convenient to introduce a time-independent density–density correlation function via

$$p(\mathbf{r}-\mathbf{r}')\equiv\int d\mathbf{v}\int d\mathbf{v}'\,G(X,X'). \tag{9.12}$$

Equation (9.11) then becomes

$$S(\mathbf{k})=1+n\int_V d\mathbf{r}\,p(\mathbf{r})\exp(-i\mathbf{k}\cdot\mathbf{r}). \tag{9.13}$$

Physically, $p(\mathbf{r}-\mathbf{r}')+1$ represents the conditional probability of finding a particle in a characteristic volume $1/n$ around the point $\mathbf{r}$ provided that there is a particle at $\mathbf{r}'$. When the particles are not correlated, such a probability would be one; hence, $p(\mathbf{r}-\mathbf{r}')$ describes the net correlation effects. The static form factor is directly connected to this correlation function via a Fourier transformation.

B. Correlation Energy

The static form factor is useful in describing the static properties of the system. For example, various thermodynamic quantities can be calculated from the static form factor in the following way.

Consider the Hamiltonian of the system with binary interactions

$$
\begin{aligned}
H &= \sum_{i=1}^{N} \frac{p_i^2}{2m} + \frac{1}{2} \sum_{i\neq j}^{N} \Phi(\mathbf{r}_i - \mathbf{r}_j) \\
&= \sum_{i=1}^{N} \frac{p_i^2}{2m} + \frac{1}{2V} \sum_{i\neq j}^{N} \sum_{\mathbf{k}}{}' \Phi_{\mathbf{k}} \exp[i\mathbf{k}\cdot(\mathbf{r}_i - \mathbf{r}_j)]
\end{aligned}
\tag{9.14}
$$

where $\mathbf{p}_i$ is the momentum of the ith particle and $\Phi_{\mathbf{k}}$ is the Fourier transformation of the binary interaction potential $\Phi(\mathbf{r})$, that is,

$$
\Phi_{\mathbf{k}} = \int_V d\mathbf{r}\, \Phi(\mathbf{r}) \exp(-i\mathbf{k}\cdot\mathbf{r}). \tag{9.15}
$$

We have assumed that no potential fields remain in the uniform state so that the $\mathbf{k}=0$ term has been omitted in the $\mathbf{k}$ summation; for a system of charged particles, this assumption corresponds to taking into account the neutralizing background of smeared-out charges.

The energy density U of the system is given by the statistical average of the Hamiltonian density; thus, we write

$$
\begin{aligned}
U &= \frac{\langle H \rangle}{V} \\
&= \frac{1}{V} \left\langle \sum_{i=1}^{N} \frac{p_i^2}{2m} \right\rangle + \frac{1}{2V^2} \sum_{i\neq j}^{N} \sum_{\mathbf{k}}{}' \Phi_{\mathbf{k}} \langle \exp[i\mathbf{k}\cdot(\mathbf{r}_i - \mathbf{r}_j)] \rangle.
\end{aligned}
$$

The first term on the right-hand side of this equation yields the kinetic energy density $\frac{3}{2}nT$; the second term may be expressed in terms of $S(\mathbf{k})$ with the aid of Eq. (9.6). Since $S(\mathbf{k})=S(-\mathbf{k})$, we find

$$U=\tfrac{3}{2}nT+(n/2V)\sum_{\mathbf{k}}{}'\Phi_{\mathbf{k}}[S(\mathbf{k})-1]. \tag{9.16a}$$

Alternatively, with the aid of (9.13) and (9.15), we write

$$U=\tfrac{3}{2}nT+(n^2/2)\int d\mathbf{r}\,\Phi(\mathbf{r})p(\mathbf{r}). \tag{9.16b}$$

The last term of Eq. (9.16a) or (9.16b) depends on the net correlations between particles. It may thus be regarded as the correlation energy of the system. From such a calculation of the energy density, we can derive various thermodynamic quantities.

C. The Dynamic Form Factor and Time-Dependent Correlation Function

The static form factor introduced in Section 9.1A does not exhaust all the information contained in the random variables $\delta\rho_{\mathbf{k}}(t)$; significantly, it does not describe the time-dependent behavior or the dynamical structure of the random variables. Such information may be retrieved if we consider the *dynamic form factor* defined by

$$S(\mathbf{k},\omega)=\frac{1}{2\pi}\int_{-\infty}^{\infty}dt\,\langle\delta\rho_{\mathbf{k}}(t'+t)\delta\rho_{-\mathbf{k}}(t')\rangle\exp(i\omega t). \tag{9.17}$$

Since the system under consideration is stationary, the entire expression of Eq. (9.17) is independent of t'. Equivalently, the dynamic form factor can also be calculated from

$$S(\mathbf{k},\omega)\delta(\omega-\omega')=(2\pi)^{-2}\langle\delta\rho(\mathbf{k},\omega)\delta\rho(-\mathbf{k},-\omega')\rangle \tag{9.18}$$

where

$$\delta\rho(\mathbf{k},\omega)=\int_{-\infty}^{\infty}dt\,\delta\rho_{\mathbf{k}}(t)\exp(i\omega t). \tag{9.19}$$

It follows from Eq. (9.18) that the dynamic form factor is a real and nonnegative definite function. This function satisfies a sum rule

$$\int_{-\infty}^{\infty}d\omega\,S(\mathbf{k},\omega)=NS(\mathbf{k}), \tag{9.20}$$

which may be proved from a straightforward calculation.

Physically, the dynamic form factor represents the power spectrum of the density fluctuations in the frequency and wave vector space. One way of seeing this is to note that Eq. (9.17) is a version of the standard Wiener–Khinchine theorem in the theory of random noise. We can also convince ourselves of that statement by deriving the relationship

$$\sum_{\mathbf{k}} \int_{-\infty}^{\infty} d\omega S(\mathbf{k},\omega) = V^2 \langle \{\delta\rho(\mathbf{r},t)\}^2 \rangle, \tag{9.21}$$

an analogue of Parseval's theorem.

It is again instructive to express the dynamic form factor in terms of the Klimontovich functions. Substituting Eq. (9.8b) into Eq. (9.17), we obtain

$$S(\mathbf{k},\omega) = \frac{N^2}{2\pi V} \int_V d(\mathbf{r}-\mathbf{r}') \int_{-\infty}^{\infty} dt \int d\mathbf{v} \int d\mathbf{v}' \langle \delta N(X;t'+t)\delta N(X';t') \rangle$$
$$\times \exp\{-i[\mathbf{k}\cdot(\mathbf{r}-\mathbf{r}') - \omega t]\}. \tag{9.22}$$

Analogous to Eq. (9.12), we may now define two time-dependent correlation functions according to [compare Eq. (9.10)]

$$\int d\mathbf{v} \int d\mathbf{v}' \langle \delta N(X;t'+t)\delta N(X';t') \rangle = (1/n) p_s(\mathbf{r}-\mathbf{r}',t) + p(\mathbf{r}-\mathbf{r}',t). \tag{9.23}$$

Here, $p_s(\mathbf{r},t)$ describes the self motion of a particle; thus $p_s(\mathbf{r},t)d\mathbf{r}$ represents the conditional probability that, when a particle was located at the origin at $t=0$, the same particle is found in the volume $d\mathbf{r}$ around $\mathbf{r}$ at time t. The second function $p(\mathbf{r},t)$ represents an extension of Eq. (9.12) into a time-dependent scheme; $p(\mathbf{r},t)+1$ gives the conditional probability of finding a particle in a characteristic volume $1/n$ around $\mathbf{r}$ at time t provided that another particle was located at the origin at $t=0$. Substitution of Eq. (9.23) into Eq. (9.22) yields

$$S(\mathbf{k},\omega) = \frac{N}{2\pi} \int_V d\mathbf{r} \int_{-\infty}^{\infty} dt [p_s(\mathbf{r},t) + np(\mathbf{r},t)] \exp[-i(\mathbf{k}\cdot\mathbf{r} - \omega t)]. \tag{9.24}$$

The first term on the right-hand side of Eq. (9.24) represents that part of the dynamic form factor stemming from the self-motion of the particles; the second term describes the effects of dynamic correlations between two different particles. For a system of uncorrelated particles, the first term only remains.

The dynamic form factor plays the central part in formulating a theory of a many-particle system, be it classical or quantum mechanical [6]. In the following sections, we shall study these functions in various cases of plasmas.

9.2 DENSITY FLUCTUATIONS IN EQUILIBRIUM PLASMAS

Let us consider the dynamic form factor or the spectral function of the electron-density fluctuation in those plasmas under thermodynamic equilibrium. The fluctuation-dissipation theorem derived in Appendix D is applicable for the calculation of the dynamic form factor in these circumstances. We begin with the simple cases of an electron gas, a single-component plasma consisting of electrons.

A. Electron Gas

The dynamic form factor of the many-particle system can be calculated with the aid of the fluctuation-dissipation theorem as soon as the linear response function relevant to the problem under consideration is found. To find such a response function, we note that the dynamic form factor defined by Eq. (9.17) may be obtained if we replace both $\mathcal{A}$ and $\mathcal{B}$ in (D.25) by the electron density fluctuations $\delta\rho$. In order that we may find $\delta\rho$ in place of $\mathcal{A}$ in the extra Hamiltonian (D.6), we must thus choose an external disturbance which couples with the electron density. Let us, therefore, apply a potential field $\varphi_{\text{ext}}(\mathbf{r},t)$ to the plasma. We then have

$$\mathcal{H}' = -e\int d\mathbf{r}\,\delta\rho(\mathbf{r})\varphi_{\text{ext}}(\mathbf{r},t) \tag{9.25}$$

so that $a(\mathbf{r},t)$ in (D.6) should be given by $e\varphi_{\text{ext}}(\mathbf{r},t)$. The response function (D.4) in this case is

$$K(\mathbf{k},\omega) = \frac{\delta\rho(\mathbf{k},\omega)}{e\varphi_{\text{ext}}(\mathbf{k},\omega)} \tag{9.26}$$

where $\delta\rho(\mathbf{k},\omega)$ is the Fourier component of the induced density fluctuation of the electrons.

Such density fluctuations in turn induce potential fields in the plasma; these may be calculated from the Poisson equation as

$$\varphi_{\text{ind}}(\mathbf{k},\omega) = -\frac{4\pi e}{k^2}\delta\rho(\mathbf{k},\omega). \tag{9.27}$$

Recalling the definition of the dielectric response function as in Eq. (6.2), we can express Eq. (9.26) as

$$K(\mathbf{k},\omega) = \frac{k^2}{4\pi e^2}\left[1 - \frac{1}{\epsilon(\mathbf{k},\omega)}\right]. \tag{9.28}$$

Application of Eq. (D.29) thus yields†

$$S(\mathbf{k},\omega)=-\frac{N}{\pi\omega}\frac{k^2}{k_\mathrm{D}^2}\,\mathrm{Im}\frac{1}{\epsilon(\mathbf{k},\omega)}. \tag{9.29}$$

The static form factor is calculated from Eq. (9.29) with the aid of the sum rule (9.20):

$$S(\mathbf{k})=-\frac{k^2}{\pi k_\mathrm{D}^2}\int_{-\infty}^{\infty}\frac{d\omega}{\omega}\,\mathrm{Im}\frac{1}{\epsilon(\mathbf{k},\omega)}. \tag{9.30}$$

This integration can be carried out explicitly by evoking the Kramers-Kronig relation (C.5) in which we let $\omega=0$; we thus have

$$S(\mathbf{k})=\frac{k^2}{k_\mathrm{D}^2}\left[1-\mathrm{Re}\frac{1}{\epsilon(\mathbf{k},0)}\right]$$

$$=\frac{k^2}{k^2+k_\mathrm{D}^2}. \tag{9.31}$$

The final expression may be obtained from either (4.1) or (4.68); the result is not affected by the presence of a magnetic field.

The time-independent pair correlation function (9.12) may then be calculated from the inverse Fourier transformation of $S(\mathbf{k})-1$ as Eq. (9.13) indicates; we thus have

$$p(\mathbf{r})=-\frac{k_\mathrm{D}^2}{4\pi n r}\exp(-k_\mathrm{D}r)$$

$$=-\frac{g}{4\pi}\left(\frac{\lambda_\mathrm{D}}{r}\right)\exp\left(-\frac{r}{\lambda_\mathrm{D}}\right) \tag{9.32}$$

where g is the plasma parameter (1.22). This repulsive correlation has the same physical origin and character as the Debye–Hückel screening. It is quite important to note here that Eq. (9.32) predicts physically unacceptable values for $r<g\lambda_\mathrm{D}/4\pi$; the probability of finding other electrons, $p(\mathbf{r})+1$, becomes

†All the results in Appendix D have been obtained with the periodic boundary conditions appropriate to a unit volume. The dynamic form factor defined in Section 9.1C is in fact proportional to the volume of the system. This may be seen most clearly from Eq. (9.24): By the definition of the probability functions $p_s(\mathbf{r},t)$ and $p(\mathbf{r},t)$, the integration in (9.24) is independent of the volume; $S(\mathbf{k},\omega)$ is thus proportional to the volume V through N. This volume dependence is recovered in Eq. (9.29).

negative at such short distances. The origin of this failure may be traced to the limit of applicability for the Vlasov equation, upon which the calculations of the dielectric functions, (4.1) and (4.68), are based. For correlations at such short distances, the effects of the particle discreteness are much more important than the effects coming from the average distribution of all the other particles. The factorization (2.7) or assumption (2.12), for example, is no longer justified. A theory relying on such an expansion with respect to the discreteness parameters breaks down there.

The thermodynamic energy density (9.16) can likewise be calculated from (9.31) or (9.32). Since $\Phi(\mathbf{r})=e^2/r$ or $\Phi_{\mathbf{k}}=4\pi e^2/k^2$, we find

$$U=\tfrac{3}{2}nT(1-(g/12\pi)). \tag{9.33}$$

The last term of Eq. (9.33) represents the correlation energy arising from the Debye–Hückel screening. In particular, we have here reconfirmed the estimate made in Eq. (1.15).

Equation (9.33) offers the correct expression of the thermodynamic energy density to the first order in the plasma parameter. For the classical electron gas, rigorous calculations of the thermodynamic quantities are available up to the second-order terms in the expansion with respect to the plasma parameter [7–10]. To this order, the energy density is written as

$$U=\tfrac{3}{2}nT\left\{1-\tfrac{1}{3}(g/4\pi)-\tfrac{1}{3}(g/4\pi)^2\left[\tfrac{1}{2}\ln(g/4\pi)+\left(C-\tfrac{2}{3}+\tfrac{1}{2}\ln 3\right)\right]\right\} \tag{9.34}$$

where $C=0.57721\ldots$ is Euler's constant. In obtaining these results, we must take proper account of strong correlations at short distances [11].

B. Two-Component Plasma

We now proceed to consider the dynamic form factor $S_e(\mathbf{k},\omega)$ of the electrons in the two-component plasma consisting of electrons and ions. In this case, we must be a little careful about selecting a disturbance to the plasma, for an ordinary potential field as considered in (9.25) couples not only with the electrons but also with the ions. The resulting response function could not be properly used for the calculation of the dynamic form factor of the electrons through the fluctuation-dissipation theorem. Such a situation can be easily avoided, however, if we introduce a fictitious potential field which would couple only with the electrons.

We thus find that the appropriate response function for the problem under consideration is given by (see Problem 9.4)

$$K_e(\mathbf{k},\omega)=\frac{k^2}{4\pi e^2}\chi_e(\mathbf{k},\omega)\frac{1+\chi_i(\mathbf{k},\omega)}{\epsilon(\mathbf{k},\omega)}. \tag{9.35}$$

Here, $\chi_\sigma(\mathbf{k},\omega)$ represents the polarizability associated with the particles of the σ species, so that the dielectric response function of the plasma is written generally as

$$\epsilon(\mathbf{k},\omega)=1+\sum_\sigma \chi_\sigma(\mathbf{k},\omega). \tag{9.36}$$

The dynamic form factor is thereby calculated from the fluctuation-dissipation theorem as

$$S_e(\mathbf{k},\omega)=\frac{N}{\pi\omega}\frac{k^2}{k_e^2}\,\mathrm{Im}\left[\chi_e(\mathbf{k},\omega)\frac{1+\chi_i(\mathbf{k},\omega)}{\epsilon(\mathbf{k},\omega)}\right] \tag{9.37}$$

where $k_e^2=k_i^2$.

The static form factor can be calculated from Eq. (9.37) with the aid of the same technique as that employed in Eq. (9.31). The polarizabilities individually satisfy the analytic properties of the retarded response functions; hence, we have

$$\begin{aligned} S_e(\mathbf{k}) &= \frac{k^2}{k_e^2}\,\mathrm{Re}\left[\chi_e(\mathbf{k},0)\frac{1+\chi_i(\mathbf{k},0)}{\epsilon(\mathbf{k},0)}\right] \\ &= \frac{k^2+k_e^2}{k^2+2k_e^2}. \end{aligned} \tag{9.38}$$

The results obtained in this section are again applicable to plasmas with and without an external magnetic field.

9.3 DENSITY FLUCTUATIONS IN NONEQUILIBRIUM PLASMAS

We now extend our considerations of the dynamic form factor to those cases of stationary plasmas which are not in the state of thermodynamic equilibrium. These considerations are particularly important because almost all of the plasmas occurring in nature are in such nonequilibrium states. The fluctuation-dissipation theorem is no longer applicable to such a plasma. We must deal directly with the plasma kinetic equations to obtain a proper formalism of fluctuations under these conditions. We shall thereby find a general scheme of calculating the spectral functions which is applicable to a wide class of nonequilibrium plasmas [12–15]; this scheme reduces to the fluctuation-dissipation theorem when the plasma is in thermodynamic equilibrium.

A. Kinetic-Theoretical Calculation of the Static Form Factor

Let us first recall our calculations in Chapter 2, leading to Eq. (2.42). Within the validity of the ordering (2.12) and (2.13), as well as the Bogoliubov ansatz (2.25), those calculations gave us a correct expression for the pair correlation function $G_{\mathbf{k}}(\mathbf{v}_1, \mathbf{v}_2)$ to the first order in the plasma parameter. Let us also recall that this pair correlation function is related directly to the static form factor via Eq. (9.11). Consequently, we may use the results of Sections 2.4 and 2.5 for the calculation of the static form factor.

We therefore integrate Eq. (2.42) with respect to $\mathbf{v}_1$ and $\mathbf{v}_2$, and then substitute Eq. (2.36) in it. Equation (9.11) is thus calculated as

$$S(\mathbf{k}) = 1 + i\int d\mathbf{v}\int_{-\infty}^{\infty}\frac{d\omega}{2\pi}\frac{[\epsilon(\mathbf{k},\omega)-1]f(\mathbf{v})}{(\omega-\mathbf{k}\cdot\mathbf{v}-i\eta)|\epsilon(\mathbf{k},\omega)|^2}$$

$$-i\int d\mathbf{v}\int_{-\infty}^{\infty}\frac{d\omega}{2\pi}\frac{[\epsilon(-\mathbf{k},-\omega)-1]f(\mathbf{v})}{(\omega-\mathbf{k}\cdot\mathbf{v}+i\eta)|\epsilon(\mathbf{k},\omega)|^2}.$$

In view of the analytic properties of $\epsilon(\pm\mathbf{k},\pm\omega)$ on the complex ω plane, we prove the identities

$$i\int_{-\infty}^{\infty}\frac{d\omega}{2\pi}[(\omega-\mathbf{k}\cdot\mathbf{v}\pm i\eta)\epsilon(\pm\mathbf{k},\pm\omega)]^{-1} = \pm\frac{1}{2}$$

by closing the integration contour with an infinite semicircle in the upper or lower half of the complex plane. The foregoing equation thus reduces to

$$S(\mathbf{k}) = \frac{1}{N}\int_{-\infty}^{\infty} d\omega\frac{S^{(0)}(\mathbf{k},\omega)}{|\epsilon(\mathbf{k},\omega)|^2} \tag{9.39}$$

where

$$S^{(0)}(\mathbf{k},\omega) = N\int d\mathbf{v}f(\mathbf{v})\delta(\omega-\mathbf{k}\cdot\mathbf{v}). \tag{9.40}$$

For those nonequilibrium stationary plasmas in which we expect the ordering of (2.12) and (2.13) to be valid, Eq. (9.39) offers a correct calculation of the static form factor to the first order in the plasma parameter.

Comparison between Eqs. (9.20) and (9.39) may suggest that the dynamic form factor be given by

$$S(\mathbf{k},\omega) = \frac{S^{(0)}(\mathbf{k},\omega)}{|\epsilon(\mathbf{k},\omega)|^2}. \tag{9.41}$$

This conjecture in fact turns out to be the case. We shall outline a proof of Eq. (9.41) in the following section.

When the plasma is in thermodynamic equilibrium, the velocity distribution is a Maxwellian; we have the relationship

$$S^{(0)}(\mathbf{k},\omega)=\frac{N}{\pi\omega}\frac{k^2}{k_D^2}\,\mathrm{Im}\,\epsilon(\mathbf{k},\omega). \tag{9.42}$$

Equation (9.41) thus reduces to the fluctuation-dissipation theorem (9.29) in those equilibrium situations.

B. Kinetic-Theoretical Calculation of the Dynamic Form Factor

We begin with expression (9.24) for the dynamic form factor, which has been obtained by writing the statistical average of the product of two Klimontovich functions as Eq. (9.23). We wish first to establish the ordering of the probability functions $p_s(\mathbf{r},t)$ and $p(\mathbf{r},t)$ with respect to the plasma-parameter expansions, so that we may treat the calculation of $S(\mathbf{k},\omega)$ on the same footing as that of the pair correlation function in Chapter 2. In order to do so, it may be useful to note that those probability functions remain physically well-defined as one approaches the fluid limit of Section 1.3A. When each particle is cut into ν finer pieces, we can still define the self-correlation function $p_s(\mathbf{r},t)$ for the resulting smaller particle. The probability of finding other particles in a given volume will increase by a factor of ν, but the characteristic volume itself decreases by a factor of $1/\nu$; hence, $p(\mathbf{r},t)$ should remain unchanged. It thus follows from Eq. (9.24) that $p(\mathbf{r},t)$ must be calculated up to the first-order terms in the plasma-parameter expansion while $p_s(\mathbf{r},t)$ need be calculated only to the zeroth-order terms.

We now extend the definition of these probability functions into the entire domain of the phase space and write

$$\langle\delta N(X;t)\delta N(X';0)\rangle=(1/n)f(\mathbf{v})\delta(\mathbf{r}-\mathbf{r}'-\mathbf{v}t)\delta(\mathbf{v}-\mathbf{v}')+P(X,X';t). \tag{9.43}$$

We have been assuming the space–time translational invariance of the system. The first term on the right-hand side of Eq. (9.43) describes the self-motion of a particle during a time interval t; to the lowest order in the plasma parameter, the trajectories of the particles are those of rectilinear motion. Probability $p_s(\mathbf{r},t)$ is then given by

$$p_s(\mathbf{r},t)=\delta(\mathbf{r}-\mathbf{v}t). \tag{9.44}$$

We thus find that Eq. (9.40) can be expressed as

$$S^{(0)}(\mathbf{k},\omega)=(N/2\pi)\int_V d\mathbf{r}\int_{-\infty}^{\infty}dt\,p_s(\mathbf{r},t)\exp[-i(\mathbf{k}\cdot\mathbf{r}-\omega t)]. \quad (9.45)$$

We shall later establish the validity of this relationship for more general cases of $p_s(\mathbf{r},t)$.

The second term on the right-hand side of Eq. (9.43) describes time-dependent correlations in the phase space between two different particles. Substitution of Eq. (9.43) into Eq. (9.22) yields

$$S(\mathbf{k},\omega)=S^{(0)}(\mathbf{k},\omega)+\frac{N^2}{2\pi V}\int_V d(\mathbf{r}-\mathbf{r}')\int_{-\infty}^{\infty}dt\int d\mathbf{v}\int d\mathbf{v}'P(X,X';t)$$

$$\times\ \exp\{-i[\mathbf{k}\cdot(\mathbf{r}-\mathbf{r}')-\omega t]\}. \quad (9.46)$$

The dynamic form factor may thus be obtained as soon as $P(X,X';t)$ is calculated.

Since the function $P(X,X';t)$ satisfies the initial condition

$$P(X,X';0)=G(X,X'), \quad (9.47)$$

its values at later times may be determined when its equation of motion is known. To obtain such an equation, we consider the Klimontovich equation (A.17) expressed in terms of the fluctuations (9.7):

$$\left[\frac{\partial}{\partial t}+\mathrm{L}(X)-\int \mathrm{V}(X,X'')\delta N(X'';t)\,dX''\right]\delta N(X;t)$$

$$-\int \mathrm{V}(X,X'')f(\mathbf{v})\delta N(X'';t)\,dX''=0. \quad (9.48)$$

We multiply this equation by $\delta N(X';0)$ and carry out the statistical average. We then note that as long as the scaling (2.13) is valid, the average $\langle\delta N(X;t)\delta N(X';0)\delta N(X'';t)\rangle$ should likewise scale as g^2 (Problem 9.5); the term involving this average may be neglected. We thus have

$$\left[\frac{\partial}{\partial t}+\mathrm{L}(X)\right]\langle\delta N(X;t)\delta N(X';0)\rangle$$

$$-\int \mathrm{V}(X,X'')f(\mathbf{v})\langle\delta N(X'';t)\delta N(X';0)\rangle dX''=0, \quad (9.49)$$

or substituting (9.43) into this equation, we find

$$\left[\frac{\partial}{\partial t}+L(X)\right]P(X,X';t)-\int V(X,X'')f(\mathbf{v})P(X'',X';t)\,dX''$$

$$=\frac{1}{n}\int V(X,X'')f(\mathbf{v})f(\mathbf{v}')\delta(\mathbf{r}''-\mathbf{r}'-\mathbf{v}'t)\delta(\mathbf{v}'-\mathbf{v}'')\,dX''. \tag{9.50}$$

Starting from the initial condition (9.47), the time-dependent correlation function develops in time according to this equation. Its left-hand side is a linearized version of the Vlasov equation; the right-hand side represents a driving term arising from the free rectilinear motion of the individual particles.

Equation (9.50) may be solved in much the same way as Eq. (2.29). We carry out the Fourier–Laplace transformations so that

$$P_{\mathbf{k}}(\mathbf{v},\mathbf{v}';t)=\int_V d(\mathbf{r}-\mathbf{r}')P(X,X';t)\exp[-i\mathbf{k}\cdot(\mathbf{r}-\mathbf{r}')], \tag{9.51a}$$

$$\mathrm{P}_{\mathbf{k}}(\mathbf{v},\mathbf{v}';\omega)=\int_0^\infty dt P_{\mathbf{k}}(\mathbf{v},\mathbf{v}';t)\exp(i\omega t). \tag{9.51b}$$

With the initial condition (9.47), Eq. (9.50) is transformed as

$$\mathrm{P}_{\mathbf{k}}(\mathbf{v},\mathbf{v}';\omega)+\frac{\omega_p^2}{\omega-\mathbf{k}\cdot\mathbf{v}+i\eta}\frac{\mathbf{k}}{k^2}\cdot\frac{\partial}{\partial\mathbf{v}}f(\mathbf{v})\int d\mathbf{v}\,\mathrm{P}_{\mathbf{k}}(\mathbf{v},\mathbf{v}';\omega)$$

$$=i\frac{G_{\mathbf{k}}(\mathbf{v},\mathbf{v}')}{\omega-\mathbf{k}\cdot\mathbf{v}+i\eta}-i\frac{\omega_p^2}{n\,(\omega-\mathbf{k}\cdot\mathbf{v}+i\eta)(\omega-\mathbf{k}\cdot\mathbf{v}'+i\eta)}\frac{\mathbf{k}}{k^2}\cdot\frac{\partial}{\partial\mathbf{v}}[f(\mathbf{v})f(\mathbf{v}')]. \tag{9.52}$$

The explicit expression for $G_{\mathbf{k}}(\mathbf{v},\mathbf{v}')$ is available in Eq. (2.42), where we may substitute Eq. (2.38).

We now wish to calculate from Eq. (9.52) the function

$$\mathrm{P}(\mathbf{k},\omega)\equiv\int d\mathbf{v}\int d\mathbf{v}'\mathrm{P}_{\mathbf{k}}(\mathbf{v},\mathbf{v}';\omega), \tag{9.53}$$

which is closely related to the last term of (9.46). In fact, the only difference between Eq. (9.53) and the function

$$P(\mathbf{k},\omega)\equiv\int d\mathbf{v}\int d\mathbf{v}'\int_V d(\mathbf{r}-\mathbf{r}')\int_{-\infty}^{\infty} dt P(X,X';t)\exp\{-i[\mathbf{k}\cdot(\mathbf{r}-\mathbf{r}')-\omega t]\}$$

(9.54)

arises from the fact that $\mathsf{P}(\mathbf{k},\omega)$ is defined through a one-sided Fourier transformation (9.51b) while $P(\mathbf{k},\omega)$ involves the ordinary Fourier transformation. These functions are related to each other by

$$\mathsf{P}(\mathbf{k},\omega)=i\int_{-\infty}^{\infty}\frac{d\omega'}{2\pi}\frac{1}{\omega-\omega'+i\eta}P(\mathbf{k},\omega') \tag{9.55a}$$

or, since $P(\mathbf{k},\omega)$ is a real function, we have

$$P(\mathbf{k},\omega)=2\,\mathrm{Re}\,\mathsf{P}(\mathbf{k},\omega). \tag{9.55b}$$

The dynamic form factor (9.46) can thus be obtained from the real part of Eq. (9.53).

Calculations of Eq. (9.53) with the aid of Eq. (9.52) proceed in a way mathematically quite similar to those described in Appendix B. The intermediate steps are, therefore, left to the reader as an exercise (Problem 9.6); the result is

$$\mathsf{P}(\mathbf{k},\omega)=\frac{i}{n}\int d\mathbf{v}\frac{f(\mathbf{v})}{(\omega-\mathbf{k}\cdot\mathbf{v}+i\eta)|\epsilon(\mathbf{k},\mathbf{k}\cdot\mathbf{v})|^2}-\frac{i}{n}\int d\mathbf{v}\frac{f(\mathbf{v})}{\omega-\mathbf{k}\cdot\mathbf{v}+i\eta}. \tag{9.56}$$

The dynamic form factor thus obtains from (9.46) as Eq. (9.41).

C. Superposition of Dressed Test Particles

The result obtained in Eq. (9.41) contains an important physical concept; the dynamic form factor of a nonequilibrium stationary plasma may be calculated by superposing the fields of dynamically screened or "dressed" test particles without further consideration of the correlations among them. The theoretical scheme for calculating the fluctuation spectrum through this superposition principle has been extensively investigated [13, 16–18]; the method is applicable to a stable plasma in a stationary state.

The idea of the superposition principle can be explained in the following way: We pick each particle (electron) in the plasma and regard it as a test charge. Suppose that such a particle is located initially at $\mathbf{r}_j$ and has a velocity

$\mathbf{v}_j$; in the absence of collisions, the Fourier components of its density field are

$$\rho_j^{(0)}(\mathbf{k},\omega)=2\pi\exp(-i\mathbf{k}\cdot\mathbf{r}_j)\delta(\omega-\mathbf{k}\cdot\mathbf{v}_j). \tag{9.57}$$

The field of the dressed particle can be constructed by taking additional account of the induced screening cloud; hence, we define

$$\rho_j^{(s)}(\mathbf{k},\omega)\equiv\frac{\rho_j^{(0)}(\mathbf{k},\omega)}{\epsilon(\mathbf{k},\omega)}. \tag{9.58}$$

The dressed particle, therefore, consists of the bare particle field (9.57) plus the induced screening cloud. The total density fields of the dressed particles are

$$\rho^{(s)}(\mathbf{k},\omega)=\sum_{j=1}^{N}\rho_j^{(s)}(\mathbf{k},\omega)=2\pi\sum_{j=1}^{N}\exp(-i\mathbf{k}\cdot\mathbf{r}_j)\delta(\omega-\mathbf{k}\cdot\mathbf{v}_j)/\epsilon(\mathbf{k},\omega). \tag{9.59}$$

We may now use the expressions (9.59) for the calculation of the dynamic form factor in accord with Eq. (9.18). In these processes we neglect the correlations between two different dressed particles; the major effects of correlations have been already taken into account in the structure of each dressed particle (9.58). We thus have

$$\begin{aligned}S(\mathbf{k},\omega)\delta(\omega-\omega')&=(2\pi)^{-2}\langle\rho^{(s)}(\mathbf{k},\omega)\rho^{(s)}(-\mathbf{k},-\omega')\rangle\\&=N\frac{\langle\delta(\omega-\mathbf{k}\cdot\mathbf{v})\rangle}{|\epsilon(\mathbf{k},\mathbf{k}\cdot\mathbf{v})|^2}\delta(\omega-\omega').\end{aligned}$$

Introducing the single-particle velocity distribution $f(\mathbf{v})$ to carry out the required average, we obtain

$$S(\mathbf{k},\omega)=\frac{N}{|\epsilon(\mathbf{k},\omega)|^2}\int d\mathbf{v}f(\mathbf{v})\delta(\omega-\mathbf{k}\cdot\mathbf{v}), \tag{9.60}$$

a result identical to Eq. (9.41). The direct kinetic-theoretical calculation of the dynamic form factor in the previous section then offers a proof for the validity of this superposition principle.

In terms of the Bogoliubov hierarchy of the characteristic time scales involved in the plasma (Section 2.3), we may describe the superposition principle in the following way. Let us again fix on a charged particle in the plasma which we regard as a moving test charge. It will act to polarize the medium and so carry a screening cloud with it. This action corresponds to the establishment of a pair correlation; the characteristic time associated with such a process has been denoted by τ_2. In the course of its motion, however, the test charge may be involved in a short-range collision and suffer an abrupt change of its energy and momentum. The polarization cloud originally associated with the test charge will no longer represent the appropriate screening cloud; the polarization cloud must adjust itself to its new circumstances. In such an event, if the mean free time τ_1 of the collisions is much larger than τ_2, the adjustment will take place so quickly that, for most of the time between two successive short-range collisions, the test charge can be considered as carrying a well-established cloud. It is clear that the superposition calculation of Eq. (9.60) amounts to assuming that each particle, not *almost* always, but *always* carries a well-established screening cloud. We may therefore argue that the superposition principle represents a good approximation as long as $\tau_1 \gg \tau_2$. For a stable plasma with $g \ll 1$, this condition is reasonably well satisfied.

It is instructive to consider yet another argument, which exhibits quantum-theoretically the validity of the superposition principle. For this purpose, we first note that for each many-particle system consisting of charged particles, we can imagine a fictitious neutral counterpart which may be constructed by adiabatically turning off the long-range part of the Coulomb interaction between the particles in the real system [6, 16, 18]. The turning off can be achieved by subtracting the average self-consistent field of each particle, and thus the screened short-range Coulomb forces remain in the resulting fictitious system. The matrix element $(\rho_{-\mathbf{k}}^{(0)})_{n0}$ of the density fluctuation excitations in the fictitious system are then given by the product between those $(\rho_{-\mathbf{k}})_{n0}$ in the real charged system and the dielectric response function $\epsilon(\mathbf{k},\omega)$, that is

$$(\rho_{-\mathbf{k}}{}^{(0)})_{n0} = (\rho_{-\mathbf{k}})_{n0}\epsilon(-\mathbf{k},\omega_{n0}) \tag{9.61}$$

where ω_{n0} are the excitation frequencies of the system from the ground state 0 [see Eqs. (D.19), (D.23), and (D.24)]. We remark that both the density fluctuations in the real system and the dielectric response function are quantities that physically have unambiguous definitions; Eq. (9.61) may thus be regarded as a mathematical definition of the fictitious system.

The dynamic form factor may now be calculated in a way similar to Eq. (E.10); since the initial state has been fixed to 0, we have

$$S(\mathbf{k},\omega) = \frac{S^{(0)}(\mathbf{k},\omega)}{|\epsilon(-\mathbf{k},\omega)|^2} \tag{9.62}$$

where

$$S^{(0)}(\mathbf{k},\omega)=\sum_n |(\rho_{-\mathbf{k}}{}^{(0)})_{n0}|^2\delta(\omega-\omega_{n0}) \tag{9.63}$$

is the dynamic form factor of the fictitious neutral system. Equation (9.62) reduces to the form of Eq. (9.41) when the system is reflectionally invariant in either space or time.

Equation (9.62) has permitted us to replace the problem of calculating $S(\mathbf{k},\omega)$ with that of $S^{(0)}(\mathbf{k},\omega)$. For a neutral system, we may then argue that the most important contribution to the dynamic form factor comes from the self-motion of the particles. We may therefore write to a good approximation

$$S^{(0)}(\mathbf{k},\omega)=\frac{N}{2\pi}\int_V d\mathbf{r}\int_{-\infty}^{\infty} dt\, p_s(\mathbf{r},t)\exp[-i(\mathbf{k}\cdot\mathbf{r}-\omega t)] \tag{9.64}$$

where $p_s(\mathbf{r},t)$, defined through Eq. (9.23), is now the *true* probability function describing the self-motion of a particle in the system. A sum rule

$$(1/N)\int_{-\infty}^{\infty} d\omega\, S^{(0)}(\mathbf{k},\omega)=1 \tag{9.65}$$

follows from Eq. (9.64).

The superposition principle as expressed by Eqs. (9.41) and (9.64) thus makes it possible to treat some of those problems that are outside the scope of Eq. (9.60). As we have seen, the derivation of Eq. (9.60) is based on the solution of the kinetic equations with the aid of a plasma-parameter expansion. The trajectories of the dressed particles are hence restricted to those obtained from a solution of unperturbed Vlasov equations; in the absence of a magnetic field, we are led to Eq. (9.40) corresponding to the rectilinear self-motion (9.44). According to Eq. (9.64), however, we may take additional account of the collisional and diffusive aspects of the particle self-motion through, for example, a hydrodynamic approach [19]. We shall call Eqs. (9.41) and (9.64) the *dielectric superposition principle*.

Before closing this section, let us briefly describe an application of the superposition technique to a two-component plasma. We denote by $\rho_e^{(0)}(\mathbf{k},\omega)$ and $\rho_i^{(0)}(\mathbf{k},\omega)$ the density field associated with the bare electrons and ions, respectively. The fields of the dressed electrons and ions are then calculated as

$$\rho_e{}^{(s)}(\mathbf{k},\omega)=\left[1-\frac{\chi_e(\mathbf{k},\omega)}{\epsilon(\mathbf{k},\omega)}\right]\rho_e{}^{(0)}(\mathbf{k},\omega)+\frac{\chi_i(\mathbf{k},\omega)}{\epsilon(\mathbf{k},\omega)}\rho_e{}^{(0)}(\mathbf{k},\omega), \tag{9.66a}$$

$$\rho_i{}^{(s)}(\mathbf{k},\omega)=\frac{\chi_e(\mathbf{k},\omega)}{\epsilon(\mathbf{k},\omega)}\rho_i{}^{(0)}(\mathbf{k},\omega)+\left[1-\frac{\chi_i(\mathbf{k},\omega)}{\epsilon(\mathbf{k},\omega)}\right]\rho_i{}^{(0)}(\mathbf{k},\omega). \tag{9.66b}$$

We note that the first terms on the right-hand sides of Eqs. (9.66) consist of the electron density fluctuations, while the last terms consist of the ion density fluctuations. Such a mixing of the individual electron and ion coordinates to form a correlated screening cloud of a test charge is a characteristic feature of correlations in a two-component plasma. Because of this mixing, the electron density fluctuations contained in the dressed ion (9.66b), for example, may have to be regarded as moving with the ion velocity. The dynamic form factor of the electrons may now be calculated by superposing the first terms of Eqs. (9.66) in accord with Eq. (9.18). Neglecting the correlations between the dressed particles, we thus find

$$S_e(\mathbf{k},\omega)=N_e\left|1-\frac{\chi_e(\mathbf{k},\omega)}{\epsilon(\mathbf{k},\omega)}\right|^2\int d\mathbf{v}f_e(\mathbf{v})\delta(\omega-\mathbf{k}\cdot\mathbf{v})$$

$$+N_i\left|\frac{\chi_e(\mathbf{k},\omega)}{\epsilon(\mathbf{k},\omega)}\right|^2\int d\mathbf{v}f_i(\mathbf{v})\delta(\omega-\mathbf{k}\cdot\mathbf{v}). \tag{9.67}$$

Again we remark that the last term exhibits the influence of the ion velocity distribution on the frequency spectrum of the electron density fluctuations.

D. Fluctuations Associated with the Collective Modes

The dynamic form factor (9.41) clearly demonstrates both the collective and individual-particle aspects of fluctuations in the plasma. As we argued in connection with Eq. (9.64), the function $S^{(0)}(\mathbf{k},\omega)$ is essentially related to the self-motion of the individual particles. This function, therefore, describes the fluctuation spectrum associated with the motion of the individual particles; Eq. (9.40) offers the simplest example of such a spectrum.

The collective mode, determined from the dispersion relation $\epsilon(\mathbf{k},\omega)=0$, greatly affects the shape of the spectral function. Generally, it gives rise to a sharp peak in the frequency spectrum, as illustrated in Fig. 4.3. Integration across such a resonance pole should then yield a measure of the total energy contained in the fluctuations associated with the given collective mode.

Such an integration may be carried out by following the method described in Section 4.2B. We first note that the dynamic form factor is generally written in a form proportional to $|\epsilon(\mathbf{k},\omega)|^{-2}$; we then rewrite

$$\frac{1}{|\epsilon(\mathbf{k},\omega)|^2}=-\frac{\mathrm{Im}[\epsilon(\mathbf{k},\omega)]^{-1}}{\mathrm{Im}\,\epsilon(\mathbf{k},\omega)}.$$

In the vicinity of the collective mode $\omega=\omega_{\mathbf{k}}$, we may use the expansion (4.16), and hence

$$-\mathrm{Im}[\epsilon(\mathbf{k},\omega)]^{-1}=\frac{\pi}{(\partial\epsilon_1/\partial\omega_\mathbf{k})}\delta(\omega-\omega_\mathbf{k}) \tag{9.68}$$

where

$$\frac{\partial\epsilon_1}{\partial\omega_\mathbf{k}}=\left[\frac{\partial\epsilon_1(\mathbf{k},\omega)}{\partial\omega}\right]_{\omega=\omega_\mathbf{k}} \tag{9.69}$$

Again with the aid of (4.16), we have

$$\frac{1}{|\epsilon(\mathbf{k},\omega)|^2}=-\pi\left(\frac{\partial\epsilon_1}{\partial\omega_\mathbf{k}}\right)^{-2}\frac{\delta(\omega-\omega_\mathbf{k})}{\gamma_\mathbf{k}}. \tag{9.70}$$

Since we have been dealing with a stable plasma, $\gamma_\mathbf{k}<0$.

That part of the dynamic form factor (9.41) arising from the collective mode $\omega=\omega_\mathbf{k}$ is thus expressed as

$$S_{\mathrm{res}}(\mathbf{k},\omega)=-\pi\left(\frac{\partial\epsilon_1}{\partial\omega_\mathbf{k}}\right)^{-2}\frac{S^{(0)}(\mathbf{k},\omega_\mathbf{k})}{\gamma_\mathbf{k}}\delta(\omega-\omega_\mathbf{k}). \tag{9.71}$$

Integration of this spectrum across $\omega_\mathbf{k}$ then yields the static form factor associated with this particular collective mode

$$S_{\mathrm{res}}(\mathbf{k})=\frac{1}{N}\int d\omega\, S_{\mathrm{res}}(\mathbf{k},\omega)$$

$$=-\frac{\pi}{N}\left(\frac{\partial\epsilon_1}{\partial\omega_\mathbf{k}}\right)^{-2}\frac{S^{(0)}(\mathbf{k},\omega_\mathbf{k})}{\gamma_\mathbf{k}}. \tag{9.72}$$

The mean square value of the density fluctuations in a given collective mode is therefore proportional to $|\gamma_\mathbf{k}|^{-1}$; the fluctuations would increase as the damping rate decreases. We may extend Eqs. (9.71) and (9.72) similarly to the cases of a two-component plasma, starting from Eq. (9.67).

For a Maxwellian plasma, we calculate the contribution of the two plasma-wave solutions $\omega=\pm\omega_\mathrm{p}$ from Eq. (9.72) as

$$S_{\mathrm{pl}}(\mathbf{k})=k^2/k_\mathrm{D}^2 \qquad (k^2\ll k_\mathrm{D}^2). \tag{9.73}$$

The density fluctuations associated with the plasma oscillation thus exhaust the entire strength of the density fluctuations (9.31) in the limit of long wavelengths.

We can similarly investigate the fluctuations associated with the ion-acoustic waves, starting from Eq. (9.67). As we saw in Section 4.4D, the ion-acoustic wave becomes a well-defined, relatively undamped excitation when $T_e/T_i \gg 1$. The fluctuations associated with the ion-acoustic waves, $\omega = \pm sk$, in these circumstances are then calculated to be

$$S_{ac}(\mathbf{k}) = 1. \tag{9.74}$$

This strength thus represents an enhancement by a factor of two over the static form factor (9.38) with $T_e = T_i$ in the long-wavelength domain.

E. Plasma Critical Fluctuations

In Eq. (9.72), we observed the possibility that the level of the fluctuations in the collective mode grows up enormously as the rate of damping approaches zero. Such a phenomenon may be looked upon as the critical fluctuations associated with the onset of a plasma-wave instability [20], quite analogous to the critical opalescence in the vicinity of the critical point for the liquid–gas phase transition.

As a concrete example, let us consider the onset of the ion-acoustic wave instability discussed in Section 7.3. With the aid of Eq. (7.22), we obtain in the vicinity of the critical point

$$S_{crit}(\mathbf{k}, \omega) = (N\omega_k/2k)[v_d(k) - v_d]^{-1}\delta(\omega - \omega_k) \qquad (k^2 \ll k_e^2). \tag{9.75}$$

Integration of this expression across $\omega = \omega_k$ then yields

$$S_{crit}(\mathbf{k}) = \frac{1}{2}\frac{\omega_k/k}{v_d(k) - v_d} \qquad (k^2 \ll k_e^2). \tag{9.76}$$

As v_d approaches v_c, the critical velocity, the density fluctuations increase enormously in the vicinity of the critical wave number k_c.

9.4 ELECTROMAGNETIC FLUCTUATIONS

The theories developed in the preceding sections on the density fluctuations in plasmas are equally applicable to the description of fluctuations associated with electromagnetic fields. Here, we are dealing with tensorial spectral functions arising from space–time correlations of vectorial quantities. In calculating the spectral tensors of electromagnetic fluctuations, we make extensive use of the superposition principle discussed in Section 9.3C. The fundamental quantities involved in such a calculation are the spectral tensor of the individual-particle currents and the dielectric tensor; the former plays the part of $S^{(0)}(\mathbf{k},\omega)$ in (9.41) and the dielectric tensor describes the "dressing" of those currents consisting of the flows of individual test charges. We begin this section with the definition of the spectral tensors for fluctuating fields.

A. Spectral Tensors of Fluctuating Field Quantities

Consider the N-particle system as specified at the beginning of Section 2.1; the field of the electric-current density is then given by

$$\mathbf{J}(\mathbf{r},t)=q\sum_{i=1}^{N}\mathbf{v}_i\delta(\mathbf{r}-\mathbf{r}_i) \tag{9.77}$$

where $\mathbf{v}_i$ and $\mathbf{r}_i$ are the velocity and the position of the ith particle at time t. Its spatial Fourier components are

$$J_{\mathbf{k}}(t)=q\sum_{i=1}^{N}\mathbf{v}_i\exp(-i\mathbf{k}\cdot\mathbf{r}_i). \tag{9.78}$$

Analogous to Eq. (9.17), we define the spectral tensor of the current fluctuation by

$$\langle\mathbf{JJ}^*(\mathbf{k},\omega)\rangle=\frac{1}{2\pi}\int_{-\infty}^{\infty}dt\langle\mathbf{J}_{\mathbf{k}}(t'+t)\mathbf{J}_{-\mathbf{k}}(t')\rangle\exp(i\omega t). \tag{9.79}$$

Equivalently, this tensor may also be expressed as

$$\langle\mathbf{JJ}^*(\mathbf{k},\omega)\rangle=(V/2\pi)\int_V d\mathbf{r}\int_{-\infty}^{\infty}dt\langle\mathbf{J}(\mathbf{r}'+\mathbf{r},t'+t)\mathbf{J}(\mathbf{r}',t')\rangle\exp[-i(\mathbf{k}\cdot\mathbf{r}-\omega t)] \tag{9.80}$$

or

$$\langle\mathbf{JJ}^*(\mathbf{k},\omega)\rangle\delta(\omega-\omega')=(2\pi)^{-2}\langle\mathbf{J}(\mathbf{k},\omega)\mathbf{J}(-\mathbf{k},-\omega')\rangle. \tag{9.81}$$

In Eq. (9.81), $\mathbf{J}(\mathbf{k},\omega)$ are the Fourier components of the current fluctuations in the wave number and frequency domain.

The spectral tensors of other field quantities, such as $\mathbf{E}(\mathbf{r},t)$ and $\mathbf{B}(\mathbf{r},t)$, may be defined similarly.

B. Spectral Tensor of Current Fluctuations in the System of Uncorrelated Charge Particles

Various spectral tensors of the field fluctuations in the plasma may be calculated through the superposition technique of the dressed test particles. We thus pick each particle in the plasma and regard it as a test charge; the motion of the test charge produces a current field. These current fields are then dressed by the screening currents produced by all the other particles. The dressed current fields so obtained can now be superposed without further consideration of correlations among the dressed particles. The spectral tensor of the current fluctuations may thus be calculated; the expressions for the spectral tensors for other field quantities will then follow immediately with the aid of the macroscopic laws of electrodynamics.

In order to carry through all these calculations, it is essential first of all to obtain the spectral tensor of current fluctuations for an assembly of charged particles that are moving freely without mutual correlations. We thus substitute Eq. (9.78) into Eq. (9.79) and carry out the statistical average with the aid of the velocity distribution $f(\mathbf{v})$:

$$\langle \mathbf{J}\mathbf{J}^*(\mathbf{k},\omega)\rangle^{(0)} = \frac{Nq^2}{2\pi}\int_{-\infty}^{\infty} dt \int d\mathbf{v} f(\mathbf{v})\mathbf{v}(t)\mathbf{v}(0)\exp\{-i\mathbf{k}\cdot[\mathbf{r}(t)-\mathbf{r}(0)]+i\omega t\}. \tag{9.82}$$

Here, the superscript (0) reminds us that Eq. (9.82) has been evaluated with neglect of correlations between different particles; the average over the velocities may be performed with respect to either the initial velocities $\mathbf{v}(0)$ or the velocities $\mathbf{v}(t)$ at t.

In the absence of an external magnetic field, we have $\mathbf{v}(t)=\mathbf{v}(0)$ and $\mathbf{r}(t)=\mathbf{r}(0)+\mathbf{v}t$; hence,

$$\langle \mathbf{J}\mathbf{J}^*(\mathbf{k},\omega)\rangle^{(0)} = Nq^2\int d\mathbf{v} f(\mathbf{v})\mathbf{v}\mathbf{v}\delta(\omega-\mathbf{k}\cdot\mathbf{v}). \tag{9.83}$$

When a uniform magnetic field is applied in the z direction, the particles perform a spiral motion, as described by Eqs. (3.60) and (3.61). Thus,

$$\mathbf{v}(t)=\mathsf{B}(t)\cdot\mathbf{v}(0), \qquad \exp\{-i\mathbf{k}\cdot[\mathbf{r}(t)-\mathbf{r}(0)]+i\omega t\}=\exp[-i\phi(t)],$$

which may be expanded as Eq. (3.70). We may choose the wave vector $\mathbf{k}$ on the x–z plane, as illustrated in Fig. 3.1. Comparison between Eqs. (3.66) and (3.71) then shows that Eq. (9.82) is calculated as

$$\langle \mathbf{J}\mathbf{J}^*(\mathbf{k},\omega)\rangle^{(0)} = Nq^2\sum_{n=-\infty}^{\infty}\int d\mathbf{v} f(v_\perp,v_\parallel)\boldsymbol{\Pi}(v_\perp,v_\parallel;n)\delta(\omega-n\Omega-k_\parallel v_\parallel), \tag{9.84}$$

where the velocity integration has been defined by (3.73).

C. Fluctuations of the Electric Field

The microscopic electric field in the plasma fluctuates as the charged particles move randomly. The spectral tensor of such fluctuations can be calculated through a combination of the macroscopic laws of electrodynamics and the superposition principle of the dressed particles.

To describe the dressing, we introduce a tensorial response function $\mathbf{Z}(\mathbf{k},\omega)$ of the electric field against the external current density through

$$\mathbf{E}(\mathbf{k},\omega) = (4\pi/i\omega)\mathbf{Z}(\mathbf{k},\omega)\cdot\mathbf{J}_{\text{ext}}(\mathbf{k},\omega). \tag{9.85}$$

In view of Eq. (3.22), we find that the tensor $\mathbf{Z}$ is related to the dispersion tensor (3.21) via

$$\mathbf{Z}\cdot\boldsymbol{\Delta} = \mathbf{I}. \tag{9.86}$$

According to the superposition principle, the spectral tensor of the electric field fluctuation is then calculated as

$$\langle \mathbf{EE}^*(\mathbf{k},\omega)\rangle = (16\pi^2/\omega^2)\mathbf{Z}(\mathbf{k},\omega)\cdot\langle \mathbf{JJ}^*(\mathbf{k},\omega)\rangle^{(0)}\cdot\mathbf{Z}^+(\mathbf{k},\omega). \tag{9.87}$$

Each solution of the dispersion relation (3.24) thereby contributes as a pole of the spectral tensor through $\mathbf{Z}$.

When the plasma is isotropic, Eq. (9.87) can be simplified substantially. Since the dielectric tensor is now written as Eq. (3.26), we have

$$\mathbf{Z}(\mathbf{k},\omega) = \frac{\mathbf{I}_{\text{L}}}{\epsilon_{\text{L}}(k,\omega)} + \frac{\mathbf{I}_{\text{T}}}{\epsilon_{\text{T}}(k,\omega) - (kc/\omega)^2}. \tag{9.88}$$

Similarly, Eq. (9.83) can be decomposed into its longitudinal and transverse parts as

$$\langle \mathbf{JJ}^*(\mathbf{k},\omega)\rangle^{(0)} = \langle |J|^2(k,\omega)\rangle_{\text{L}}^{(0)}\mathbf{I}_{\text{L}} + \langle |J^2|(k,\omega)\rangle_{\text{T}}^{(0)}\mathbf{I}_{\text{T}} \tag{9.89}$$

where

$$\langle |J|^2(k,\omega)\rangle_{\text{L}}^{(0)} = \frac{Nq^2\omega^2}{k^2}\int d\mathbf{v} f(\mathbf{v})\delta(\omega - \mathbf{k}\cdot\mathbf{v}), \tag{9.90a}$$

$$\langle |J|^2(k,\omega)\rangle_{\text{T}}^{(0)} = \frac{Nq^2}{2}\int d\mathbf{v} f(\mathbf{v})\frac{|\mathbf{k}\times\mathbf{v}|^2}{k^2}\delta(\omega - \mathbf{k}\cdot\mathbf{v}). \tag{9.90b}$$

Substituting these formulas into Eq. (9.87) and extending the result to cases of a multicomponent plasma, we obtain

$$\langle \mathbf{EE}^*(\mathbf{k},\omega)\rangle = \sum_\sigma \frac{16\pi^2 N_\sigma q_\sigma^2}{k^2} \int d\mathbf{v}\, f_\sigma(\mathbf{v})\delta(\omega - \mathbf{k}\cdot\mathbf{v})$$

$$\times \left\{ \frac{\mathbf{I}_L}{|\epsilon_L(k,\omega)|^2} + \frac{1}{2}\frac{|\mathbf{k}\times\mathbf{v}|^2}{\omega^2}\frac{\mathbf{I}_T}{|\epsilon_T(k,\omega) - (kc/\omega)^2|^2} \right\}. \tag{9.91}$$

The longitudinal and transverse fluctuations are clearly separated here.

D. Energy Spectrum of Fluctuations

Each electromagnetic fluctuation carries an electromagnetic energy with it. We can therefore define an energy spectrum of fluctuations associated with a spectral distribution of such electromagnetic fluctuations.

The energy density $\mathcal{E}(\mathbf{r},t)$ of the electromagnetic field in a dispersive medium is generally determined from the energy conservation law of the Maxwell equations as [21]

$$\frac{\partial \mathcal{E}}{\partial t} = \frac{1}{4\pi}\left(\mathbf{E}\cdot\frac{\partial \mathbf{D}}{\partial t} + \mathbf{H}\cdot\frac{\partial \mathbf{B}}{\partial t}\right) \tag{9.92}$$

where $\mathbf{D}(\mathbf{r},t)$ and $\mathbf{H}(\mathbf{r},t)$ satisfy the field equations equivalent to Eqs. (3.13); these may be related to the other pair of the field quantities via (3.14) and (3.15), for example.† The right-hand side of Eq. (9.92) can be shown to be equal to $-\nabla\cdot\mathcal{S}$ from the Maxwell equations in the absence of external perturbations, where

$$\mathcal{S} = (c/4\pi)(\mathbf{E}\times\mathbf{H}) \tag{9.93}$$

is Poynting's vector; Eq. (9.92) thus represents the continuity equation for energy density.

The energy spectrum may be obtained from a statistical analysis of Eq. (9.92) since we are dealing with the energies contained in fluctuating fields. In so doing, we shall confine ourselves to a calculation of the energy spectrum associated with the electrostatic fluctuations in this section. Energy contained in a collective mode can then be readily obtained from such a calculation. More general treatments, including the energy in the magnetic fluctuations, become important in connection with the calculation of the radiation spectrum from the plasma; we shall treat such a problem later from a statistical analysis of Poynting's vector.

† In Eqs. (3.13)–(3.15), the field quantities have been expressed in terms of their Fourier components.

For electrostatic fluctuations we may expand

$$\mathbf{E}(\mathbf{r},t)=(2\pi V)^{-1}\sum_{\mathbf{k}}\int_{-\infty}^{\infty}d\omega\,\mathbf{E}(\mathbf{k},\omega)\exp[i(\mathbf{k}\cdot\mathbf{r}-\omega t)], \tag{9.94a}$$

$$\mathbf{D}(\mathbf{r},t)=(2\pi V)^{-1}\sum_{\mathbf{k}}\int_{-\infty}^{\infty}d\omega\,\epsilon(\mathbf{k},\omega)\mathbf{E}(\mathbf{k},\omega)\exp[i(\mathbf{k}\cdot\mathbf{r}-\omega t)], \tag{9.94b}$$

where $\epsilon(\mathbf{k},\omega)$ is the dielectric response function defined in Section 3.2. The energy density associated with such fluctuations can be calculated from the statistical average of Eq. (9.92) as

$$\langle\mathcal{E}\rangle=\frac{1}{4\pi V}\int^{t}dt'\int_{V}d\mathbf{r}\left\langle\mathbf{E}(\mathbf{r},t')\cdot\frac{\partial\mathbf{D}(\mathbf{r},t')}{\partial t'}\right\rangle. \tag{9.95}$$

We assume that the system is uniform and stationary; $\langle\mathcal{E}\rangle$ will be independent of space–time coordinates.

We now substitute (9.94) into (9.95); after a simple manipulation involving symmetrization with respect to ω and ω', we have

$$\langle\mathcal{E}\rangle=-\frac{i}{32\pi^3V^2}\int^{t}dt'\sum_{\mathbf{k}}\int_{-\infty}^{\infty}d\omega\int_{-\infty}^{\infty}d\omega'\,[\omega\epsilon(\mathbf{k},\omega)-\omega'\epsilon(-\mathbf{k},-\omega')]$$
$$\times\langle\mathbf{E}(\mathbf{k},\omega)\cdot\mathbf{E}(-\mathbf{k},-\omega')\rangle\exp[-i(\omega-\omega')t'].$$

The statistical average $\langle\mathbf{E}(\mathbf{k},\omega)\cdot\mathbf{E}(-\mathbf{k},-\omega')\rangle$ has a sharply peaked structure at $\omega=\omega'$ [compare Eq. (9.81)]. We may therefore expand $\omega\epsilon(\mathbf{k},\omega)$ around $\omega=\omega'$ so that

$$\omega\epsilon(\mathbf{k},\omega)=\omega'\epsilon(\mathbf{k},\omega')+(\omega-\omega')\frac{\partial[\omega'\epsilon(\mathbf{k},\omega')]}{\partial\omega'}.$$

When the dissipation in the medium is negligibly small, we can ignore the imaginary part of the dielectric function, and thus, $\epsilon(\mathbf{k},\omega)=\epsilon(-\mathbf{k},-\omega)$; we shall henceforth assume such a situation. For a relatively undamped collective mode, for example, this is certainly a good approximation. Carrying out the t' integration, we obtain

$$\langle\mathcal{E}\rangle=\frac{1}{32\pi^3V^2}\sum_{\mathbf{k}}\int_{-\infty}^{\infty}d\omega\int_{-\infty}^{\infty}d\omega'\frac{\partial[\omega'\epsilon(\mathbf{k},\omega')]}{\partial\omega'}$$
$$\times\langle\mathbf{E}(\mathbf{k},\omega)\cdot\mathbf{E}(-\mathbf{k},-\omega')\rangle\exp[-i(\omega-\omega')t].$$

Analogous to Eq. (9.18) or (9.81), we may now introduce the spectral function according to

$$\langle \mathbf{E}(\mathbf{k},\omega)\cdot\mathbf{E}(-\mathbf{k},-\omega')\rangle = (2\pi)^2 \operatorname{Tr}\langle \mathbf{E}\mathbf{E}^*(\mathbf{k},\omega)\rangle\delta(\omega-\omega')$$

$$= (2\pi)^2\langle |E^2|(\mathbf{k},\omega)\rangle\delta(\omega-\omega'). \tag{9.96}$$

We thus have the final expression

$$\langle \mathcal{E}\rangle = \frac{1}{8\pi V^2}\sum_{\mathbf{k}}\int_{-\infty}^{\infty} d\omega \frac{\partial\,[\omega\epsilon(\mathbf{k},\omega)]}{\partial\omega}\langle |E^2|(\mathbf{k},\omega)\rangle \tag{9.97}$$

for the energy density associated with the electrostatic fluctuations. Only the real part of the dielectric response function need be considered in Eq. (9.97) and in the subsequent equations obtained therefrom; its imaginary part does not contribute to Eq. (9.97) because of its symmetry.†

Equation (9.97) describes a total energy density resulting from a summation of the contributions of various field fluctuations with different **k** and ω. This expression thus enables us to decompose the energy density into its spectral components

$$\mathcal{E}(\mathbf{k},\omega) = \frac{1}{8\pi V}\frac{\partial\,[\omega\epsilon(\mathbf{k},\omega)]}{\partial\omega}\langle |E^2|(\mathbf{k},\omega)\rangle, \tag{9.98}$$

so that

$$\langle \mathcal{E}\rangle = \frac{1}{V}\sum_{\mathbf{k}}\int_{-\infty}^{\infty} d\omega\, \mathcal{E}(\mathbf{k},\omega)$$

$$= \int \frac{d\mathbf{k}}{(2\pi)^3}\int_{-\infty}^{\infty} d\omega\, \mathcal{E}(\mathbf{k},\omega). \tag{9.99}$$

The total energy contained in the **k** mode of the fluctuations is then given by

$$\mathcal{E}_{\mathbf{k}} = \int_{-\infty}^{\infty} d\omega\, \mathcal{E}(\mathbf{k},\omega) = \frac{1}{8\pi V}\int_{-\infty}^{\infty} d\omega \frac{\partial\,[\omega\epsilon(\mathbf{k},\omega)]}{\partial\omega}\langle |E^2|(\mathbf{k},\omega)\rangle. \tag{9.100}$$

These are the general formulas relating the electrostatic fluctuations with their energy spectrum. Since $\mathcal{E}_{\mathbf{k}}$ has been defined as the energy per unit volume in real space and per unit volume in the **k** space [see Eq. (9.99)], it has the dimension of energy.

†The spectral function $\langle |E^2|(\mathbf{k},\omega)\rangle$ has the symmetry $\langle |E^2|(\mathbf{k},\omega)\rangle = \langle |E^2|(-\mathbf{k},-\omega)\rangle$.

A particularly important application of these formulas occurs in the fluctuations associated with a relatively undamped collective mode. As we see in the expression of Eq. (9.71), the frequency spectrum of the electric field fluctuations is also expected to have a peak at $\omega=\omega_{\mathbf{k}}$; we may correspondingly write

$$\langle|E^2|(\mathbf{k},\omega)\rangle_{\text{res}}=\langle|E^2|(\mathbf{k})\rangle_{\text{res}}\delta(\omega-\omega_{\mathbf{k}}). \tag{9.101}$$

Let us recall that the frequency $\omega_{\mathbf{k}}$ has been determined from the dispersion relation $\epsilon(\mathbf{k},\omega)=0$. Thus substituting (9.101) into (9.100), we find that the energy density in the collective mode with the wave vector $\mathbf{k}$ is given by

$$\mathcal{E}_{\mathbf{k}}=\frac{1}{8\pi V}\omega_{\mathbf{k}}\left(\frac{\partial\epsilon_1}{\partial\omega_{\mathbf{k}}}\right)\langle|E^2|(\mathbf{k})\rangle_{\text{res}}. \tag{9.102}$$

Here, we have explicitly singled out the real part ϵ_1 of the dielectric function.

For a single component plasma, we prove the relation

$$\langle|E^2|(\mathbf{k})\rangle_{\text{res}}=\frac{16\pi^2Ne^2}{k^2}S_{\text{res}}(\mathbf{k}) \tag{9.103}$$

from the Poisson equation. Thus, the energy spectrum associated with the static form factor (9.73) for the plasma oscillation is found to be

$$\mathcal{E}_{\mathbf{k}}=T, \tag{9.104a}$$

a value anticipated from the equipartition theorem. We can similarly show that

$$\mathcal{E}_{\mathbf{k}}=T_e \tag{9.104b}$$

for the ion-acoustic wave with $T_e \gg T_i$ (Problem 9.9).

9.5 INTERACTIONS BETWEEN PARTICLES AND FIELDS

As the calculations in the previous section have demonstrated, there exist electromagnetic fields produced by the charged particles in a plasma when viewed in a microscopic scale. In addition, we may apply a macroscopic electromagnetic field externally to the plasma. The motion of each charged particle is then affected by the presence of these macroscopic or microscopic electromagnetic fields. Such mutual interactions between the particles and the fields account for a number of interesting phenomena in a plasma.

Consider a monochromatic electromagnetic wave applied to a plasma. The motion of the charged particles is modulated by the wave; the particles thereby reradiate electromagnetic waves in directions and at frequencies different from those of the incident wave. The entire phenomenon may then be looked upon as the incoherent scattering of the electromagnetic radiation by the plasma [22–24]. Such a scattering experiment directly measures the dynamic form factor of the electrons in the plasma; it thus provides an important diagnostic technique.

We may also consider the processes in which the particles are scattered by the fluctuating microscopic fields and then radiate; such processes correspond to the bremsstrahlung of a plasma [25, 26]. Similarly, if we single out the conversion processes of an incident transverse wave into longitudinal excitations through particle scattering, such a calculation would amount to a consideration of the so-called free–free absorption processes. The purpose of the present section is to present a general formalism pertaining to those problems, and then to apply it to explicit calculations of the incoherent scattering cross section and the bremsstrahlung spectrum.

Scattering of particles by microscopic field fluctuations is also closely related to the relaxation processes in the plasma; these problems will be considered in Chapter 10.

A. Basic Formulas

We begin with the generalized Klimontovich equation, which can be obtained by substituting Eq. (A.4) into Eq. (A.3). The fluctuation components (9.7) of the Klimontovich distribution for the paticles of σ species are then determined as

$$\delta N_\sigma(\mathbf{k},\omega;\mathbf{v}) = -\frac{q_\sigma}{im_\sigma}\frac{1}{\mathbf{k}\cdot\mathbf{v}-\omega-i\eta}\frac{\partial f_\sigma(\mathbf{v})}{\partial \mathbf{v}}\cdot\left[\left(1-\frac{\mathbf{k}\cdot\mathbf{v}}{\omega}\right)\mathsf{I}+\frac{\mathbf{k}\mathbf{v}}{\omega}\right]\cdot\mathbf{E}(\mathbf{k},\omega)$$

$$-\frac{q_\sigma}{im_\sigma V}\sum_{\mathbf{k}'}\int_{-\infty}^{\infty}\frac{d\omega'}{2\pi}\frac{1}{\mathbf{k}\cdot\mathbf{v}-\omega-i\eta}\frac{\partial\delta N_\sigma(\mathbf{k}-\mathbf{k}',\omega-\omega';\mathbf{v})}{\partial\mathbf{v}}$$

$$\cdot\left[\left(1-\frac{\mathbf{k}'\cdot\mathbf{v}}{\omega'}\right)\mathsf{I}+\frac{\mathbf{k}'\mathbf{v}}{\omega'}\right]\cdot\mathbf{E}(\mathbf{k}',\omega'). \qquad (9.105)$$

Correspondingly, the Fourier components of the current-density fluctuations are calculated to be

$$\delta\mathbf{J}(\mathbf{k},\omega)=\sum_\sigma q_\sigma n_\sigma\int d\mathbf{v}\,\mathbf{v}\,\delta N_\sigma(\mathbf{k},\omega;\mathbf{v})$$

$$=-\frac{i\omega}{4\pi}[\epsilon(\mathbf{k},\omega)-\mathsf{I}]\cdot\mathbf{E}(\mathbf{k},\omega)$$

$$-\sum_\sigma\frac{q_\sigma^2 n_\sigma}{im_\sigma V}\sum_{\mathbf{k}'}\int_{-\infty}^{\infty}\frac{d\omega'}{2\pi}\int d\mathbf{v}\frac{\mathbf{v}}{\mathbf{k}\cdot\mathbf{v}-\omega-i\eta}\frac{\partial\delta N_\sigma(\mathbf{k}-\mathbf{k}',\omega-\omega';\mathbf{v})}{\partial\mathbf{v}}$$

$$\cdot\left[\left(1-\frac{\mathbf{k}'\cdot\mathbf{v}}{\omega'}\right)\mathsf{I}+\frac{\mathbf{k}'\mathbf{v}}{\omega'}\right]\cdot\mathbf{E}(\mathbf{k}',\omega'). \qquad (9.106)$$

The first terms on the right-hand sides of Eqs. (9.105) and (9.106) represent the linear response relationships between the electric field and the induced fluctuations; in particular, the first term in the latter equation is the same as the conductivity relation (3.19). The major physical consequences arising from such linear electromagnetic relationships have been well investigated in the foregoing chapters. The second terms, on the other hand, describe the nonlinear coupling between the electric fields and and the fluctuations of the microscopic distribution. The origin of these electric fields may be either macroscopic or microscopic. Equations (9.105) and (9.106) thus serve as the basic equations describing the interactions between the particles and the fields in a plasma.

B. Incoherent Scattering of Electromagnetic Radiation

Consider a monochromatic electromagnetic wave

$$\mathbf{E}_1(\mathbf{r},t)=\boldsymbol{\eta}_1 E_1\cos(\mathbf{k}_1\cdot\mathbf{r}-\omega_1 t) \qquad (9.107)$$

incident into the plasma; $\boldsymbol{\eta}_1$ is a unit polarization vector perpendicular to $\mathbf{k}$. The frequency ω_1 is chosen so that the wave can travel through the plasma. The Fourier components of (9.107) are

$$\mathbf{E}_1(\mathbf{k},\omega)=\boldsymbol{\eta}_1 E_1 V\pi\{\delta(\mathbf{k},\mathbf{k}_1)\delta(\omega-\omega_1)+\delta(\mathbf{k},-\mathbf{k}_1)\delta(\omega+\omega_1)\} \qquad (9.108)$$

where $\delta(\mathbf{k},\mathbf{k}_1)$ is the three-dimensional extension of Kronecker's delta (5.25).

The current density induced in the plasma through coupling between this monochromatic wave and the density fluctuations in the plasma is obtained from the second term of Eq. (9.106) in which (9.108) is substituted in place of $\mathbf{E}(\mathbf{k}',\omega')$. The contribution from a given particle species is inversely proportional to the particle mass; the electrons thus provide the major contribution to the induced current. We also note that for a nonrelativistic plasma $v \ll c = \omega_1/k_1$. Hence, the nonlinear term of Eq. (9.106) is simplified as

$$\delta\mathbf{J}(\mathbf{k},\omega) = i\frac{ne^2}{2m_e}E_1\int d\mathbf{v}\frac{\mathbf{v}}{\mathbf{k}\cdot\mathbf{v}-\omega-i\eta}\boldsymbol{\eta}_1\cdot\frac{\partial}{\partial\mathbf{v}}[\delta N(\mathbf{k}-\mathbf{k}_1,\omega-\omega_1;\mathbf{v})$$
$$+\delta N(\mathbf{k}+\mathbf{k}_1,\omega+\omega_1;\mathbf{v})]. \qquad (9.109)$$

We now perform a partial integration with respect to **v**. To do so, we note

$$\frac{\partial}{\partial\mathbf{v}}\left[\frac{\mathbf{v}}{\mathbf{k}\cdot\mathbf{v}-\omega-i\eta}\right] = -\frac{\omega}{(\mathbf{k}\cdot\mathbf{v}-\omega-i\eta)^2}\left[\left(1-\frac{\mathbf{k}\cdot\mathbf{v}}{\omega}\right)\mathbf{I}+\frac{\mathbf{k}\mathbf{v}}{\omega}\right]. \qquad (9.110)$$

We also remark that **k** and ω in Eq. (9.109) will eventually be identified as those of the outgoing scattered electromagnetic waves; hence, we may assume that $\omega/k = c \gg v$ in (9.109). We thereby obtain

$$\delta\mathbf{J}(\mathbf{k},\omega) = i\frac{ne^2}{2m_e\omega}E_1\boldsymbol{\eta}_1[\delta N(\mathbf{k}-\mathbf{k}_1,\omega-\omega_1)+\delta N(\mathbf{k}+\mathbf{k}_1,\omega+\omega_1)] \qquad (9.111)$$

where

$$\delta N(\mathbf{k},\omega) \equiv \int d\mathbf{v}\,\delta N(\mathbf{k},\omega;\mathbf{v}). \qquad (9.112)$$

Inverse Fourier transformations of (9.111) yield

$$\delta\mathbf{J}(\mathbf{r},t) = \frac{ne^2}{2m_eV}E_1\boldsymbol{\eta}_1\sum_{\mathbf{k}}\int_{-\infty}^{\infty}\frac{d\omega}{2\pi}\frac{i}{\omega}[\delta N(\mathbf{k}-\mathbf{k}_1,\omega-\omega_1)+\delta N(\mathbf{k}+\mathbf{k}_1,\omega+\omega_1)]$$
$$\times\exp[i(\mathbf{k}\cdot\mathbf{r}-\omega t)]. \qquad (9.113)$$

The radiation field produced by this current-density fluctuation at a distance R_0 from the plasma may be described by a retarded vector potential. Choosing R_0 sufficiently large as compared with the size of the plasma, we write the retarded potential for such scattered waves as

$$\mathbf{A}(t) = \frac{1}{cR_0}\int_V d\mathbf{r}\,\delta\mathbf{J}\left(\mathbf{r},t-\frac{R_0}{c}+\frac{\hat{k}\cdot\mathbf{r}}{c}\right) \qquad (9.114)$$

where $\hat{k}$ is the unit vector in the direction of $\mathbf{k}$. Substituting (9.113) here, we obtain

$$\mathbf{A}(t)=\frac{nr_e c}{2R_0}E_1\boldsymbol{\eta}_1\sum_{\mathbf{k}}\int_{-\infty}^{\infty}\frac{d\omega}{2\pi}\frac{i}{\omega}[\delta N(\mathbf{k}-\mathbf{k}_1,\omega-\omega_1)+\delta N(\mathbf{k}+\mathbf{k}_1,\omega+\omega_1)]$$

$$\times\ \delta\left(\mathbf{k},\frac{\hat{k}\omega}{c}\right)\exp\left[-i\omega\left(t-\frac{R_0}{c}\right)\right], \qquad (9.115)$$

where $r_e=e^2/m_e c^2=2.818\times10^{-13}$ [cm] is the classical electron radius. The electric and magnetic fields of the scattered waves are then given by

$$\mathbf{E}_2(t)=\frac{1}{c}(\dot{\mathbf{A}}\times\hat{k})\times\hat{k}, \qquad (9.116a)$$

$$\mathbf{B}_2(t)=\frac{1}{c}(\dot{\mathbf{A}}\times\hat{k}). \qquad (9.116b)$$

Let us now suppose that the observation of the scattered wave is made in a fixed polarization direction $\boldsymbol{\eta}_2$; the intensity of the observed electric field then is $\boldsymbol{\eta}_2\cdot\mathbf{E}_2(t)$. The energy flux associated the scattered wave is calculated from Poynting's vector (9.93) in which we set $\mathbf{H}=\mathbf{B}$. Since $\mathbf{E}_2(t)$ and $\mathbf{B}_2(t)$ are stochastic variables reflecting the behavior of the density fluctuations δN, we carry out a statistical average for $\mathfrak{S}(t)$. The energy flux observed at the distance R_0 is thus calculated as

$$\langle\mathfrak{S}\rangle=\frac{c\hat{k}}{4\pi}(\boldsymbol{\eta}_1\cdot\boldsymbol{\eta}_2)^2\left(\frac{nr_eE_1}{2R_0}\right)^2\sum_{\mathbf{k},\mathbf{k}'}\int_{-\infty}^{\infty}\frac{d\omega}{2\pi}\int_{-\infty}^{\infty}\frac{d\omega'}{2\pi}$$

$$\times\langle[\delta N(\mathbf{k}-\mathbf{k}_1,\omega-\omega_1)+\delta N(\mathbf{k}+\mathbf{k}_1,\omega+\omega_1)]$$

$$\times[\delta N(\mathbf{k}'-\mathbf{k}_1,\omega'-\omega_1)+\delta N(\mathbf{k}'+\mathbf{k}_1,\omega'+\omega_1)]\rangle$$

$$\times\delta\left(\mathbf{k},\frac{\hat{k}\omega}{c}\right)\delta\left(\mathbf{k}',\frac{\hat{k}\omega'}{c}\right)\exp\left[-i(\omega+\omega')\left(t-\frac{R_0}{c}\right)\right].$$

The statistical averages involved here may be expressed in terms of the dynamic form factors defined by Eq. (9.18); in view of Eq. (9.8a), we note that $\delta\rho(\mathbf{k},\omega)=n\delta N(\mathbf{k},\omega)$. After a straightforward algebra, we find

$$\langle S\rangle=\hat{k}\left(\frac{cE_1^2}{8\pi}\right)(\eta_1\cdot\eta_2)^2\frac{r_e^2}{R_0^2}\sum_{\mathbf{k}}\int_{-\infty}^{\infty}d\omega\,S(\mathbf{k}-\mathbf{k}_1,\omega-\omega_1)\delta\left(\mathbf{k},\frac{\hat{k}\omega}{c}\right). \quad (9.117)$$

In this expression, the quantity $(cE_1{}^2/8\pi)$ is the time-averaged energy flux of the incident wave. Hence, we may obtain from Eq. (9.177) the differential cross section for the transfer of momentum $\hbar\mathbf{k}$ (corresponding to scattering into a solid angle do) and energy $\hbar\omega$ from the plasma to the electromagnetic wave as

$$\frac{d^2Q}{do\,d\omega}=(r_e)^2(\eta_1\cdot\eta_2)^2S(\mathbf{k},\omega) \quad (9.118)$$

where

$$\mathbf{k}=\mathbf{k}_2-\mathbf{k}_1, \qquad \omega=\omega_2-\omega_1 \quad (9.119)$$

are the wave vector and frequency differences between the scattered and incident waves.

If the experiment is carried out in such a way that no attention is paid to the polarization directions, we average Eq. (9.118) over the incident polarizations and sum over the polarizations of the scattered waves. The result is

$$\frac{d^2Q}{do\,d\omega}=(r_e)^2(1-\tfrac{1}{2}\sin^2\theta)S(\mathbf{k},\omega) \quad (9.120)$$

where θ is the angle between $\mathbf{k}_1$ and $\mathbf{k}_2$. Finally, if no resolutions are made on the frequencies of the scattered waves, we integrate Eq. (9.120) over the frequencies to obtain

$$\frac{dQ}{do}=Nr_e^2(1-\tfrac{1}{2}\sin^2\theta)S(\mathbf{k}). \quad (9.121)$$

C. Bremsstrahlung

Let us now consider the radiation arising from the scattering of charged particles by the microscopic fluctuations of the electrostatic fields in the plasma. Since these microscopic fields originate from the electrostatic fields of the individual particles, the problem under consideration corresponds to a treatment of the bremsstrahlung from the plasma.

The electric field entering in Eq. (9.106) is now longitudinal; we may substitute

$$\mathbf{E}(\mathbf{k},\omega)=i\hat{k}E_l(\mathbf{k},\omega) \tag{9.122}$$

in the second term of Eq. (9.106), which describes the scattering of the particles by the field fluctuations. Again the major contribution to the current density comes from the electrons; the remark made after Eq. (9.110) applies in the present case also. Hence, carrying out the partial integration with respect to $\mathbf{v}$ in the second term of Eq. (9.106), we obtain the induced current density of our interest as

$$\delta\mathbf{J}(\mathbf{k},\omega)=-\frac{ne^2}{m_eV}\sum_{\mathbf{q}}\int_{-\infty}^{\infty}\frac{dx}{2\pi}\frac{\hat{q}}{\omega}\delta N(\mathbf{k}-\mathbf{q},\omega-x)E_l(\mathbf{q},x)$$

or

$$\delta\mathbf{J}(\mathbf{r},t)=-\frac{ne^2}{m_eV^2}\sum_{\mathbf{k},\mathbf{q}}\int_{-\infty}^{\infty}\frac{d\omega}{2\pi}\int_{-\infty}^{\infty}\frac{dx}{2\pi}\frac{\hat{q}}{\omega}\delta N(\mathbf{k}-\mathbf{q},\omega-x)E_l(\mathbf{q},x)$$

$$\times\exp[i(\mathbf{k}\cdot\mathbf{r}-\omega t)]. \tag{9.123}$$

To evaluate the intensity of radiation emitted by these current fluctuations, we calculate the retarded vector potential (9.114) at the observation point R_0 sufficiently far away from the plasma. The energy flux may then be obtained from the statistical average of Poynting's vector (9.93) with the aid of Eqs. (9.116). Since E_l as well as δN are now stochastic variables in Eq. (9.123), the resulting expression for $\langle\mathfrak{S}\rangle$ involves a statistical average

$$\langle\delta N(\mathbf{k}-\mathbf{q},\omega-x)\delta N(\mathbf{k}'-\mathbf{q}',\omega'-x')E_l(\mathbf{q},x)E_l(\mathbf{q}',x')\rangle$$

in its integrand. Neglecting the quadruple correlations altogether, we evaluate this statistical average by means of factorization, $\langle ABCD\rangle=\langle AB\rangle\langle CD\rangle+\langle AC\rangle\langle BD\rangle+\langle AD\rangle\langle BC\rangle$. We thus find

$$\langle\mathfrak{S}\rangle=(cr_e^2/4\pi V^2)\sum_{\mathbf{k},\mathbf{q}}\int_{-\infty}^{\infty}d\omega\int_{-\infty}^{\infty}dx\left(\hat{q}\cdot\hat{k}\hat{k}-\hat{q}\right)$$

$$\times\left\{\left(\hat{q}\times\hat{k}\right)S(\mathbf{k}-\mathbf{q},\omega-x)\langle|E_l^2|(\mathbf{q},x)\rangle\right.$$

$$+\left[\left(\widehat{k-q}\right)\times\hat{k}\right]n^2\langle\delta NE_l^*(\mathbf{k}-\mathbf{q},\omega-x)\rangle$$

$$\left.\times\langle E_l\delta N^*(\mathbf{q},x)\rangle\right\}\delta\left(\mathbf{k},\hat{k}\omega/c\right) \tag{9.124}$$

where, analogous to Eq. (9.81), we define

$$\langle \delta N E_l^*(\mathbf{k},\omega)\rangle \delta(\omega-\omega') \equiv (2\pi)^{-2} \langle \delta N(\mathbf{k},\omega) E_l(-\mathbf{k},-\omega')\rangle. \tag{9.125}$$

The average $\langle E_l \delta N^*(\mathbf{k},\omega)\rangle$ may be defined similarly; this quantity is complex conjugate to $\langle \delta N E_l^*(\mathbf{k},\omega)\rangle$.

The frequency spectrum of the electromagnetic energy emitted per unit time into a unit solid angle in the direction of $\hat{k}$ is then obtained from Eq. (9.124) as

$$\begin{aligned}\frac{d^2P}{do\,d\omega} = \frac{cr_e^2}{4\pi V^2} \sum_{\mathbf{q}} \int_{-\infty}^{\infty} dx \left(\hat{q}\cdot\hat{k}\hat{k} - \hat{q}\right) \\ \times \Big\{ \left(\hat{q}\times\hat{k}\right) S(\mathbf{k}-\mathbf{q},\omega-x) \langle |E_l^2|(\mathbf{q},x)\rangle \\ + \left[\left(\widehat{k-q}\right)\times\hat{k}\right] n^2 \langle \delta N E_l^*(\mathbf{k}-\mathbf{q},\omega-x)\rangle \langle E_l \delta N^*(\mathbf{q},x)\rangle \Big\}_{k=\omega/c}.\end{aligned} \tag{9.126}$$

For a two-component plasma consisting of N_e electrons and N_i ions with charge Ze (hence, $ZN_i = N_e$ for macroscopic charge neutrality), we determine the field fluctuations from the Poisson equation as

$$E_l(\mathbf{k},\omega) = \frac{4\pi e N_e}{kV} \delta N_e(\mathbf{k},\omega) - \frac{4\pi Z e N_i}{kV} \delta N_i(\mathbf{k},\omega). \tag{9.127}$$

Substituting (9.127) into (9.126) and recovering the usual subscripts e and i for electron and ion, we have

$$\begin{aligned}\frac{d^2P}{do\,d\omega} = \frac{4\pi e^2 r_e^2 c}{V^2} \sum_{\mathbf{q}} \int_{-\infty}^{\infty} dx \frac{\left(\hat{q}\cdot\hat{k}\hat{k} - \hat{q}\right)}{q^2} \\ \times \Big\{ \left(\hat{q}\times\hat{k}\right) S_e(\mathbf{k}-\mathbf{q},\omega-x) [S_e(\mathbf{q},x) + Z^2 S_i(\mathbf{q},x) - Z S_{ei}(\mathbf{q},x) - Z S_{ie}(\mathbf{q},x)] \\ + \frac{q}{|\mathbf{k}-\mathbf{q}|} \left[\left(\widehat{k-q}\right)\times\hat{k}\right] [S_e(\mathbf{k}-\mathbf{q},\omega-x) S_e(\mathbf{q},x) \\ + Z^2 S_{ei}(\mathbf{k}-\mathbf{q},\omega-x) S_{ie}(\mathbf{q},x) - Z S_e(\mathbf{k}-\mathbf{q},\omega-x) S_{ie}(\mathbf{q},x) \\ - Z S_{ei}(\mathbf{k}-\mathbf{q},\omega-x) S_e(\mathbf{q},x)] \Big\}_{k=\omega/c}.\end{aligned} \tag{9.128}$$

Here, $S_{ei}(\mathbf{k},\omega)$ and $S_{ie}(\mathbf{k},\omega)$ describe the cross-correlations between the electrons and the ions; they are defined by

$$S_{ei}(\mathbf{k},\omega)\delta(\omega-\omega') = [S_{ie}(\mathbf{k},\omega)]^{*}\delta(\omega-\omega')$$

$$= \frac{N_e N_i}{(2\pi)^2 V^2}\langle \delta N_e(\mathbf{k},\omega)\delta N_i(-\mathbf{k},-\omega')\rangle. \qquad (9.129)$$

Equation (9.126) or (9.128) represents the general expression for the radiation spectrum arising from the electrostatic interactions between particles in the plasma.

Among various wave vectors involved in Eq. (9.128), $\mathbf{k}$ is associated with emitted radiation, while $\mathbf{q}$ is related to a longitudinal fluctuation; hence, we can assume that $|k/q| \ll 1$. We may thus expand Eq. (9.128) with respect to $|k/q|$; to the lowest order, we have the radiation spectrum within the dipole approximation:

$$\frac{d^2P}{do\,d\omega} = \frac{4\pi(Ze)^2 r_e^2 c}{V^2}\sum_{\mathbf{q}}\int_{-\infty}^{\infty} dx \frac{|\hat{q}\times\hat{k}|^2}{q^2} \qquad (9.130)$$

$$\times[S_e(-\mathbf{q},\omega-x)S_i(\mathbf{q},x) - S_{ei}(-\mathbf{q},\omega-x)S_{ie}(\mathbf{q},x)].$$

Note that the terms involving the product $S_e(\mathbf{k}-\mathbf{q},\omega-x)S_e(\mathbf{q},x)$ in Eq. (9.128) do not contribute to the radiation spectrum within the dipole approximation; an electron–electron scattering keeps the total dipole moment of the electrons unchanged.

In order to carry through the calculation of (9.130) further, we assume an isotropic Maxwellian plasma; for simplicity, we consider the limit $m_i \to \infty$. The various dynamic form factors appearing in Eq. (9.130) may be obtained by superposing the fields of the dressed particles as

$$S_e(\mathbf{k},\omega) = \frac{N_e}{(2\pi T_e/m_e)^{1/2} k|\epsilon(\mathbf{k},\omega)|^2}\exp\left[-\frac{\omega^2}{2k^2(T_e/m_e)}\right]$$

$$+ N_e Z\frac{k_e^4}{(k_e^2+k^2)^2}\delta(\omega), \qquad (9.131a)$$

$$S_i(\mathbf{k},\omega) = N_i\delta(\omega), \qquad (9.131b)$$

$$S_{ei}(\mathbf{k},\omega) = S_{ie}(\mathbf{k},\omega) = N_e\frac{k_e^2}{k_e^2+k^2}\delta(\omega) \qquad (9.131c)$$

where $\epsilon(\mathbf{k},\omega)$ is given by Eq. (4.1) for the electrons. Substitution of these expressions into Eq. (9.130) yields

$$\frac{d^2P}{do\,d\omega}=\frac{4\pi c(Ze)^2 r_e^2 N_e N_i}{(2\pi T_e/m_e)^{1/2}V^2}\sum_{\mathbf{q}}\frac{|\hat{q}\times\hat{k}|^2}{q^3|\epsilon(\mathbf{q},\omega)|^2}\exp\left[-\frac{\omega^2}{2q^2(T_e/m_e)}\right]. \quad (9.132)$$

The term under the $\mathbf{q}$ summation takes on an appreciable value only when $q>|\omega|/(T_e/m_e)^{1/2}$. Since ω is the frequency of the emitted radiation out of the plasma, it must be greater than ω_p. Hence, the bulk of the effects comes from the $\mathbf{q}$ domain such that $q>k_e$. In such a domain, we may then approximate $\epsilon(\mathbf{q},\omega)=1$; the collective effects are not important in the bremsstrahlung processes. Converting the $\mathbf{q}$ summation into the integration via (2.31) and carrying out the angular integrations, we calculate (9.132) as

$$\frac{d^2P}{do\,d\omega}=\frac{4c(Ze)^2 r_e^2 N_e N_i}{3\pi(2\pi T_e/m_e)^{1/2}V}\int_0^{k_m}dq\frac{1}{q}\exp\left[-\frac{\omega^2}{2q^2(T_e/m_e)}\right]. \quad (9.133)$$

Here, we have set the upper limit of the q integration to be a finite value k_m; we thus avoid the logarithmic divergence. The physical origin of this maximum wave number is the same as that introduced in the calculation of the stopping power in Eq. (4.36); we may choose (4.37) or (4.39) with $v_0 \cong (3T_e/m_e)^{1/2}$ for the estimation of this quantity. Equation (9.133) may be calculated explicitly in terms of the exponential integral function (C is Euler's constant),

$$Ei(-x)=-\int_x^{\infty}dt\,(1/t)\exp(-t)$$

$$=\ln x+C-x+\frac{x^2}{2\cdot 2!}-\cdots \qquad (x>0), \quad (9.134)$$

as

$$\frac{d^2P}{do\,d\omega}=\frac{2c(Ze)^2 r_e^2 N_e N_i}{3\pi(2\pi T_e/m_e)^{1/2}V}\left\{-Ei\left[-\frac{\omega^2}{2k_m^2(T_e/m_e)}\right]\right\}. \quad (9.135)$$

For $\omega^2<k_m^2(T_e/m_e)$, the radiation spectrum may be approximated by

$$\frac{d^2P}{do\,d\omega}=\frac{4c(Ze)^2 r_e^2 N_e N_i}{3\pi(2\pi T_e/m_e)^{1/2}V}\left\{\ln\frac{\sqrt{2}\,k_m(T_e/m_e)^{1/2}}{\omega}-\frac{C}{2}\right\}. \quad (9.136)$$

In this domain, the spectrum thus shows only a weak logarithmic dependence on the frequency.

References

1. H. B. Callen and T. A. Welton, *Phys. Rev.* **83**, 34 (1951).
2. R. Kubo, *J. Phys. Soc. Japan* **12**, 570 (1957).
3. L. D. Landau and E. M. Lifshitz, *Statistical Physics*, 2nd ed. (J. B. Sykes and M. J. Kearsley, *transls.*) Chap. XII (Addison-Wesley, Reading, Mass. ,1969).
4. A. G. Sitenko, *Electromagnetic Fluctuations in Plasma* (M. D. Friedman, *transl.*)(Academic, New York, 1967).
5. L. Van Hove, *Phys. Rev.* **95**, 249 (1954).
6. D. Pines and P. Nozières, *The Theory of Quantum Liquids*, Vol. I (W. A. Benjamin, New York, 1966).
7. R. Abe, *Prog. Theoret. Phys. (Kyoto)* **22**, 213 (1959).
8. D. L. Bowers and E. E. Salpeter, *Phys. Rev.* **119**, 1180 (1960).
9. T. O'Neil and N. Rostoker, *Phys. Fluids* **8**, 1109 (1965).
10. T. J. Lie and Y. H. Ichikawa, *Revs. Mod. Phys.* **38**, 680 (1966).
11. S. Ichimaru, *Phys. Rev.* **A2**, 494 (1970).
12. E. E. Salpeter, *Phys. Rev.* **120**, 1528 (1960); *Phys. Rev.* **122**, 1663 (1961).
13. W. B. Thompson and J. Hubbard, *Revs. Mod. Phys.* **32**, 714 (1960); J. Hubbard, *Proc. Roy. Soc. (London)* **A260**, 114 (1961).
14. N. Rostoker, *Nucl. Fusion* **1**, 101 (1961).
15. S. Ichimaru, *Ann. Phys. (N.Y.)* **20**, 78 (1962); in *Fluctuation Phenomena in Solids* (R. E. Burgess, ed.), p. 125 (Academic, New York, 1965).
16. P. Nozières and D. Pines, *Phys. Rev.* **109**, 762 (1958); *Nuovo Cimento* **9**, 470 (1958).
17. N. Rostoker, *Phys. Fluids* **7**, 479, 491 (1964).
18. S. Ichimaru, *Phys. Rev.* **140**, B226 (1965).
19. S. Ichimaru, *J. Phys. Soc. Japan* **19**, 1207 (1964); *J. Phys. Soc. Japan* **21**, 996 (1966).
20. S. Ichimaru, D. Pines, and N. Rostoker, *Phys. Rev. Letters* **8**, 231 (1962).
21. L. D. Landau and E. M. Lifshitz, *Electrodynamics of Continuous Media* (J. B. Sykes and J. S. Bell, *transls.*), Sec. 61. (Addison-Wesley, Reading, Mass., 1960).
22. A. I. Akhiezer, I. G. Prokhoda, and A. G. Sitenko, *Zhur. Eksptl. Teoret. Fiz.* **33**, 750 (1957) [*Soviet Phys. JETP* **6**, 576 (1958)].
23. W. E. Gordon, *Proc. IRE* **46**, 1824 (1958).
24. M. N. Rosenbluth and N. Rostoker, *Phys. Fluids* **5**, 776 (1962).
25. J. Dawson and C. Oberman, *Phys. Fluids* **5**, 517 (1962); *Phys. Fluids* **6**, 394 (1963); T. Birmingham, J. Dawson, and C. Oberman, *Phys. Fluids* **8**, 297 (1965).
26. T. H. Dupree, *Phys. Fluids* **6**, 1714 (1963); *Phys. Fluids* **7**, 923 (1964); D. A. Tidman and T. H. Dupree, *Phys. Fluids* **8**, 1860 (1965).

Problems

9.1. Show the equivalence between Eqs. (9.17) and (9.18).

9.2. Derive Eq. (9.21).

9.3. Calculate the thermodynamic pressure of the electron gas from the energy density (9.33) and (9.34).

9.4. Consider a fictitious potential field $\varphi_{-}(\mathbf{r},t)$ which couples only with electrons. When such a potential field is applied to a two-component plasma, the Vlasov equations for the electrons and ions may be written, within the electrostatic approximation, as

$$\frac{\partial f_e}{\partial t}+\mathbf{v}\cdot\frac{\partial f_e}{\partial \mathbf{r}}+\frac{e}{m_e}\frac{\partial}{\partial \mathbf{r}}(\varphi_- + \varphi_{\text{ind}})\cdot\frac{\partial f_e}{\partial \mathbf{v}}=0,$$

$$\frac{\partial f_i}{\partial t}+\mathbf{v}\cdot\frac{\partial f_i}{\partial \mathbf{r}}-\frac{e}{m_i}\frac{\partial \varphi_{\text{ind}}}{\partial \mathbf{r}}\cdot\frac{\partial f_i}{\partial \mathbf{v}}=0.$$

The induced potential field $\varphi_{\text{ind}}(\mathbf{r},t)$ is to be determined from the Poisson equation. Carry out linear response calculations from these equations and show that Eq. (9.35) is the appropriate response function for the calculation of the dynamic form factor of the electrons.

9.5. Express the statistical average $\langle \delta N(X;t)\delta N(X';t)\delta N(X'';t)\rangle$ in terms of the functions F, G, and H of Section 2.2; and show that it scales as g^2 if (2.13) is assumed.

9.6. Carry out the calculation of Eq. (9.53) with the aid of Eq. (9.52), and show that Eq. (9.56) results.

9.7. When the velocity distributions are Maxwellian and $N \equiv N_e = N_i$, show that Eq. (9.67) reduces to Eq. (9.37).

9.8. Derive Eqs. (9.73) and (9.74).

9.9. For the ion-acoustic wave with $T_e \gg T_i$, we can calculate $\langle |E^2|(\mathbf{k})\rangle_{\text{res}}$ from the longitudinal part of Eq. (9.91) with the aid of the technique explained in Section 9.3D. Carry out such a calculation and show Eq. (9.104b).

9.10. Derive Eq. (9.120) from Eq. (9.118).

9.11. Derive Eqs. (9.131).

CHAPTER 10

RELAXATION PROCESSES

In Chapter 2, we studied the structure and properties of the collision term based on the solution of the plasma kinetic equations. In the light of the fluctuation analyses described in Chapter 9, we can recognize another possible approach by which to investigate the relaxation processes in a plasma, based on the theory of Brownian motion. Microscopically, each particle feels the fluctuating electric field produced by all the other particles in the plasma; its motion in the phase space should therefore resemble that of a Brownian particle. The behavior of such a particle may then be described by the Fokker–Planck equation, which contains the fluctuation spectrum of the electric field in its coefficients. The collision term of the plasma can be determined, and the rates of various relaxation processes calculated, from this equation. This approach thus makes it possible to express various transport coefficients in terms of the spectral function of the electric-field fluctuations.

This method of analyzing relaxation processes from a fluctuation-theoretical point of view has a definite advantage over the Balescu–Lenard type of kinetic-theoretical approach. The results of the former theory may be applicable both to a plasma near thermodynamic equilibrium and to one in a turbulent state, as long as the relevant fluctuation spectrum is known either experimentally or theoretically. In addition, through such a formulation we can investigate the extent to which each different frequency or wave vector regime of the fluctuation spectrum contributes to a given relaxation process.

In this chapter, we therefore adopt such a fluctuation-theoretical approach and calculate the Fokker–Planck coefficients in terms of the spectral function of the electric-field fluctuations and the dielectric response function. As concrete examples of the relaxation phenomena, we shall consider the temperature relaxation and the plasma diffusion across the magnetic field.

10.1 THE FOKKER–PLANCK EQUATION [1]

In the theory of Brownian motion, it is essential to recognize the stochastic nature of the phenomenon and thereby to seek a description in terms of the transition probability in the phase space starting from a given initial distribution. In this connection, it is considered as of the essence of Brownian motion that there exist time intervals τ during which the phase-space coordinates of the Brownian particle change by infinitesimal amounts while there occur a very large number of fluctuations characteristic of the motion and arising from the interaction with the particles of the surrounding medium. In terms of the various time scales considered in Section 2.3, we thus assume the existence of those τ which satisfy

$$\tau_1 \gg \tau \gg \tau_2. \tag{10.1}$$

In the comparison leading to (2.25), we noted that the foregoing is a reasonable assumption in many cases of a quiescent plasma.

The time evolution of the distribution function $f(X;t)$ for the Brownian particles may then be described by the integral equation

$$f(X;t+\tau) = \int f(X-\Delta X;t) W_\tau(X-\Delta X;\Delta X)\, d(\Delta X). \tag{10.2}$$

Here, $X \equiv (\mathbf{r}, \mathbf{v})$ are the phase-space coordinates of a particle, and $W_\tau(X;\Delta X)$ represents the *transition probability* that X changes by an increment ΔX during the time interval τ. The probability function is normalized so that

$$\int W_\tau(X;\Delta X)\, d(\Delta X) = 1. \tag{10.3}$$

In expecting the integral equation (10.2) to be true, we are actually supposing that the course which a Brownian particle will take depends only on the instantaneous values of its physical parameters and is entirely independent of its whole previous history. In probability theory, a stochastic process which has this characteristic, namely, that what happens at a given instant of time t depends only on the state of the system at time t, is said to be a *Markoff process.* That we should be able to idealize the motion of charged particles in a plasma as a Markoff process appears fairly reasonable, but far from obvious. In fact, we know from the discussion in Section 4.2A that electron plasma oscillations can have an infinite lifetime in the limit of long wavelengths; the behavior of these particles coupled with such long-lived excitations is expected to show a non-Markoffian character. Let us note, however, that the intensities of plasma oscillations become infinitesimally small as $k \to 0$ [see Eq. (9.73)]; their presence may thus have only a minor and negligible effect on the overall properties of the relaxation processes in the plasma. To the extent that the effects of long-lived collective oscillations are negligible in a given relaxation process,† we can adopt a Markoffian hypothesis of Eq. (10.2). We expect this assumption to be valid for plasmas near

†Some of these collective effects are investigated in [2] and are shown to be indeed negligible in the cases treated therein.

thermodynamic equilibrium; the assumption becomes dubious as one approaches the onset conditions of a plasma wave instability.

Equation (10.2) is now expanded in a Taylor series with respect to τ and ΔX:

$$f(X;t)+\frac{\partial f}{\partial t}\tau+O\left(\frac{\tau^2}{\tau_1^2}\right)$$

$$=\int d(\Delta X)\left[f(X;t)-\Delta X\cdot\frac{\partial f}{\partial X}+\frac{1}{2}(\Delta X\Delta X):\frac{\partial^2 f}{\partial X\partial X}-\cdots\right]$$

$$\times\left[W_\tau(X;\Delta X)-\Delta X\cdot\frac{\partial W_\tau}{\partial X}+\frac{1}{2}(\Delta X\Delta X):\frac{\partial^2 W_\tau}{\partial X\partial X}-\cdots\right]. \quad (10.4)$$

Recalling the normalization (10.3), we may define and calculate the average values of the increments according to

$$\langle\Delta X\rangle=\int \Delta X W_\tau(X;\Delta X)\,d(\Delta X) \quad (10.5a)$$

$$\langle\Delta X\Delta X\rangle=\int \Delta X\Delta X\, W_\tau(X;\Delta X)\,d(\Delta X). \quad (10.5b)$$

In terms of these averages, we can rewrite Eq. (10.4); neglecting the third term on its left-hand side, which is smaller by a factor of τ/τ_1 than the second term, we obtain

$$\frac{\partial f}{\partial t}=-\frac{\partial}{\partial X}\cdot\left[\frac{\langle\Delta X\rangle}{\tau}f\right]+\frac{1}{2}\frac{\partial^2}{\partial X\partial X}:\left[\frac{\langle\Delta X\Delta X\rangle}{\tau}f\right]+O\left(\frac{\langle\Delta X\Delta X\Delta X\rangle}{\tau}\right). \quad (10.6)$$

This is the *Fokker–Planck equation* in its general form; the factors $\langle\Delta X\rangle/\tau$ and $\langle\Delta X\Delta X\rangle/\tau$ are called the Fokker–Planck coefficients.

In a treatment of the relaxation problems through the Fokker–Planck equation, the Fokker–Planck coefficients are calculated by considering the effects of the microscopic fluctuating forces on the motion of a particle. In a homogeneous plasma, those coefficients as well as the distribution function are independent of the spatial coordinates; we thus simplify Eq. (10.6) as

$$\frac{\partial f}{\partial t}=-\frac{\partial}{\partial \mathbf{v}}\cdot\left[\frac{\langle\Delta\mathbf{v}\rangle}{\tau}f\right]+\frac{1}{2}\frac{\partial^2}{\partial\mathbf{v}\partial\mathbf{v}}:\left[\frac{\langle\Delta\mathbf{v}\Delta\mathbf{v}\rangle}{\tau}f\right]. \quad (10.7)$$

In obtaining this formula, we have truncated the series expansion by ignoring the terms with the coefficients $\langle \Delta\mathbf{v}\Delta\mathbf{v}\Delta\mathbf{v}\rangle/\tau$ and those of higher order in $\Delta\mathbf{v}$. For a plasma near thermodynamic equilibrium with the plasma parameter $g \ll 1$, such a truncation may be justified [3]. The essential point is that in a Coulomb-interacting system the averages $\langle \Delta\mathbf{v}\rangle$ and $\langle \Delta\mathbf{v}\Delta\mathbf{v}\rangle$ contain terms that diverge logarithmically and thus lead to a Coulomb logarithm when an appropriate cutoff is adopted for the integration; the averages $\langle \Delta\mathbf{v}\Delta\mathbf{v}\Delta\mathbf{v}\rangle$ and higher order, on the other hand, do not involve such a divergent term. (See Problem 10.1.) The logarithmically divergent terms leading to Coulomb logarithms have been called the "dominant" terms by Chandrasekhar [4]; these represent the major effects for relaxations in a plasma with $g \ll 1$. We shall consider an explicit example of such a dominant-term approximation in Section 10.3.

The right-hand side of Eq. (10.7) should correspond to the collision term of Section 2.2. The Fokker–Planck coefficients in the velocity space,

$$\mathbf{F}(\mathbf{v}) = \frac{1}{\tau}\langle \Delta\mathbf{v}\rangle, \tag{10.8}$$

$$\mathsf{D}(\mathbf{v}) = \frac{1}{\tau}\langle \Delta\mathbf{v}\Delta\mathbf{v}\rangle, \tag{10.9}$$

are then the basic quantities which determine the collision term of the plasma within the Fokker–Planck formalism. $\mathbf{F}(\mathbf{v})$ and $\mathsf{D}(\mathbf{v})$ are, respectively, the friction and diffusion coefficients of the Brownian particle in the velocity space.

10.2 FOKKER–PLANCK COEFFICIENTS

The effects of fluctuating electric fields on the motion of a charged particle in the absence of an external magnetic field have been investigated extensively by Hubbard and Thompson [5]. They were concerned with the first-order effects of the fluctuation spectrum on friction and diffusion of the particle in the velocity space. Within such a framework their calculations are rigorous and provide an alternative method of deriving the Balescu–Lenard collision term when a theoretical expression of the fluctuation spectrum for a stable plasma is substituted into the resulting Fokker–Planck coefficients. Let us therefore follow their approach rather closely in this section and calculate the Fokker–Planck coefficients.

A. Particle Motion in Fluctuating Fields

Consider the motion of a particle with electric charge q and mass m in the fluctuating electric field $\mathbf{E}(\mathbf{r}, t)$. The spectral distribution of the fluctuations is assumed to be uniform in space and stationary in time. The equation of motion is

$$\frac{d\mathbf{v}(t)}{dt}=\frac{q}{m}\mathbf{E}[\mathbf{r}(t),t] \tag{10.10}$$

where $\mathbf{r}(t)$ and $\mathbf{v}(t)[=d\mathbf{r}(t)/dt]$ represent the trajectory of the test particle in the phase space. The trajectories may be determined by integrating Eq. (10.10) as

$$\mathbf{r}(t)=\mathbf{r}(0)+\mathbf{v}(0)t+\frac{q}{m}\int_0^t dt'(t-t')\mathbf{E}[\mathbf{r}(t'),t'], \tag{10.11}$$

$$\mathbf{v}(t)=\mathbf{v}(0)+\frac{q}{m}\int_0^t dt'\mathbf{E}[\mathbf{r}(t'),t']. \tag{10.12}$$

The increments $\Delta\mathbf{r}(\tau)$ and $\Delta\mathbf{v}(\tau)$ of the particle coordinates during a time interval τ are readily calculated from Eqs. (10.11) and (10.12) as $\Delta\mathbf{r}(\tau)=\mathbf{r}(\tau)-\mathbf{r}(0)$ and $\Delta\mathbf{v}(\tau)=\mathbf{v}(\tau)-\mathbf{v}(0)$. We expand these quantities with respect to the fluctuating electric fields; we thus find

$$\Delta\mathbf{r}(\tau)=\mathbf{v}\tau+\frac{q}{m}\int_0^\tau ds\,s\mathbf{E}[\mathbf{r}_0(\tau-s),\tau-s]$$
$$+\frac{q^2}{m^2}\int_0^\tau ds\,s\int_0^{\tau-s}ds'\,s'\mathbf{E}[\mathbf{r}_0(\tau-s-s'),\tau-s-s']\cdot\frac{\partial}{\partial\mathbf{r}_0}\mathbf{E}[\mathbf{r}_0(\tau-s),\tau-s]+\cdots, \tag{10.13}$$

$$\Delta\mathbf{v}(\tau)=\frac{q}{m}\int_0^\tau ds\,\mathbf{E}[\mathbf{r}_0(s),s]$$
$$+\frac{q^2}{m^2}\int_0^\tau ds\int_0^s ds'\,s'\mathbf{E}[\mathbf{r}_0(s-s'),s-s']\cdot\frac{\partial}{\partial\mathbf{r}_0}\mathbf{E}[\mathbf{r}_0(s),s]+\cdots \tag{10.14}$$

where $\mathbf{v}\equiv\mathbf{v}(0)$ is the initial velocity, and

$$\mathbf{r}_0(t)=\mathbf{r}(0)+\mathbf{v}t \tag{10.15}$$

is the unperturbed orbit of the test particle.

When a uniform magnetic field is applied in the z direction, we take additional account of the spiral motion of the particle as described by Eqs.

(3.60) and (3.61). Equations (10.13)–(10.15) are now replaced by

$$\Delta \mathbf{r}(\tau)=\frac{1}{\Omega}\mathsf{H}(\tau)\cdot\mathbf{v}+\frac{c}{B}\int_0^\tau ds\,\mathsf{H}(s)\cdot\mathbf{E}[\mathbf{r}_0(\tau-s),\tau-s]$$

$$+\frac{c^2}{B^2}\int_0^\tau ds\,\mathsf{H}(s)\cdot\left[\left\{\int_0^{\tau-s}ds'\,\mathsf{H}(s')\cdot\mathbf{E}[\mathbf{r}_0(\tau-s-s'),\tau-s-s']\right\}\right.$$

$$\left.\cdot\frac{\partial}{\partial\mathbf{r}_0}\mathbf{E}[\mathbf{r}_0(\tau-s),\tau-s]\right]+\cdots, \tag{10.16}$$

$$\Delta\mathbf{v}(\tau)=[\mathsf{B}(\tau)-\mathsf{I}]\cdot\mathbf{v}+\frac{q}{m}\int_0^\tau ds\,\mathsf{B}(s)\cdot\mathbf{E}[\mathbf{r}_0(\tau-s),\tau-s]$$

$$+\frac{qc}{mB}\int_0^\tau ds\,\mathsf{B}(s)\cdot\left[\left\{\int_0^{\tau-s}ds'\,\mathsf{H}(s')\cdot\mathbf{E}[\mathbf{r}_0(\tau-s-s'),\tau-s-s']\right\}\right.$$

$$\left.\cdot\frac{\partial}{\partial\mathbf{r}_0}\mathbf{E}[\mathbf{r}_0(\tau-s),\tau-s]\right]+\cdots \tag{10.17}$$

$$\mathbf{r}_0(t)=\mathbf{r}(0)+\frac{1}{\Omega}\mathsf{H}(t)\cdot\mathbf{v}. \tag{10.18}$$

Those equations form the basis for calculating various relaxation processes in the following sections.

B. Diffusion Coefficients in Velocity Space

According to Eq. (10.10), the position and velocity of the test particle fluctuate as the microscopic electric fields; their instantaneous displacements are given by Eqs. (10.13) and (10.14). The diffusion tensor in the velocity space may then be calculated through definition (10.9) where the time interval τ must be chosen in such a way as to satisfy (10.1).

We thus substitute Eq. (10.14) in Eq. (10.9); to the lowest order in the fluctuation spectrum, we have

$$\mathsf{D}(\mathbf{v})=\frac{q^2}{\tau m^2}\int_0^\tau ds\int_0^\tau ds'\langle\mathbf{E}[\mathbf{r}_0(s),s]\mathbf{E}[\mathbf{r}_0(s'),s']\rangle. \tag{10.19}$$

We find it useful to change one of the integration variables to t via $t=s-s'$.

Then, Eq. (10.19) becomes

$$\mathsf{D}(\mathbf{v})=\frac{q^2}{\tau m^2}\int_0^{\tau}ds\int_{s-\tau}^{s}dt\langle\mathbf{E}[\mathbf{r}_0(s),s]\mathbf{E}[\mathbf{r}_0(s-t),s-t]\rangle. \qquad (10.20)$$

The domain of the double integration in Eq. (10.20) is illustrated in Fig. 10.1 as the parallelogram $ABCO$. By the definition of the correlation time τ_2, the integrand of Eq. (10.20) takes on significant values only in the shaded domain

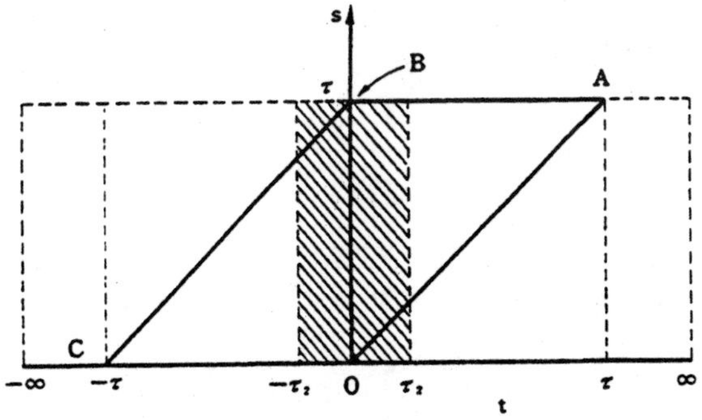

Fig. 10.1 Domain of integration in Eq. (10.20).

of Fig. 10.1. Hence, neglecting the errors coming from the contributions of the small triangular domains outside the parallelogram in the vicinity of O and B, we may extend the upper and lower limits of t integration in Eq. (10.20) to $+\infty$ and $-\infty$, respectively; the errors involved are of the order of τ_2/τ and negligible in the light of (10.1). Equation (10.20) thus reduces to

$$\mathsf{D}(\mathbf{v})=\frac{q^2}{\tau m^2}\int_0^{\tau}ds\int_{-\infty}^{\infty}dt\langle\mathbf{E}[\mathbf{r}_0(s),s]\mathbf{E}[\mathbf{r}_0(s-t),s-t]\rangle. \qquad (10.21)$$

We now introduce a spectral tensor of the electric-field fluctuations according to the scheme of Section 9.4A; we have

$$\langle\mathbf{E}(\mathbf{r}'+\mathbf{r},t'+t)\mathbf{E}(\mathbf{r}',t')\rangle=\frac{1}{V^2}\sum_{\mathbf{k}}\int_{-\infty}^{\infty}d\omega\langle\mathbf{E}\mathbf{E}^*(\mathbf{k},\omega)\rangle\exp[i(\mathbf{k}\cdot\mathbf{r}-\omega t)].$$

(10.22)

Substitution of Eqs. (10.15) and (10.22) into Eq. (10.21) then yields

$$\mathsf{D}(\mathbf{v}) = \frac{2\pi q^2}{m^2V^2}\sum_{\mathbf{k}}\int_{-\infty}^{\infty} d\omega \langle \mathbf{EE}^*(\mathbf{k},\omega)\rangle \delta(\omega - \mathbf{k}\cdot\mathbf{v})$$

$$= \frac{2\pi q^2}{m^2V^2}\sum_{\mathbf{k}} \langle \mathbf{EE}^*(\mathbf{k},\mathbf{k}\cdot\mathbf{v})\rangle. \qquad (10.23)$$

This equation gives a general relationship between the diffusion tensor and the spectral tensor of the electric-field fluctuations.

When the fluctuations under consideration are primarily of a longitudinal character, a condition which we shall frequently assume in the following calculations, the tensor of the electric-field fluctuation spectrum has a simplified form,

$$\langle \mathbf{EE}^*(\mathbf{k},\omega)\rangle = (\mathbf{kk}/k^2)\langle |E^2|(\mathbf{k},\omega)\rangle, \qquad (10.24)$$

where $\langle |E^2|(\mathbf{k},\omega)\rangle$ is a power spectral function of the longitudinal electric-field fluctuations. In these circumstances, Eq. (10.23) takes the form

$$\mathsf{D}(\mathbf{v}) = \frac{2\pi q^2}{m^2V^2}\sum_{\mathbf{k}} \frac{\mathbf{kk}}{k^2}\langle |E^2|(\mathbf{k},\mathbf{k}\cdot\mathbf{v})\rangle. \qquad (10.25)$$

The explicit expression for the spectral function is available in the first term on the right-hand side of Eq. (9.91); substituting this expression into Eq. (10.25), we obtain

$$\mathsf{D}(\mathbf{v}) = \frac{2\pi q^2}{m^2V}\sum_{\sigma}\sum_{\mathbf{k}}\int d\mathbf{v}'\left(\frac{4\pi q_\sigma}{k^2}\right)^2 \mathbf{kk}\frac{n_\sigma f_\sigma(\mathbf{v}')}{|\epsilon(\mathbf{k},\mathbf{k}\cdot\mathbf{v})|^2}\delta[\mathbf{k}\cdot(\mathbf{v}-\mathbf{v}')]. \qquad (10.26)$$

Here we have replaced the function $\epsilon_L(k,\mathbf{k}\cdot\mathbf{v})$ by the dielectric response function $\epsilon(\mathbf{k},\mathbf{k}\cdot\mathbf{v})$ defined in Section 3.2; such a generalization may be justified within the electrostatic approximation (10.24).

In the presence of a uniform magnetic field in the z direction, we can calculate the diffusion tensor in a way quite similar to the foregoing, starting from Eq. (10.17) [2]. Within the approximation (10.24), we express in Cartesian coordinates as

$$\mathsf{D}(\mathbf{v}) = \begin{bmatrix} D_\perp(v_\perp,v_\parallel) & 0 & 0 \\ 0 & D_\perp(v_\perp,v_\parallel) & 0 \\ 0 & 0 & D_\parallel(v_\perp,v_\parallel) \end{bmatrix} \qquad (10.27)$$

where

$$D_\perp(v_\perp,v_\parallel)=\frac{2\pi q^2}{m^2V^2}\sum_{\mathbf{k}}\sum_{n=-\infty}^{\infty}\frac{k_\perp^2}{4k^2}[J_{n-1}^2(\mathfrak{z})+J_{n+1}^2(\mathfrak{z})]\langle|E^2|(\mathbf{k},n\Omega+k_\parallel v_\parallel)\rangle,$$

(10.28)

$$D_\parallel(v_\perp,v_\parallel)=\frac{2\pi q^2}{m^2V^2}\sum_{\mathbf{k}}\sum_{n=-\infty}^{\infty}\frac{k_\parallel^2}{k^2}J_n^2(\mathfrak{z})\langle|E^2|(\mathbf{k},n\Omega+k_\parallel v_\parallel)\rangle \qquad \left(\mathfrak{z}=\frac{k_\perp v_\perp}{\Omega}\right).$$

(10.29)

With the knowledge of the fluctuation spectrum, these formulas enable us to calculate the diffusion coefficients in the magnetic field.

C. Induced Polarization Field of a Test Particle

When we single out a particle in the plasma and regard it as a test charge, an induced electric field arises due to polarization of the medium. Such an effect has been studied in Section 4.3B; according to Eq. (4.30), the induced field acting on the test particle with velocity $\mathbf{v}$ and charge q is†

$$\mathbf{E}_{\text{ind}}(\mathbf{v})=\frac{4\pi q}{V}\sum_{\mathbf{k}}\frac{\mathbf{k}}{k^2}\operatorname{Im}[\epsilon(\mathbf{k},\mathbf{k}\cdot\mathbf{v})]^{-1}. \qquad (10.30)$$

The test charge therefore feels both the induced field (10.30) and the spontaneously fluctuating fields $\mathbf{E}(\mathbf{r},t)$ considered in Eq. (10.10). We would then argue that the diffusion coefficients should have been calculated not from $\mathbf{E}(\mathbf{r},t)$ alone, but from the total field

$$\mathbf{E}_{\text{tot}}=\mathbf{E}_{\text{ind}}(\mathbf{v})+\mathbf{E}(\mathbf{r},t).$$

We would then have

$$\mathsf{D}(\mathbf{v})=(q^2/m^2)\mathbf{E}_{\text{ind}}(\mathbf{v})\mathbf{E}_{\text{ind}}(\mathbf{v})\tau+\text{Eq. }(10.23), \qquad (10.31)$$

assuming the translational invariance so that $\langle\mathbf{E}[\mathbf{r}_0(t),t]\rangle=0$ along the unperturbed orbit (10.15) of the particle.

It is possible to show, however, that the contribution of the first term on the right-hand side of Eq. (10.31) is in fact negligibly small, so that the results of the previous section represent correct evaluations. The arguments, due to

†Here, we recover the factor due to the finiteness of the plasma volume V.

Hubbard, proceed as follows: Since the velocity of the test charge undergoes a change of order v in a time of order τ_1, we must have

$$(q/m)|\mathbf{E}_{\text{tot}}| \sim v/\tau_1,$$

so that

$$\frac{q^2}{m^2}|\mathbf{E}_{\text{ind}}(\mathbf{v})\mathbf{E}_{\text{ind}}(\mathbf{v})|\tau \lesssim \frac{q^2}{m^2}|\mathbf{E}_{\text{tot}}|^2\tau \sim \frac{v^2\tau}{\tau_1^2}. \tag{10.32}$$

On the other hand, $\mathsf{D}(\mathbf{v})$ must be of the order v^2/τ_1; the collision terms should describe the effects pertaining to the relaxation time τ_1. In the light of (10.1) and (10.32), we find that the first term on the right-hand side of Eq. (10.31) is indeed negligible.

D. Friction Coefficients in Velocity Space

The friction coefficients may be calculated in accord with definition (10.8). Since the statistical average of the first term on the right-hand side of Eq. (10.14) vanishes, the second term represents the major contribution of the fluctuating fields. We must then retain the effects of the induced polarization field in the calculation of the friction coefficients; we thus have

$$\mathbf{F}(\mathbf{v}) = \frac{q}{m}\mathbf{E}_{\text{ind}}(\mathbf{v}) + \frac{q^2}{\tau m^2}\int_0^\tau ds \int_0^s ds' s' \langle \mathbf{E}[\mathbf{r}_0(s-s'),s-s'] \cdot \frac{\partial}{\partial \mathbf{r}_0}\mathbf{E}[\mathbf{r}_0(s),s]\rangle. \tag{10.33}$$

Since $\tau \gg \tau_2$, we may extend the upper limit of the s' integration to $+\infty$; substituting (10.15) in (10.33), we have

$$\begin{aligned}\mathbf{F}(\mathbf{v}) &= \frac{q}{m}\mathbf{E}_{\text{ind}}(\mathbf{v}) + \frac{q^2}{m^2}\int_0^\infty ds' s' \langle \mathbf{E}[\mathbf{r}_0(s),s] \cdot \frac{\partial}{\partial \mathbf{r}_0}\mathbf{E}[\mathbf{r}_0(s)+\mathbf{v}s',s+s']\rangle \\ &= \frac{q}{m}\mathbf{E}_{\text{ind}}(\mathbf{v}) + \frac{q^2}{m^2}\frac{\partial}{\partial \mathbf{v}} \cdot \int_0^\infty ds' \langle \mathbf{E}[\mathbf{r}_0(s),s]\mathbf{E}[\mathbf{r}_0(s)+\mathbf{v}s',s+s']\rangle.\end{aligned}$$

The statistical average may be expressed in terms of the spectral tensor of the electric-field fluctuations according to Eq. (10.22); hence

$$\mathbf{F}(\mathbf{v}) = \frac{q}{m}\mathbf{E}_{\text{ind}}(\mathbf{v}) - i\frac{q^2}{m^2 V^2}\frac{\partial}{\partial \mathbf{v}} \cdot \sum_{\mathbf{k}} \int_{-\infty}^{\infty} d\omega \frac{\langle \mathbf{E}\mathbf{E}^*(\mathbf{k},\omega)\rangle}{\mathbf{k}\cdot\mathbf{v}-\omega-i\eta}.$$

For reasons of symmetry in **k** and ω space, the imaginary part of the last term vanishes. In the light of Eq. (10.30), we thus find

$$\mathbf{F}(\mathbf{v}) = \frac{4\pi q^2}{mV}\sum_{\mathbf{k}}\frac{\mathbf{k}}{k^2}\operatorname{Im}[\epsilon(\mathbf{k},\mathbf{k}\cdot\mathbf{v})]^{-1} + \frac{\pi q^2}{m^2V^2}\frac{\partial}{\partial\mathbf{v}}\cdot\sum_{\mathbf{k}}\langle\mathbf{E}\mathbf{E}^*(\mathbf{k},\mathbf{k}\cdot\mathbf{v})\rangle. \quad (10.34)$$

With the aid of (10.23), the friction coefficients may also be expressed as

$$\mathbf{F}(\mathbf{v}) = \frac{4\pi q^2}{mV}\sum_{\mathbf{k}}\frac{\mathbf{k}}{k^2}\operatorname{Im}[\epsilon(\mathbf{k},\mathbf{k}\cdot\mathbf{v})]^{-1} + \frac{1}{2}\frac{\partial}{\partial\mathbf{v}}\cdot\mathsf{D}(\mathbf{v}). \quad (10.35)$$

For a plasma in a magnetic field, the calculation of the friction coefficients proceeds quite similarly to that described above. Expressing $\mathbf{F}(\mathbf{v})$ as a summation of two terms, polarization term $\mathbf{F}_p(\mathbf{v})$ and the fluctuation term $\mathbf{F}_f(\mathbf{v})$, we find in the electrostatic approximation [2]

$$\mathbf{F}_p(\mathbf{v}) = \frac{2q^2}{mV}\sum_{\mathbf{k}}\int_{-\infty}^{\infty}d\omega\int_{-\infty}^{\infty}dt\,\frac{\mathbf{k}}{k^2}\operatorname{Im}[\epsilon(\mathbf{k},\omega)]^{-1}\exp\{-i[\mathbf{k}\cdot\Delta\mathbf{r}_0(t)-\omega t]\}, \quad (10.36)$$

$$\mathbf{F}_f(\mathbf{v}) = i\frac{qc}{2mBV^2}\sum_{\mathbf{k}}\int_{-\infty}^{\infty}d\omega\int_{-\infty}^{\infty}dt\,[\mathbf{k}\cdot\mathsf{H}(t)\cdot\mathbf{k}]\frac{\mathbf{k}}{k^2}\langle|E^2|(\mathbf{k},\omega)\rangle \times\exp\{i[\mathbf{k}\cdot\Delta\mathbf{r}_0(t)-\omega t]\} \quad (10.37)$$

where $\Delta\mathbf{r}_0(t)$ has been defined in terms of the orbit (10.18).

E. Collision Term

Now that the Fokker–Planck coefficients have been evaluated in Eqs. (10.26) and (10.35), it is not difficult to obtain an explicit form of the collision term through Eq. (10.7). We note that Eq. (10.7) may be written as

$$\frac{\partial f}{\partial t} = \frac{\partial}{\partial\mathbf{v}}\cdot\left\{\frac{1}{2}\mathsf{D}(\mathbf{v})\cdot\frac{\partial}{\partial\mathbf{v}}f(\mathbf{v}) - \frac{4\pi q^2}{mV}\sum_{\mathbf{k}}\frac{\mathbf{k}}{k^2}\operatorname{Im}[\epsilon(\mathbf{k},\mathbf{k}\cdot\mathbf{v})]^{-1}f(\mathbf{v})\right\}. \quad (10.38)$$

Substitution of Eqs. (2.45) and (10.26) into Eq. (10.38) then yields the Balescu–Lenard collision term (2.44).

10.3 FOKKER–PLANCK COEFFICIENTS FOR PLASMAS IN THERMODYNAMIC EQUILIBRIUM

In many cases, the relaxation processes are considered for plasmas near thermodynamic equilibrium. For this reason, it is useful to evaluate the Fokker–Planck coefficient for a plasma with an isotropic Maxwellian distribution. We shall assume that no magnetic field is applied to the plasma. Based on an argument similar to the one leading to Eq. (3.26), we note that the diffusion tensor $\mathbf{D}(\mathbf{v})$ must be expressible as

$$\mathbf{D}(\mathbf{v}) = D_{\parallel}(v)\frac{\mathbf{vv}}{v^2} + D_{\perp}(v)\left[\mathbf{I} - \frac{\mathbf{vv}}{v^2}\right]. \tag{10.39}$$

We refer $D_{\parallel}(v)$ and $D_{\perp}(v)$, respectively, to the longitudinal and transverse diffusion coefficients; these depend on the magnitude of $\mathbf{v}$ only.

A. Longitudinal Diffusion Coefficient

We first obtain from Eqs. (10.26) and (10.39)

$$D_{\parallel}(v) = \frac{2\pi q^2}{m^2 v^2 V}\sum_{\sigma}\sum_{\mathbf{k}}\int d\mathbf{v}'\left(\frac{4\pi q_\sigma}{k^2}\right)^2 \frac{(\mathbf{k}\cdot\mathbf{v})^2 n_\sigma f_\sigma(\mathbf{v}')}{|\epsilon(\mathbf{k},\mathbf{k}\cdot\mathbf{v})|^2}\delta[\mathbf{k}\cdot(\mathbf{v}-\mathbf{v}')]. \tag{10.40}$$

The distribution function is given by Eq. (4.43). For such a Maxwellian plasma, the dielectric response function is written in terms of the W function as

$$\begin{aligned}\epsilon(\mathbf{k},\mathbf{k}\cdot\mathbf{v}) &= 1 + \sum_{\sigma}\frac{k_\sigma^2}{k^2}W\left(\frac{\mathbf{k}\cdot\mathbf{v}}{k(T_\sigma/m_\sigma)^{1/2}}\right)\\ &= 1 + \frac{k_{\mathrm{D}}^2}{k^2}(X+iY)\end{aligned} \tag{10.41}$$

where

$$\begin{aligned}k_{\mathrm{D}}^2 &= \sum_{\sigma}k_\sigma^2 = \sum_{\sigma}\frac{4\pi n_\sigma q_\sigma^2}{T_\sigma},\\ X &= \frac{1}{k_{\mathrm{D}}^2}\sum_{\sigma}k_\sigma^2 \,\mathrm{Re}\, W\left(\frac{\mathbf{k}\cdot\mathbf{v}}{k(T_\sigma/m_\sigma)^{1/2}}\right),\\ Y &= \frac{1}{k_{\mathrm{D}}^2}\sum_{\sigma}k_\sigma^2 \,\mathrm{Im}\, W\left(\frac{\mathbf{k}\cdot\mathbf{v}}{k(T_\sigma/m_\sigma)^{1/2}}\right).\end{aligned} \tag{10.42}$$

Note that X and Y are independent of the magnitude of the wave vector $\mathbf{k}$ and are functions of only its direction cosine, $\mu \equiv \cos\theta$, where θ is the angle between $\mathbf{k}$ and $\mathbf{v}$.

Transforming the $\mathbf{k}$ summation into integration via (2.31), we find that Eq. (10.40) may be calculated as

$$D_{\parallel}(v) = \left(\frac{2}{\pi}\right)^{1/2} \frac{q^2}{m^2} \sum_\sigma \frac{4\pi n_\sigma q_\sigma^2}{(T_\sigma/m_\sigma)^{1/2}} \int_{-1}^{1} d\mu\, \mu^2 \exp\left(-\frac{v^2}{2(T_\sigma/m_\sigma)}\mu^2\right)$$

$$\times \int_0^{k_m} dk \frac{k^3}{(k^2 + k_{\mathrm{D}}^2 X)^2 + k_{\mathrm{D}}^4 Y^2}$$

$$= \left(\frac{2}{\pi}\right)^{1/2} \frac{q^2}{m^2} \sum_\sigma \frac{4\pi n_\sigma q_\sigma^2}{(T_\sigma/m_\sigma)^{1/2}} \int_{-1}^{1} d\mu\, \mu^2 \exp\left(-\frac{v^2}{2(T_\sigma/m_\sigma)}\mu^2\right)$$

$$\times \left\{ \ln\frac{k_{\mathrm{m}}}{k_{\mathrm{D}}} + \frac{1}{4}\ln\frac{\left[1 + (k_{\mathrm{D}}/k_{\mathrm{m}})^2 X\right]^2 + (k_{\mathrm{D}}/k_{\mathrm{m}})^4 Y^2}{X^2 + Y^2} - \frac{X}{2|Y|}\left[\frac{\pi}{2} - \tan^{-1}\frac{X}{|Y|}\right]\right\}. \tag{10.43}$$

The first term, $\ln(k_{\mathrm{m}}/k_{\mathrm{D}})$, in the curly bracket of Eq. (10.43) diverges logarithmically as $k_{\mathrm{m}} \to \infty$, while the other two terms remain finite. The former term corresponds to the "dominant" term of Chandrasekhar [4]. Since the value of k_{m} may be estimated as the inverse of the average distance of the closest approach between two colliding particles [see Eq. (4.37)], the $\ln(k_{\mathrm{m}}/k_{\mathrm{D}})$ term represents the major effects as long as the plasma parameter $g \ll 1$. We may then identify

$$\ln(k_{\mathrm{m}}/k_{\mathrm{D}}) = \ln\Lambda, \tag{10.44}$$

the Coulomb logarithm. It must be remarked, however, that such a dominant-term approximation would break down if both X and Y take on infinitesimally small values at certain μ and v.

Within the validity of the dominant-term approximation, we thus have

$$D_{\parallel}(v) = \left(\frac{8}{\pi}\right)^{1/2} \frac{q^2}{m^2} \sum_\sigma \frac{4\pi n_\sigma q_\sigma^2}{(T_\sigma/m_\sigma)^{1/2}} \ln\Lambda \int_0^1 d\mu\, \mu^2 \exp\left(-\frac{v^2}{2(T_\sigma/m_\sigma)}\mu^2\right).$$

Changing the integration variable from μ to

$$y = \frac{v}{(2T_\sigma/m_\sigma)^{1/2}}\mu$$

and carrying out a partial integration, we find

$$D_\parallel(v) = \frac{16\pi^{1/2}q^2}{m^2v^3}\sum_\sigma n_\sigma q_\sigma^2\left(\frac{T_\sigma}{m_\sigma}\right)\ln\Lambda$$

$$\times\left[\int_0^{v/(2T_\sigma/m_\sigma)^{1/2}} dy\exp(-y^2) - \frac{v}{(2T_\sigma/m_\sigma)^{1/2}}\exp\left(-\frac{v^2}{2T_\sigma/m_\sigma}\right)\right]$$

$$= \frac{8\pi q^2}{m^2 v}\sum_\sigma n_\sigma q_\sigma^2 \ln\Lambda\, G\left[\frac{v}{(2T_\sigma/m_\sigma)^{1/2}}\right]. \qquad (10.45)$$

Here, the function $G(x)$ is defined in terms of the error function

$$\Phi(x) = (2/\pi^{1/2})\int_0^x dy\exp(-y^2) \qquad (10.46)$$

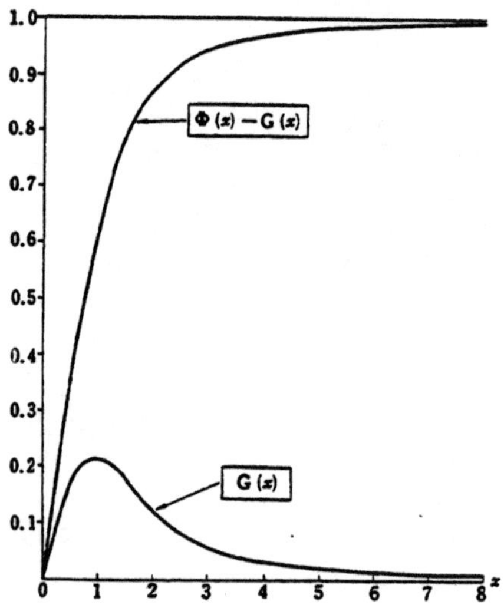

Fig. 10.2 Graphs of $G(x)$ and $\Phi(x) - G(x)$.

as

$$G(x)=\frac{\Phi(x)-x\Phi'(x)}{2x^2}. \tag{10.47}$$

The function $G(x)$ is plotted in Fig. 10.2.

B. Transverse Diffusion Coefficient

The transverse diffusion coefficient $D_\perp(v)$ is determined from (10.39) as

$$D_\perp(v)=\tfrac{1}{2}[\mathrm{Tr}\,\mathbf{D}(\mathbf{v})-D_\parallel(v)]. \tag{10.48}$$

From Eq. (10.26), we calculate

$$\mathrm{Tr}\,\mathbf{D}(\mathbf{v})=\frac{2\pi q^2}{m^2}\sum_\sigma\sum_{\mathbf{k}}\int d\mathbf{v}'\frac{(4\pi q_\sigma)^2}{k^2}\frac{n_\sigma f_\sigma(\mathbf{v}')}{|\epsilon(\mathbf{k},\mathbf{k}\cdot\mathbf{v})|^2}\delta[\mathbf{k}\cdot(\mathbf{v}-\mathbf{v}')]$$

$$=\left(\frac{2}{\pi}\right)^{1/2}\frac{q^2}{m^2}\sum_\sigma\frac{4\pi n_\sigma q_\sigma^2}{(T_\sigma/m_\sigma)^{1/2}}\int_{-1}^{1}d\mu\exp\left[-\frac{v^2}{2(T_\sigma/m_\sigma)}\mu^2\right]$$

$$\times\int_0^{k_m}dk\frac{k^3}{(k^2+k_D^2X)^2+k_D^4Y^2}.$$

Retaining the dominant term only, we have

$$\mathrm{Tr}\,\mathbf{D}(\mathbf{v})=\frac{8\pi q^2}{m^2 v}\sum_\sigma n_\sigma q_\sigma^2\ln\Lambda\,\Phi\left(\frac{v}{(2T_\sigma/m_\sigma)^{1/2}}\right). \tag{10.49}$$

Using the results of (10.45) and (10.49) in (10.48), we find

$$D_\perp(v)=\frac{4\pi q^2}{m^2 v}\sum_\sigma n_\sigma q_\sigma^2\ln\Lambda\left[\Phi\left(\frac{v}{(2T_\sigma/m_\sigma)^{1/2}}\right)-G\left(\frac{v}{(2T_\sigma/m_\sigma)^{1/2}}\right)\right]. \tag{10.50}$$

The function $\Phi(x)-G(x)$ is also plotted in Fig. 10.2.

C. Friction Coefficient

The friction coefficient has been given by Eq. (10.35). The second term on its right-hand side may be calculated directly from Eq. (10.49) as

$$\frac{1}{2}\frac{\partial}{\partial \mathbf{v}}\cdot\mathbf{D}(\mathbf{v})=\frac{1}{2}\frac{\partial}{\partial \mathbf{v}}[\mathrm{Tr}\mathbf{D}(\mathbf{v})]$$

$$=-\left(\frac{\mathbf{v}}{v}\right)\frac{4\pi q^2}{m^2}\sum_\sigma \frac{n_\sigma q_\sigma^2}{(T_\sigma/m_\sigma)}\ln\Lambda G\left[\frac{v}{(2T_\sigma/m_\sigma)^{1/2}}\right]. \tag{10.51}$$

The calculation of the first term proceeds in much the same way as those leading to (10.45) and (10.49); within the dominant-term approximation, we find

$$\frac{q}{m}\mathbf{E}_{\text{ind}}(\mathbf{v})=-\left(\frac{\mathbf{v}}{v}\right)\frac{4\pi q^2}{m}\sum_\sigma \frac{n_\sigma q_\sigma^2}{T_\sigma}\ln\Lambda G\left[\frac{v}{(2T_\sigma/m_\sigma)^{1/2}}\right]. \tag{10.52}$$

Collecting Eqs. (10.51) and (10.52), we finally obtain

$$\mathbf{F}(\mathbf{v})=-\left(\frac{\mathbf{v}}{v}\right)\frac{4\pi q^2}{m^2}\sum_\sigma \frac{n_\sigma q_\sigma^2}{(T_\sigma/m_\sigma)}\left(1+\frac{m}{m_\sigma}\right)\ln\Lambda G\left[\frac{v}{(2T_\sigma/m_\sigma)^{1/2}}\right]. \tag{10.53}$$

We remark in passing that the fluctuation term (10.51) and the polarization term (10.52) are of the same order of magnitude.

10.4 TEMPERATURE RELAXATION

As a simple applicational example of the Fokker–Planck coefficients calculated in the previous sections, let us consider the problem of temperature relaxation between the electrons and the ions [2]. Such a problem in the absence of an external magnetic field was considered by Spitzer [6]; several theoretical attempts [7–9] exist for the cases with a magnetic field.

A. Relaxation of Particle Energy

Consider the kinetic energy of a test particle, $w=\frac{1}{2}mv^2$. As in the cases of (10.8) and (10.9), its time rate of change may be defined and calculated as

$$\frac{dw}{dt}=\frac{\langle\Delta w(\tau)\rangle}{\tau}=\frac{m}{2\tau}\langle|\mathbf{v}+\Delta\mathbf{v}(\tau)|^2-|\mathbf{v}|^2\rangle$$

$$=m\mathbf{v}\cdot\mathbf{F}(\mathbf{v})+\frac{m}{2}\mathrm{Tr}\mathbf{D}(\mathbf{v}). \tag{10.54}$$

Each term in Eq. (10.54) can readily be calculated with the aid of Eqs. (10.28), (10.29), (10.36), and (10.37):

$$\frac{m}{2}\operatorname{Tr}\mathbf{D}(\mathbf{v})=\frac{\pi q^2}{mV^2}\sum_{\mathbf{k}}\sum_{n}\int_{-\infty}^{\infty}d\omega\left\{\frac{k_\perp^2}{2k^2}[J_{n-1}^2(\mathfrak{z})+J_{n+1}^2(\mathfrak{z})]+\frac{k_\parallel^2}{k^2}J_n^2(\mathfrak{z})\right\}$$

$$\times\langle|E^2|(\mathbf{k},\omega)\rangle\delta(n\Omega+k_\parallel v_\parallel-\omega) \tag{10.55}$$

$$m\mathbf{v}\cdot\mathbf{F}_{\mathrm{f}}(\mathbf{v})=-\frac{\pi q^2}{mV^2}\sum_{\mathbf{k}}\sum_{n}\int_{-\infty}^{\infty}d\omega\left[\frac{k_\perp^2}{2k^2}\left\{[J_{n-1}^2(\mathfrak{z})+J_{n+1}^2(\mathfrak{z})]\right.\right.$$

$$\left.-\frac{\omega}{\Omega}[J_{n-1}^2(\mathfrak{z})-J_{n+1}^2(\mathfrak{z})]\right\}\langle|E^2|(\mathbf{k},\omega)\rangle$$

$$\left.-\frac{k_\parallel^2}{k^2}J_n^2(\mathfrak{z})\,\omega\frac{\partial}{\partial\omega}\langle|E^2|(\mathbf{k},\omega)\rangle\right]\delta(n\Omega+k_\parallel v_\parallel-\omega), \tag{10.56}$$

$$m\mathbf{v}\cdot\mathbf{F}_{\mathrm{p}}(\mathbf{v})=\frac{4\pi q^2}{V}\sum_{\mathbf{k}}\sum_{n}\int_{-\infty}^{\infty}d\omega\frac{\omega}{k^2}J_n^2(\mathfrak{z})\operatorname{Im}[\epsilon(\mathbf{k},\omega)]^{-1}\delta(n\Omega+k_\parallel v_\parallel-\omega). \tag{10.57}$$

In obtaining Eqs. (10.56) and (10.57), we have carried out the average with respect to the difference of the azimuthal angles between **k** and **v**. Summing Eqs. (10.55)–(10.57), we obtain

$$\frac{dw}{dt}=\frac{\pi q^2}{mV^2}\sum_{\mathbf{k}}\sum_{n}\int_{-\infty}^{\infty}d\omega\left\{\frac{k_\perp^2}{2k^2}[J_{n-1}^2(\mathfrak{z})-J_{n+1}^2(\mathfrak{z})]\frac{\omega}{\Omega}\langle|E^2|(\mathbf{k},\omega)\rangle\right.$$

$$\left.+\frac{k_\parallel^2}{k^2}J_n^2(\mathfrak{z})\frac{\partial}{\partial\omega}[\omega\langle|E^2|(\mathbf{k},\omega)\rangle]+4mV\frac{\omega}{k^2}J_n^2(\mathfrak{z})\operatorname{Im}[\epsilon(\mathbf{k},\omega)]^{-1}\right\}\delta(n\Omega+k_\parallel v_\parallel-\omega). \tag{10.58}$$

To calculate the rate of temperature relaxation, we multiply Eq. (10.58) by the Maxwellian (4.66) and integrate over the velocities. After lengthy but straightforward algebra with the aid of some of the well-known identities on the Bessel functions, we obtain

$$\frac{dT}{dt}=\frac{(2\pi)^{1/2}q^2}{3m(T/m)^{3/2}V^2}\sum_{\mathbf{k}}\sum_{n}\int_{-\infty}^{\infty}d\omega\frac{\omega^2}{|k_{\parallel}|k^2}\Big\{\langle|E^2|(\mathbf{k},\omega)\rangle$$

$$+V\frac{4T}{\omega}\operatorname{Im}[\epsilon(\mathbf{k},\omega)]^{-1}\Big\}\Lambda_n(\beta)\exp\left[-\frac{(n\Omega-\omega)^2}{2k_{\parallel}^2(T/m)}\right], \tag{10.59}$$

where β and $\Lambda_n(\beta)$ have been defined by Eqs. (4.69) and (4.70).

Equation (10.59) contains important physical implications: Generally, the fluctuation spectrum of a system in thermodynamic equilibrium can be calculated with the aid of the fluctuation-dissipation theorem; for a classical plasma, this theorem gives

$$\langle|E^2|(\mathbf{k},\omega)\rangle=-V(4T/\omega)\operatorname{Im}[\epsilon(\mathbf{k},\omega)]^{-1}. \tag{10.60}$$

A natural conclusion which follows from Eqs. (10.59) and (10.60) is that for a plasma in thermodynamic equilibrium $dT/dt=0$, as it should be.

Alternatively, we may choose to turn the arguments around and regard the calculations in this section as another way of deriving the fluctuation-dissipation theorem. Up to Eq. (10.59), all we did was to calculate the average rate of change in energy for the test particles when the plasma dielectric function is given by $\epsilon(\mathbf{k},\omega)$ and the spectrum of the (longitudinal) electric-field fluctuations by $\langle|E^2|(\mathbf{k},\omega)\rangle$. For thermodynamic equilibrium, we may then *demand* that this rate of energy change vanish; the relationship Eq. (10.60) then follows from this requirement.

B. Rate of Temperature Relaxation between Electrons and Ions

Let us now proceed and consider a nonequilibrium situation in which the temperatures of the electrons and the ions are different; we shall use the subscripts $\sigma=\mathrm{e}$ for the electrons and $\sigma=\mathrm{i}$ for the ions. For simplicity, we assume $q_{\mathrm{i}}=e$ and $N_{\mathrm{e}}=N_{\mathrm{i}}\equiv N$.

The relaxation time τ_{ei} of the temperatures between the electrons and the ions is defined through the equation†

$$\frac{dT_e}{dt} = -\frac{T_e - T_i}{\tau_{ei}}. \tag{10.61}$$

This time constant may thus be obtained from a detailed analysis of Eq. (10.59).

For the two-component Maxwellian plasma under consideration, the dielectric function $\epsilon(\mathbf{k},\omega)$ takes the form

$$\epsilon(\mathbf{k},\omega) = 1 + \sum_{\sigma=e,i} \frac{k_\sigma^2}{k^2}\left\{1 + \sum_n \frac{\omega}{\omega - n\Omega_\sigma}\left[W\left(\frac{\omega - n\Omega_\sigma}{|k_\parallel|(T_\sigma/m_\sigma)^{1/2}}\right) - 1\right]\Lambda_n(\beta_\sigma)\right\}. \tag{10.62}$$

The spectral function $\langle|E^2|(\mathbf{k},\omega)\rangle$ of the electric-field fluctuations can be calculated with the aid of the superposition technique of the dressed test particles. We thus find

$$\langle|E^2|(\mathbf{k},\omega)\rangle = \frac{16\pi^2 e^2}{k^2|\epsilon(\mathbf{k},\omega)|^2}\sum_{\sigma=e,i} S_\sigma^{(0)}(\mathbf{k},\omega), \tag{10.63}$$

$$S_\sigma^{(0)}(\mathbf{k},\omega) = \frac{N}{(2\pi T_\sigma/m_\sigma)^{1/2}|k_\parallel|}\sum_n \Lambda_n(\beta_\sigma)\exp\left[-\frac{(n\Omega_\sigma - \omega)^2}{2k_\parallel^2(T_\sigma/m_\sigma)}\right]. \tag{10.64}$$

We now substitute Eqs. (10.62) and (10.63) into Eq. (10.59); after a straightforward calculation, we obtain

$$\frac{1}{\tau_{ei}} = \frac{16\pi^2 N e^4 (m_e m_i)^{1/2}}{3(T_e T_i)^{3/2} V^2}\sum_{\mathbf{k}}\sum_{n,n'}\int_{-\infty}^{\infty} d\omega \frac{\omega^2}{k_\parallel^2 k^4 |\epsilon(\mathbf{k},\omega)|^2}$$

$$\times \Lambda_n(\beta_e)\Lambda_{n'}(\beta_i)\exp\left[-\frac{(\omega - n\Omega_e)^2}{2k_\parallel^2(T_e/m_e)} - \frac{(\omega - n'\Omega_i)^2}{2k_\parallel^2(T_i/m_i)}\right]. \tag{10.65}$$

†Sometimes the relaxation time is defined through the equation

$$\frac{d(T_e - T_i)}{dt} = -\frac{T_e - T_i}{\bar{\tau}_{ei}}$$

together with the energy conservation relation, $N_e T_e + N_i T_i = \text{const}$. The two relaxation times are related to each other via

$$\tau_{ei} = [1 + (N_e/N_i)]\bar{\tau}_{ei}.$$

In the limit of $B \to 0$, we may go over to the continuum (4.73) for the n and n' summation, and Eq. (10.65) reduces to the Spitzer value [6]

$$\frac{1}{\tau_{ei}^{(0)}} = \frac{8(2\pi)^{1/2} N e^4}{3 m_e m_i (T_e/m_e + T_i/m_i)^{3/2} V} \ln\Lambda, \tag{10.66}$$

when only the "dominant" terms of Chandrasekhar are retained in the resulting k integration; $\ln\Lambda$ in Eq. (10.66) is the Coulomb logarithm arising from such a treatment.

C. Anomalous Term

When the magnetic field is finite, it is generally a difficult task to rigorously carry through all the summations and integrations involved in Eq. (10.65). Instead, we find it advantageous to note a small dimensionless parameter, m_e/m_i, and investigate the most dominant contribution in Eq. (10.65) in the limit $m_e/m_i \to 0$; in this way we can secure a systematic framework in which to analyze the effects of the magnetic field on the temperature-relaxation processes. We shall thereby find that in the presence of the magnetic field the relaxation rate contains a term which diverges as $\ln(m_e/m_i)$ in addition to the usual relaxation process described by Eq. (10.66).

To begin with, we note that the domain of the $\mathbf{k}$ space such that

$$|k_\parallel|(T_e/m_e)^{1/2} > k_\perp (T_i/m_i)^{1/2}, \tag{10.67a}$$

$$|\Omega_e| > k_\perp (T_i/m_i)^{1/2} \tag{10.67b}$$

covers most of the important range in Eq. (10.65). We may then replace the electronic polarizability by its static value and write the dielectric function as

$$\epsilon(\mathbf{k},\omega) = \epsilon_e(\mathbf{k}) + \frac{k_i^2}{k^2}\left\{1 + \sum_n \frac{\omega}{\omega - n\Omega_i}\left[W\left(\frac{\omega - n\Omega_i}{|k_\parallel|(T_i/m_i)^{1/2}}\right) - 1\right]\Lambda_n(\beta_i)\right\} \tag{10.68}$$

where $\epsilon_e(\mathbf{k})$ is the static dielectric constant defined by (4.64). Equation (10.68) enables us to rewrite Eq. (10.65) as

$$\frac{1}{\tau_{ei}} = -\left(\frac{2}{\pi}\right)^{1/2} \frac{16\pi^2 N e^4 m_e^{1/2}}{3 m_i T_e^{3/2} \omega_i^2 V^2} \sum_{\mathbf{k}} \sum_n \int_{-\infty}^{\infty} d\omega \frac{\omega}{|k_\parallel| k^2} \operatorname{Im}[\epsilon(\mathbf{k},\omega)]^{-1}$$

$$\times \Lambda_n(\beta_e) \exp\left[-\frac{(\omega - n\Omega_e)^2}{2k_\parallel^2 (T_e/m_e)}\right]. \tag{10.69}$$

By virtue of (10.67) we may ignore the ω dependence in the exponent of Eq. (10.69). The ω integration can then be carried out with the aid of Kramers–Kronig relations as

$$\int_{-\infty}^{\infty} d\omega\, \omega \,\mathrm{Im}[\epsilon(\mathbf{k},\omega)]^{-1} = -\pi \lim_{\omega^2\to\infty} \omega^2\{[\epsilon(\mathbf{k},\omega)]^{-1} - [\epsilon_e(\mathbf{k})]^{-1}\}$$

$$= -\frac{\pi\omega_i^2}{[\epsilon_e(\mathbf{k})]^2}.$$

Hence, Eq. (10.69) becomes

$$\frac{1}{\tau_{ei}} = \frac{8(2\pi)^{1/2} N e^4 m_e^{1/2}}{3 m_i T_e^{3/2} V} \int_0^{\infty} dk_\perp \int_{k_\perp (m_e T_i/m_i T_e)^{1/2}}^{\infty} dk_\parallel \frac{k_\perp \Lambda_0(\beta_e)}{k_\parallel k^2 [\epsilon_e(\mathbf{k})]^2}$$

$$+ \frac{8(2\pi)^{1/2} N e^4 m_e^{1/2}}{3 m_i T_e^{3/2} V} \sum_n{}' \int_0^{\infty} dk_\perp \int_0^{\infty} dk_\parallel \frac{k_\perp \Lambda_n(\beta_e)}{k_\parallel k^2 [\epsilon_e(\mathbf{k})]^2} \exp\left[-\frac{n^2 \Omega_e^2}{2 k_\parallel^2 (T_e/m_e)} \right] \tag{10.70}$$

where the prime on the n summation means omission of the term with $n=0$ in the summation, and we have transformed the **k** summation into integration via

$$\sum_{\mathbf{k}} \to \frac{V}{2\pi^2} \int_0^{\infty} k_\perp\, dk_\perp \int_{k_\perp (m_e T_i/m_i T_e)^{1/2}}^{\infty} dk_\parallel. \tag{10.71}$$

The lower limit of the $k_\parallel$ integration in Eq. (10.71) derives from (10.67a). In the second term on the right-hand side of Eq. (10.70), this limit has been set equal to zero by going over to $m_e/m_i \to 0$, as no anomaly arises there. This term reduces to the Spitzer value, Eq. (10.66), when the continuum limit of Eq. (4.73) is applied to the n summation; such a procedure may be justified for the domain $\beta_e \gg 1$, which contributes the bulk of the effect.

The first term of Eq. (10.70), on the contrary, exhibits quite a singular behavior in the limit of $m_e/m_i \to 0$; the $k_\parallel$ integration diverges logarithmically for small $k_\parallel$. We must, therefore, keep m_e/m_i finite in that integration to avoid such a divergence. Since we may neglect the $k_\parallel$ dependence in $\epsilon_e(\mathbf{k})$, the anomalous contribution arising from the first term of Eq. (10.70) is

$$\frac{1}{\tau_{ei}^*} = \frac{4(2\pi)^{1/2} N e^4 m_e^{1/2}}{3 m_i T_e^{3/2} V} \int_0^{\infty} dk_\perp \frac{k_\perp^3 \Lambda_0(\beta_e)}{(k_\perp^2 + k_e^2)^2} \ln\left(\frac{m_i}{m_e}\right). \tag{10.72}$$

Equation (10.72) vanishes in the limit $B\to 0$; this anomalous term therefore represents a new effect arising from the presence of the magnetic field.

The integration in Eq. (10.72) can be expressed in terms of the modified Bessel functions of the second kind, K_0 and K_1; we have [2]

$$\frac{1}{\tau_{\mathrm{ei}}^*}=\frac{2(2\pi)^{1/2}Ne^4}{3m_e m_i(T_e/m_e)^{3/2}V}\exp(\alpha)[(1+\alpha)K_0(\alpha)-\alpha K_1(\alpha)]\ln\left(\frac{m_i}{m_e}\right) \tag{10.73}$$

where

$$\alpha\equiv k_e^2(T_e/m_e\Omega_e^2)=\omega_e^2/\Omega_e^2. \tag{10.74}$$

When the magnetic field is strong so that $\Omega_e^2\gg\omega_e^2$ (i.e., $\alpha\ll 1$), Eq. (10.73) reduces to

$$\frac{1}{\tau_{\mathrm{ei}}^*}=\frac{2(2\pi)^{1/2}Ne^4}{3m_e m_i(T_i/m_e)^{3/2}V}\ln\left(\frac{\Omega_e^2}{\omega_e^2}\right)\ln\left(\frac{m_i}{m_e}\right). \tag{10.75}$$

Equation (10.73) or (10.75) is the relaxation rate which appears in the presence of the magnetic field in addition to the usual term of Eq. (10.66). It, therefore, seems instructive to compare the magnitudes of those two contributions; from Eqs. (10.66) and (10.75), we have

$$\frac{1/\tau_{\mathrm{ei}}^*}{1/\tau_{\mathrm{ei}}^{(0)}}=\frac{\ln|\Omega_e/\omega_e|\ln(m_i/m_e)}{2\ln\Lambda} \tag{10.76}$$

where we have neglected T_i/m_i in comparison with T_e/m_e in Eq. (10.66). Assuming $\ln\Lambda\cong 15$ and $\Omega_i\cong\omega_i$, we find the foregoing ratio to be approximately 1.1 for a deuterium plasma; the additional temperature-relaxation process of Eq. (10.73) or (10.75) is here seen to be as effective as the ordinary process of Eq. (10.66) for a plasma in a strong magnetic field.

The physical origin of such an anomalous relaxation process may be traced to the spiral motion of the electrons along the magnetic lines of force. Those electrons naturally tend to couple strongly with long-wavelength fluctuations (i.e., small $k_\parallel$) along the magnetic field. In addition, when such fluctuations are characterized by slow variation in time (i.e., small ω), the contact time or the rate of energy exchange between the electrons and the fluctuations will be further enhanced. In a two-component plasma, such low-frequency fluctuations are provided mostly by the thermal motion of the ions; the foregoing coupling can, therefore, be an effective mechanism of energy exchange between the electrons and the ions. In the limit of $m_i\to\infty$, the frequency spectrum of the ionic thermal fluctuations consists essentially of a δ function at $\omega=0$; the contact time would become finite and the relaxation rate would diverge.

The reasoning above, emphasizing the importance of the special role which the fluctuations with small $k_\parallel$ and small ω play, may lead us to suspect that there is another significant contribution to the energy exchange processes arising from the collective modes of similar character. For a plasma in a strong magnetic field, the electron plasma waves and the ion acoustic waves have the frequencies $(k_\parallel/k)\omega_e$ and $k_\parallel(T_e/m_i)^{1/2}$, respectively; potentially, they might provide a significant energy-exchange mechanism. Let us first note, however, that the use of Eq. (10.68) for the dielectric function automatically takes care of whatever contributions the ion-acoustic waves may have on such processes. We can also calculate the spectral strengths of various collective modes in the plasma following the method of Section 9.3D; we may then evaluate the contributions of those collective modes to the energy exchange processes. Such an analysis indicates that the fluctuations in the collective modes are not strong enough to be able to affect the result, Eq. (10.73).

D. Relaxation between the Temperatures Parallel and Perpendicular to the Magnetic Field

For a plasma in a magnetic field, it is sometimes possible to define two temperatures for the particles of a given species, associated with the parallel and perpendicular degrees of freedom with respect to the magnetic field. Interaction between identical particles essentially accounts for the relaxation between those two temperatures.

We may analyze the relaxation processes in these circumstances in a way quite similar to the method used in the previous section. Here, instead of the total energy of a particle, we deal separately with $w_\perp = (m/2)v_\perp^2$ and $w_\parallel = (m/2)v_\parallel^2$. Time rates of change for these quantities are multiplied by the anisotropic Maxwellian (7.23) and integrated over the velocities. We thus obtain [2]

$$\frac{dT_\perp}{dt} = \left(\frac{\pi}{2}\right)^{1/2} \frac{q^2}{m(T_\parallel/m)^{3/2}V^2} \sum_{\mathbf{k}} \sum_{n} \int_{-\infty}^{\infty} d\omega \frac{n\Omega}{|k_\parallel| k^2} \left\{ n\Omega\left(\frac{T_\parallel}{T_\perp} - 1\right) \langle |E^2|(\mathbf{k},\omega)\rangle \right.$$
$$\left. + \omega\left[\langle |E^2|(\mathbf{k},\omega)\rangle + V\frac{4T_\parallel}{\omega}\,\mathrm{Im}[\epsilon(\mathbf{k},\omega)]^{-1}\right]\right\} \Lambda_n(\beta) \exp\left[-\frac{(n\Omega-\omega)^2}{2k_\parallel^2(T_\parallel/m)}\right], \tag{10.77}$$

$$\frac{dT_\parallel}{dt} = \frac{(2\pi)^{1/2} q^2}{m(T_\parallel/m)^{3/2}V^2} \sum_{\mathbf{k}} \sum_{n} \int_{-\infty}^{\infty} d\omega \frac{\omega - n\Omega}{|k_\parallel| k^2} \left\{ n\Omega\left(\frac{T_\parallel}{T_\perp} - 1\right) \langle |E^2|(\mathbf{k},\omega)\rangle \right.$$
$$\left. + \omega\left[\langle |E^2|(\mathbf{k},\omega)\rangle + V\frac{4T_\parallel}{\omega}\,\mathrm{Im}[\epsilon(\mathbf{k},\omega)]^{-1}\right]\right\} \Lambda_n(\beta) \exp\left[-\frac{(n\Omega-\omega)^2}{2k_\parallel^2(T_\parallel/m)}\right]. \tag{10.78}$$

The relaxation rates are again written as functionals of the spectral density of the fluctuations and the dielectric response function. In thermodynamic equilibrium $T_\perp = T_\parallel$, and both Eqs. (10.77) and (10.78) vanish by virtue of the fluctuation-dissipation theorem.

As an example of the nonequilibrium situation, we consider a case in which the ions are characterized by the anisotropic distribution, Eq. (7.23). For the sake of simplicity, however, we treat the electrons only as providing a static dielectric background to the ions; we thus concentrate on the relaxation processes between $T_\perp$ and $T_\parallel$ for the ions. The ionic relaxation time τ_i is defined through the equation

$$\frac{dT_\perp}{dt} = -\frac{1}{2}\frac{dT_\parallel}{dt} = -\frac{T_\perp - T_\parallel}{\tau_i}. \tag{10.79}$$

It can be calculated in a way quite similar to τ_{ei} in the previous sections. Within the dominant-term approximation, we find

$$\frac{1}{\tau_i} = \frac{8\pi^{1/2}Ne^4}{15m_i^{1/2}(T_{\text{eff}})^{3/2}V}\ln\Lambda, \tag{10.80}$$

where the effective ion temperature T_{eff} has been defined through the integral relation

$$\frac{1}{(T_{\text{eff}})^{3/2}} \equiv \frac{15}{4}\int_{-1}^{1} d\mu \frac{\mu^2(1-\mu^2)}{[(1-\mu^2)T_\perp + \mu^2 T_\parallel]^{3/2}}. \tag{10.81}$$

When $T_\parallel = T_\perp$, T_{eff} becomes equal to those temperatures. In this relaxation process, no anomalous term corresponding to (10.73) appears; the effects of the collective modes, such as the ion Bernstein modes of Section 4.6D, are negligible.

10.5 SPATIAL DIFFUSION ACROSS A MAGNETIC FIELD

In the previous section, we investigated the problem of temperature relaxations in a plasma with a magnetic field, based on a study of the motion of a charged particle in fluctuating electric fields. We have thereby found that the relaxation rate between the electron and ion temperatures contains, in addition to the usual Spitzer rate, an anomalous term which exhibits a different functional dependence on the magnetic field and which diverges logarithmically as the mass ratio of the electron to the ion approaches zero. The physical origin of this anomalous term has been identified as the strong coupling between the spiral motion of the electrons and the long-wavelength, low-frequency fluctuations produced by the ions.

In the present section, we wish to extend the foregoing approach and consider the problem of spatial diffusion of a plasma across a magnetic field [10]. We thus calculate the first and second moments of the guiding-center displacements of a test particle under the action of the fluctuating electric fields. Originally, such calculations were carried out by Longmire and Rosenbluth [11] based on a collision-theoretical consideration. Several theoretical developments exist in these directions [12, 13].

A. Motion of Guiding Centers and Spatial Diffusion

We consider the problem of a plasma in which the density depends on the x coordinate alone, in a uniform magnetic field $\mathbf{B}$ in the z direction. We shall calculate the flux $\Gamma_1(2)$ of the guiding centers of the particles 1 due to scattering in the fluctuating electric fields $\mathbf{E}(\mathbf{r},t)$ produced by the particles 2; the two types of particles, denoted by subscripts 1 and 2, need not be different. According to the stochastic formula (10.6), the flux may be calculated as

$$\Gamma_1(2)=N_1(X)\frac{\langle\Delta X_1\rangle}{\tau}-\frac{1}{2}\frac{\partial}{\partial X}\left[N_1(X)\frac{\langle(\Delta X_1)^2\rangle}{\tau}\right] \tag{10.82}$$

where $N_1(X)$ is the *density* of the guiding centers of particles 1. In this section, we let X denote the x coordinate of the guiding center.

The first and second moments $\langle\Delta X_1\rangle$ and $\langle(\Delta X_1)^2\rangle$ of the guiding-center displacements of a particle 1 in the x direction may be evaluated with the knowledge of the motion of the guiding center in the fluctuating electric fields. With the aid of Eqs. (10.16) and (10.17) for the instantaneous position $\mathbf{r}(t)=\mathbf{r}(0)+\Delta\mathbf{r}(t)$ and velocity $\mathbf{v}(t)=\mathbf{v}+\Delta\mathbf{v}(t)$ of a particle with charge q and mass m, the position $\mathbf{R}(t)$ of the guiding center is obtained as

$$\begin{aligned}\mathbf{R}(t)&=\mathbf{r}(t)-\frac{1}{\Omega}\hat{z}\times\mathbf{v}(t)\\&=\mathbf{R}(0)-\frac{c}{B}\hat{z}\times\int_0^t dt'\mathbf{E}[\mathbf{r}(t'),t']+\hat{z}\left\{v_\parallel(0)t+\frac{q}{m}\int_0^t dt'\int_0^{t'}dt''E_\parallel[\mathbf{r}(t''),t'']\right\}.\end{aligned} \tag{10.83}$$

We thus have

$$\frac{\langle\Delta X_1\rangle}{\tau}=\frac{c}{\tau B}\int_0^\tau dt\langle E_y[\mathbf{r}_1(t),t]\rangle, \tag{10.84}$$

$$\frac{\langle(\Delta X_1)^2\rangle}{\tau}=\frac{c^2}{\tau B^2}\int_0^\tau dt\int_0^\tau dt'\langle E_y[\mathbf{r}_1(t),t]E_y[\mathbf{r}_1(t'),t']\rangle. \tag{10.85}$$

It is clear from the definition of Eq. (10.82) that the electric-field fluctuations in Eqs. (10.84) and (10.85) must be those produced by the particles 2.

As a specific example of application, we shall consider the diffusion phenomena in an electron–ion plasma. The electron and ion fluxes calculated according to Eq. (10.82) may generally be different; the electric field will then be induced in the x direction to make the diffusion ambipolar [14]. The most important case to be considered in these circumstances is that where $1 = \mathrm{e}$ and $2 = \mathrm{i}$.

B. Calculation of the Second Moment

We begin with the calculation of the second moment, Eq. (10.85), in this section. We first note that this moment takes on a finite value for a homogeneous plasma. From symmetry considerations, we then find that the first-order contributions of the spatial inhomogeneity (i.e., density gradient) vanish; the next leading terms therefore arise from the second-order contributions of inhomogeneity. We shall not delve into such higher-order calculations. From a collision-theoretical point of view, such higher-order terms have been calculated by Longmire and Rosenbluth [11].

Assuming the longitudinal fluctuations and following the procedure of Section 10.4, we thus obtain

$$\begin{aligned}\frac{\langle(\Delta X_1)^2\rangle}{\tau} &= \frac{c^2}{B^2V^2}\sum_{\mathbf{k}}\int_{-\infty}^{\infty}d\omega\int_{-\infty}^{\infty}dt\,\frac{k_y^2}{k^2}\langle|E^2|(\mathbf{k},\omega)\rangle_2\exp\{i[\mathbf{k}\cdot\Delta\mathbf{r}_0(t)-\omega t]\}\\ &= \frac{\pi c^2}{B^2V^2}\sum_{\mathbf{k}}\sum_{n}\int_{-\infty}^{\infty}d\omega\langle|E^2|(\mathbf{k},\omega)\rangle_2\frac{k_\perp^2}{k^2}J_n^2(\mathfrak{z}_1)\delta(n\Omega_1+k_\parallel v_\parallel-\omega)\end{aligned} \quad (10.86)$$

where $\langle|E^2|(\mathbf{k},\omega)\rangle_2$ denotes that part of the fluctuation spectrum produced by the dynamically screened particles 2.

For a Maxwellian plasma, we may average Eq. (10.86) with respect to a Maxwellian distribution of particles 1; we obtain

$$\begin{aligned}\frac{\langle(\Delta X_1)^2\rangle}{\tau} = \frac{\pi c^2}{B^2V^2}\left(\frac{m_1}{2\pi T_1}\right)^{1/2}\sum_{\mathbf{k}}\sum_{n}\int_{-\infty}^{\infty}d\omega\langle|E^2|(\mathbf{k},\omega)\rangle_2\frac{k_\perp^2}{|k_\parallel|k^2}\\ \times\Lambda_n(\beta_1)\exp\left[-\frac{(n\Omega_1-\omega)^2}{2k_\parallel^2(T_1/m_1)}\right]\end{aligned} \quad (10.87)$$

where β and $\Lambda_n(\beta)$ have been defined by Eqs. (4.69) and (4.70). The fluctuation spectrum of the electric field may be written as

$$\langle |E^2|(\mathbf{k},\omega)\rangle_2 = \frac{16\pi^2 q_2^2}{k^2|\epsilon(\mathbf{k},\omega)|^2} S_2^{(0)}(\mathbf{k},\omega) \tag{10.88}$$

where

$$S_2^{(0)}(\mathbf{k},\omega) = \frac{N_2 V}{(2\pi T_2/m_2)^{1/2}|k_\parallel|} \sum_n \Lambda_n(\beta_2) \exp\left[-\frac{(n\Omega_2-\omega)^2}{2k_\parallel^2(T_2/m_2)} \right] \tag{10.89}$$

and $\epsilon(\mathbf{k},\omega)$ is the dielectric function given by Eq. (10.62); N_2 in Eq. (10.89) may contain a spatial variation. Equation (10.88) is clearly an adequate approximation for the case when particles 2 are ions with the Larmor radius larger than the Debye length.

C. Calculation of the First Moment

As may be clear from Eq. (10.84), the first moment $\langle \Delta X_1 \rangle$ is proportional to the average acceleration of a particle 1 in the y direction due to scattering in the fields produced by the particles 2. As we have remarked earlier, such an acceleration arises from two separate sources: one due to the field fluctuations and another due to the polarization. These may be calculated, respectively, as

$$\frac{\langle \Delta X_1 \rangle_f}{\tau} = i\frac{c^2}{2B^2V^2} \sum_{\mathbf{k}} \int_{-\infty}^{\infty} d\omega \int_{-\infty}^{\infty} dt \frac{k_y}{k^2} \mathbf{k}\cdot\mathbf{H}(t)\cdot\mathbf{k}\langle |E^2|(\mathbf{k},\omega)\rangle_2 \exp\{ i[\mathbf{k}\cdot\Delta\mathbf{r}(t)-\omega t]\} \tag{10.90}$$

$$\frac{\langle \Delta X_1 \rangle_p}{\tau} = -\frac{2q_1 c}{BV} \sum_{\mathbf{k}} \int_{-\infty}^{\infty} d\omega \int_{-\infty}^{\infty} dt \frac{k_y}{k^2} \frac{\mathrm{Im}\chi_2(\mathbf{k},\omega)}{|\epsilon(\mathbf{k},\omega)|^2} \exp\{ i[\mathbf{k}\cdot\Delta\mathbf{r}(t)-\omega t]\} \tag{10.91}$$

where $\mathrm{Im}\chi_2(\mathbf{k},\omega)$ denotes the imaginary part of that portion of $\epsilon(\mathbf{k},\omega)$ which arises from the polarizability of the particles 2. For a Maxwellian plasma,

$$\mathrm{Im}\chi_2(\mathbf{k},\omega) = \frac{4\pi^2 q_2^2 \omega}{T_2 k^2 V} S_2^{(0)}(\mathbf{k},\omega) \tag{10.92}$$

where $S_2^{(0)}(\mathbf{k},\omega)$ is given by Eq. (10.89).

For a homogeneous system, Eqs. (10.90) and (10.91) vanish identically because of symmetry. The leading terms should therefore be proportional to the density gradient $\partial N_2/\partial X$ of the particles 2. Physically, such a density gradient acts in two ways to produce finite contributions in $\langle \Delta X_1 \rangle$: First, it gives rise to a net drift motion of the particles 2 in the y direction with the velocity [see Eq. (8.1)]

$$v_2 = \frac{T_2}{m_2 \Omega_2} \frac{1}{N_2} \frac{\partial N_2}{\partial X}. \tag{10.93}$$

In the frame of reference comoving with the average motion of the particles 2, the test particle 1 acquires a net drift velocity $-v_2$ in the y direction superposed on its own cyclotron motion; it thereby suffers a frictional force and produces $\langle \Delta X_1 \rangle$. Second, the particle 1 gyrates with a finite Larmor radius; in the presence of $\partial N_2/\partial X$, it then perceives different densities of the particles 2 at one side of its cyclotron orbit from the other. The frictional forces at the two opposite sides of the orbit do not exactly cancel each other. The resulting net frictional force points in the y direction and thus contributes to $\langle \Delta X_1 \rangle$.

Mathematically, the former processes may be taken into consideration by substituting

$$\Delta \mathbf{r}(t) = \Delta \mathbf{r}_0(t) - \hat{y} v_2 t$$

in Eqs. (10.90) and (10.91). We may then expand the resulting expressions with respect to v_2, and retain the terms linear in v_2; we are interested only in the linear contributions of the density gradient. The latter may similarly be taken into account through the expansion,

$$N_2(X) = N_2 - \frac{v_y}{\Omega_1} \frac{\partial N_2}{\partial X}.$$

Equation (10.90) thus becomes

$$\frac{\langle \Delta X_1 \rangle_f}{\tau} = -i \frac{c^2}{2B^2 V^2} \sum_{\mathbf{k}} \int_{-\infty}^{\infty} d\omega \int_{-\infty}^{\infty} dt \frac{k_y}{k^2} \left[i\mathbf{k} \cdot \frac{\partial \Delta \mathbf{r}_0(t)}{\partial \mathbf{v}} \cdot \hat{y} v_2 + \frac{v_y}{\Omega_1} \frac{1}{N_2} \frac{\partial N_2}{\partial X} \right]$$

$$\times \mathbf{k} \cdot \mathsf{H}(t) \cdot \mathbf{k} \langle |E^2| (\mathbf{k}, \omega) \rangle_2 \exp\{ i[\mathbf{k} \cdot \Delta \mathbf{r}_0(t) - \omega t] \}. \tag{10.94}$$

Noticing that $\partial \Delta \mathbf{r}_0(t)/\partial \mathbf{v} = \mathsf{H}(t)/\Omega_1$, we carry out the t integration. We then average Eq. (10.94) with respect to a Maxwellian distribution for the particle

1; after a lengthy calculation, we obtain

$$\frac{\langle \Delta X_1 \rangle_f}{\tau} = \frac{\pi c^2}{2B^2V^2}\frac{1}{N_2}\frac{\partial N_2}{\partial X}\left(1+\frac{q_1T_2}{q_2T_1}\right)\left(\frac{m_1}{2\pi T_1}\right)^{1/2}\sum_{\mathbf{k}}\sum_n\int_{-\infty}^{\infty}d\omega\langle|E^2|(\mathbf{k},\omega)\rangle_2\,\frac{1}{|k_\parallel|k^2}$$
$$\times\left\{\frac{k_\perp^2}{2}[\Lambda_{n-1}(\beta_1)+\Lambda_{n+1}(\beta_1)]-\frac{\omega n\Omega_1}{(T_1/m_1)}\Lambda_n(\beta_1)\right\}\exp\left[-\frac{(n\Omega_1-\omega)^2}{2k_\parallel^2(T_1/m_1)}\right]. \tag{10.95}$$

Similarly, we calculate Eq. (10.91) to find

$$\frac{\langle \Delta X_1 \rangle_P}{\tau} = \frac{\pi c^2}{2B^2V^2}\frac{1}{N_2}\frac{\partial N_2}{\partial X}\left(1+\frac{q_1T_2}{q_2T_1}\right)\left(\frac{m_1}{2\pi T_1}\right)^{1/2}\sum_{\mathbf{k}}\sum_n\int_{-\infty}^{\infty}d\omega\langle|E^2|(\mathbf{k},\omega)\rangle_2$$
$$\times\frac{\omega n\Omega_1}{|k_\parallel|k^2(T_2/m_1)}\Lambda_n(\beta_1)\exp\left[-\frac{(n\Omega_1-\omega)^2}{2k_\parallel^2(T_1/m_1)}\right] \tag{10.96}$$

where we have used relations (10.88) and (10.92). Finally, collecting Eqs. (10.95) and (10.96), we obtain

$$\frac{\langle \Delta X_1 \rangle}{\tau} = \frac{\pi c^2}{2B^2V^2}\frac{1}{N_2}\frac{\partial N_2}{\partial X}\left(1+\frac{q_1T_2}{q_2T_1}\right)\left(\frac{m_1}{2\pi T_1}\right)^{1/2}\sum_{\mathbf{k}}\sum_n\int_{-\infty}^{\infty}d\omega\langle|E^2|(\mathbf{k},\omega)\rangle_2$$
$$\times\frac{1}{|k_\parallel|k^2}\left\{\frac{k_\perp^2}{2}[\Lambda_{n-1}(\beta_1)+\Lambda_{n+1}(\beta_1)]-m_1\left(\frac{1}{T_1}-\frac{1}{T_2}\right)\omega n\Omega_1\Lambda_n(\beta_1)\right\}$$
$$\times\exp\left[-\frac{(n\Omega_1-\omega)^2}{2k_\parallel^2(T_1/m_1)}\right]. \tag{10.97}$$

D. Diffusion Coefficient

A complete expression of the diffusion equation may be obtained by substituting Eqs. (10.87) and (10.97) into Eq. (10.82). In many practical cases of ambipolar diffusion, we can reasonably assume that $N_1(X)=N_2(X)\equiv N(X)$. In these circumstances, the diffusion equation simplifies substantially; we may then define a diffusion coefficient $D_\perp$ across the magnetic field through the equation

$$\Gamma_1 = -D_\perp\frac{\partial N}{\partial X}. \tag{10.98}$$

Equivalently, we shall find it useful to introduce an effective collision frequency ν_D pertaining to the diffusion process by

$$\nu_D \equiv (m_1/T_1)\Omega_1^2 D_\perp . \tag{10.99}$$

This expression finds its analogy in the problem of one-dimensional random walk [1] in which a particle suffers displacements along a straight line with an average step of $(2T_1/m_1\Omega_1^2)^{1/2}$, an average Larmor radius. With the aid of Eqs. (10.82), (10.87), (10.97), and (10.98), we obtain

$$\nu_D = \left(\frac{\pi}{2}\right)^{1/2} \frac{q_1^2}{m_1^2 V^2} \left(\frac{m_1}{T_1}\right)^{3/2} \sum_{\mathbf{k}} \sum_n \int_{-\infty}^{\infty} d\omega \langle |E^2|(\mathbf{k},\omega)\rangle_2 \frac{k_\perp^2}{|k_\parallel| k^2}$$
$$\times \left[\Lambda_n(\beta_1) - \frac{1}{2}\left(1 + \frac{q_1 T_2}{q_2 T_1}\right) \left\{ \frac{1}{2}[\Lambda_{n-1}(\beta_1) + \Lambda_{n+1}(\beta_1)] \right.\right.$$
$$\left.\left. - m_1\left(\frac{1}{T_1} - \frac{1}{T_2}\right) \frac{\omega n \Omega_1}{k_\perp^2} \Lambda_n(\beta_1) \right\} \right] \exp\left[-\frac{(n\Omega_1 - \omega)^2}{2k_\parallel^2 (T_1/m_1)} \right]. \tag{10.100}$$

E. Classical Diffusion

If the effects of the magnetic field are neglected in the calculation of the collision frequency, the treatment above should correspond to a classical theory of plasma diffusion. We thus consider the limit of $B \to 0$ in Eq. (10.100) and go over to the continuum for the n summation as prescribed by Eq. (4.73). The result of such a calculation is

$$\nu_D^{(0)} = \frac{8(2\pi)^{1/2} N q_1^2 q_2^2}{3 m_1 m_2 (T_1/m_1 + T_2/m_2)^{3/2}} \ln\Lambda \left\{ \left(\frac{m_2}{m_1}\right)\left(1 + \frac{m_1 T_2}{m_2 T_1}\right) - \frac{m_1 + m_2}{2m_1}\left(1 + \frac{q_1 T_2}{q_2 T_1}\right) \right\} \tag{10.101}$$

where $\ln\Lambda$ is the Coulomb logarithm. Note that $\nu_D^{(0)} = 0$ when $1 = 2$; the collisions between like particles do not contribute to the classical diffusion processes.

In the cases where the particles 1 are the electrons and the particles 2 the ions, we may take $q_i = -q_e = e$ for charge neutrality and assume $m_i \gg m_e$. Equation (10.101) then becomes

$$\nu_D^{(0)} = \frac{4(2\pi)^{1/2} N e^4}{3 m_e^2 (T_e/m_e + T_i/m_i)^{3/2}} \left(1 + \frac{T_i}{T_e}\right) \ln\Lambda. \tag{10.102}$$

F. Anomalous Term

As we remarked in Section 10.4C, an anomalous contribution of the magnetic field to the collisional processes exists even for a plasma in thermodynamic equilibrium. Thus, we wish to single out such a term in Eq. (10.100), and study the physical consequences arising from it.

For the sake of definiteness, we adopt the situation as described in the last paragraph of the previous section, that is, 1=e and 2=i. Following the arguments based on (10.67), we may first of all write

$$\langle |E^2|(\mathbf{k},\omega)\rangle_i = -V\frac{4T_i}{\omega}\operatorname{Im}[\epsilon(\mathbf{k},\omega)]^{-1}. \tag{10.103}$$

With the aid of the Kramers–Kronig relations as applied to the imaginary part of $1/\epsilon(\mathbf{k},\omega)$, we may then carry out the ω integration; transforming the $\mathbf{k}$ summation into integration via Eq. (10.71), we find that Eq. (10.100) reduces

$$\begin{aligned}\nu_D = {} & \frac{4(2\pi)^{1/2}Ne^4}{m_e^{1/2}T_e^{3/2}}\left(1+\frac{T_i}{T_e}\right)\int_0^\infty dk_\perp \int_{k_\perp(m_eT_i/m_iT_e)^{1/2}}^\infty dk_\parallel \frac{k_\perp^3}{k_\parallel k^2(k^2+k_D^2)} \\ & \times\left\{\Lambda_0(\beta_e)-\frac{1}{2}\left(1-\frac{T_i}{T_e}\right)\Lambda_1(\beta_e)\right\} \\ & + \frac{4(2\pi)^{1/2}Ne^4}{m_e^{1/2}T_e^{3/2}}\left(1+\frac{T_i}{T_e}\right)\sum_n{}' \int_0^\infty dk_\perp \int_0^\infty dk_\parallel \frac{k_\perp^3}{k_\parallel k^2(k^2+k_D^2)} \\ & \times\left\{\Lambda_n(\beta_e)-\frac{1}{4}\left(1-\frac{T_i}{T_e}\right)[\Lambda_{n-1}(\beta_e)+\Lambda_{n+1}(\beta_e)]\right\}\exp\left[-\frac{n^2\Omega_e^2}{2k_\parallel^2(T_e/m_e)}\right]\end{aligned} \tag{10.104}$$

where

$$k_D^2 \equiv k_e^2 + k_i^2 \tag{10.105}$$

and the prime on the n summation means omission of the term with $n=0$ in the summation. The second term on the right-hand side of Eq. (10.104) reduces to Eq. (10.102) with $T_e/m_e \gg T_i/m_i$, when the continuum limit is applied to the n summation; such a procedure may be justified for the domain $\beta_e \gg 1$, which contributes the bulk of the effect.

The first term of Eq. (10.104), on the other hand, contains a singular term which diverges logarithmically as m_e/m_i approaches zero. Such an anomalous contribution $\nu_{\rm D}^*$ may be calculated in much the same way as we obtained Eq. (10.73); we find

$$\nu_{\rm D}^* = \frac{(2\pi)^{1/2} N e^4}{m_{\rm e}^{1/2} T_{\rm e}^{3/2}} \left(1 + \frac{T_{\rm i}}{T_{\rm e}}\right) \ln\left(\frac{m_{\rm i}}{m_{\rm e}}\right)$$

$$\times \left\{ \exp(\alpha^*) K_0(\alpha^*) - \frac{1}{2}\left(1 - \frac{T_{\rm i}}{T_{\rm e}}\right)\left[\exp(\alpha^*) K_1(\alpha^*) - \frac{1}{\alpha^*}\right]\right\} \qquad (10.106)$$

where K_0 and K_1 are the modified Bessel functions of the second kind, and

$$\alpha^* \equiv k_{\rm D}^2 \left(\frac{T_{\rm e}}{m_{\rm e} \Omega_{\rm e}^2}\right) = \left(1 + \frac{T_{\rm e}}{T_{\rm i}}\right) \frac{\omega_{\rm e}^2}{\Omega_{\rm e}^2}. \qquad (10.107)$$

While the ratio $|\omega_{\rm e}/\Omega_{\rm e}|$ may take on a value much smaller than unity for a plasma with a strong magnetic field, we note that such may not always be the case with α^* in Eq. (10.107) because of the factor $(1 + T_{\rm e}/T_{\rm i})$. In fact, the ratio $|\omega_{\rm e}/\Omega_{\rm e}|$ takes on values of the order of unity for certain Tokamak and Stellarator experiments. Hence, as long as $T_{\rm e} \gg T_{\rm i}$, the parameter α^* may still have to be regarded as a quantity greater than unity even in those cases of strong magnetic fields.

The anomalous part $D_\perp^*$ of the diffusion coefficient pertaining to the collisional processes of Eq. (10.106) may be calculated with the aid of Eq. (10.99). When $\alpha^* \gg 1$, we may expand $K_0(\alpha^*)$ and $K_1(\alpha^*)$ in Eq. (10.106) in asymptotic series and keep only the leading terms; we thus obtain

$$D_\perp^* = \frac{e^2 k_{\rm e}}{8 m_{\rm e} |\Omega_{\rm e}|} \left(\frac{T_{\rm i}}{T_{\rm e}}\right)^{1/2} \left(1 + \frac{T_{\rm i}}{T_{\rm e}}\right)^{3/2} \ln\left(\frac{m_{\rm i}}{m_{\rm e}}\right) \qquad (\alpha^* \gg 1). \qquad (10.108)$$

Note that this diffusion coefficient is proportional to B^{-1}; later we shall investigate a possible connection between Eq. (10.108) and the Bohm diffusion coefficient [14, 15] in terms of enhanced fluctuations above a thermal level in a turbulent plasma.

The other limiting cases of $\alpha^* \ll 1$ may be similarly obtained by series expansions of $K_0(\alpha^*)$ and $K_1(\alpha^*)$ for small α^*; we find

$$D_\perp^* = \frac{(2\pi)^{1/2} N e^4}{m_{\rm e}^{3/2} T_{\rm e}^{1/2} \Omega_{\rm e}^2} \left(1 + \frac{T_{\rm i}}{T_{\rm e}}\right) \ln\left(\frac{m_{\rm i}}{m_{\rm e}}\right) \ln\left[\frac{\Omega_{\rm e}^2}{\omega_{\rm e}^2} \frac{T_{\rm i}}{(T_{\rm e} + T_{\rm i})}\right] \qquad (\alpha^* \ll 1).$$

$$(10.109)$$

This diffusion coefficient, which is valid only with a very strong magnetic field as long as $T_e \gg T_i$, is therefore proportional to $B^{-2}\ln B$.

The results, Eqs. (10.106), (10.108), and (10.109), which are anomalous in the sense that they exhibit different dependence on (m_e/m_i) and B from the corresponding classical evaluations, do not by themselves represent a substantial enhancement of transport, however, since they are calculated under the conditions of thermodynamic equilibrium. From Eqs. (10.102) and (10.106), we may compare

$$\frac{\nu_D^*}{\nu_D^{(0)}} = \frac{3\ln(m_i/m_e)}{4\ln\Lambda}\left\{\exp(\alpha^*)K_0(\alpha^*) - \frac{1}{2}\left(1-\frac{T_i}{T_e}\right)\left[\exp(\alpha^*)K_1(\alpha^*) - \frac{1}{\alpha^*}\right]\right\}. \tag{10.110}$$

For a deuterium plasma with $\alpha^* = 1$, $\ln\Lambda = 15$, and $T_e \gg T_i$, we compute this ratio to be 0.34; when α^* is decreased to 0.1, this ratio becomes 0.92.

We wish to emphasize that those anomalous terms arise as a consequence of electron fluctuations in the low-frequency domain such that $\omega < |\Omega_e|$, as may be clearly noted from the decomposition of Eq. (10.104). In many cases of turbulent plasmas, we expect that such low-frequency fluctuations are especially enhanced above the thermal levels. In these circumstances, those anomalous terms should grow accordingly, resulting in an enhanced diffusion; $\nu_D^{(0)}$ is not related to such an enhancement, however.

References

1. S. Chandrasekhar, *Revs. Mod. Phys.* **15**, 1 (1943).
2. S. Ichimaru and M. N. Rosenbluth, *Phys. Fluids* **13**, 2778 (1970).
3. M. N. Rosenbluth, W. M. MacDonald, and D. L. Judd, *Phys. Rev.* **107**, 1 (1957).
4. S. Chandrasekhar, *Principles of Stellar Dynamics* (University of Chicago Press, Chicago, Illinois, 1942).
5. W. B. Thompson and J. Hubbard, *Revs. Mod. Phys.* **32**, 714 (1960); J. Hubbard, *Proc. Roy. Soc. (London)* **A260**, 114 (1961); W. B. Thompson, *An Introduction to Plasma Physics* (Addison-Wesley, Reading, Mass., 1964).
6. L. Spitzer, Jr., *Physics of Fully Ionized Gases*, 2nd ed. (Wiley (Interscience), New York, 1962).
7. T. Kihara and Y. Midzuno, *Revs. Mod. Phys.* **32**, 722 (1960).
8. Y. Itikawa, *J. Phys. Soc. Japan* **19**, 748 (1964).
9. V. P. Silin, in *1968 Tokyo Summer Lectures in Theoretical Physics: Statistical Physics of Charged Particle Systems* (R. Kubo and T. Kihara, eds.), p. 13 (Syokabo, Tokyo, and W. A. Benjamin, New York, 1969).
10. S. Ichimaru and M. N. Rosenbluth, in *Plasma Physics and Controlled Nuclear Fusion Research*, Vol. II, p. 373 (IAEA, Vienna, 1971).
11. C. L. Longmire and M. N. Rosenbluth, *Phys. Rev.* **103**, 507 (1956).
12. L. Spitzer, Jr., *Phys. Fluids* **3**, 659 (1960).
13. J. B. Taylor, *Phys. Fluids* **4**, 1142 (1961).
14. F. C. Hoh, *Revs. Mod. Phys.* **34**, 267 (1962).

15. D. Bohm, in *The Characteristics of Electrical Discharges in Magnetic Fields* (A. Guthrie and R. K. Wakerling, eds.), Chapter 2, Section 5 (McGraw-Hill, New York, 1949).

Problems

10.1. Consider the events of Coulomb scattering as illustrated in Fig. 1.1; $\Delta\mathbf{v}$ denotes the velocity increment before and after the scattering. Calculate the quantities

$$\{\Delta\mathbf{v}\} \equiv \int \Delta\mathbf{v}\, dQ \qquad \text{and} \qquad \{\Delta\mathbf{v}\Delta\mathbf{v}\} \equiv \int (\Delta\mathbf{v}\Delta\mathbf{v})\, dQ$$

for the cross section (1.2), and show that these contain logarithmically divergent terms as the lower limit of the χ integration approaches zero. Also show that

$$\{\Delta\mathbf{v}\Delta\mathbf{v}\Delta\mathbf{v}\} \equiv \int (\Delta\mathbf{v}\Delta\mathbf{v}\Delta\mathbf{v})\, dQ$$

does not involve such a divergent term.

10.2. Show that Eq. (10.38) reduces to the Balescu–Lenard collision term (2.44).

10.3. Show the intermediate steps in obtaining Eq. (10.60) through the fluctuation-dissipation theorem.

10.4. Derive Eqs. (10.63) and (10.64).

10.5. Calculate Eq. (10.65) in the limit of $B \rightarrow 0$ and obtain Eq. (10.66).

CHAPTER 11

PLASMA TURBULENCE

In Chapter 9, we studied the theory of fluctuations for stationary plasmas near the state of thermodynamic equilibrium. In such a plasma, the density fluctuations associated with the collective modes are at the level of thermal fluctuations; the plasma may be said to be in a quiescent state. The motion of single particles is not affected appreciably by the presence of such weak fluctuations; the transport processes are described through a consideration of standard Coulomb collisions. The plasma parameter $g \equiv 1/n\lambda_D^3$ provides a measure of the ratio between the fluctuation energy and the kinetic energy in these circumstances. The Bogoliubov–Born–Green–Kirkwood–Yvon (BBGKY) hierarchy may be solved for such a quiescent plasma by means of a systematic expansion in powers of the plasma parameter.

In most cases, the plasmas found in nature are significantly far from thermodynamic equilibrium. A collective mode may even become unstable in some cases; the fluctuations are then greatly enhanced over the thermal level. The plasma may eventually go over to a state of turbulence. The enhanced fluctuations are expected to affect drastically the rates of relaxation processes; the so-called anomalous transport of physical quantities may take place in such a plasma. The resulting state may be either weakly or strongly turbulent, depending on the deviation of the plasma from equilibrium. Clearly, the BBGKY hierarchy should be able to produce solutions applicable to any of the three plasma states, namely, the quiescent, weakly turbulent, and strongly turbulent states, although the truncation scheme of a quiescent plasma is no longer applicable to those plasmas in a turbulent state. It is the purpose of the present chapter to elucidate some of the fundamental problems associated with the theory of plasma turbulence.

11.1 INTRODUCTORY SURVEY

In a theoretical treatment of plasma turbulence, it may be significant to distinguish the following two classes of the problems involved. One is what may be called an *initial value problem*. Here, the plasma is isolated from the external energy source; initially, however, it is characterized by physical conditions such that a certain collective mode can grow exponentially.

Experimentally, a situation corresponding to this case may, for example, be realized in the initial stage of a beam–plasma interaction experiment when a pulsed beam of charged particles is injected into a quiescent plasma. The theoretical problem then is to analyze the time development, or approach to equilibrium, of the combined system of particles and oscillations, starting from such unstable initial conditions. Since the amount of the extra free energy available in the initial state to drive the plasma-wave instability is limited, the resulting plasma is most likely to stay in a weakly turbulent state. The advent of the quasilinear theory of plasma oscillations has marked an important step toward the solution of the problems associated with weakly turbulent plasmas [1–9]. This approach takes explicit account of the feedback action of the growing oscillations upon the single-particle distribution function and treats the wave–wave interaction in a perturbation-theoretical way.

Many significant contributions have been made toward the understanding of nonlinear interactions between various collective modes in plasmas [10–14]. The treatments involved here are basically perturbation-theoretical. Iterative solutions of equations such as Eqs. (9.105) and (9.106) provide us with a description of the nonlinear coupling processes, including both the longitudinal and the transverse modes, in a plasma.

Consider now a second class of turbulence problems, which arise when a system maintains connection with an external source and a sink of energy. In these circumstances, if we wait long enough, the plasma may reach a new kind of stationary state: a *turbulent stationary state*. Experimental examples pertaining to this case may be found in various plasma phenomena, both in the laboratory and in an actual astronomical setting. The plasma may be either strongly or weakly turbulent, depending on, among other things, the ability of the external source to feed energy to the plasma turbulence. In either of these states, a steady flow of energy is established through interactions among particles and oscillations. We may then ask: What is the characteristic feature of such turbulent stationary state, particularly if the state is characterized by those physical parameters for which a conventional linear theory predicts an exponential growth of oscillations, or instability?

The general structure of the energy flow pattern established in such a turbulent stationary plasma may be understood in the following way. First of all, we consider a plasma which is kept in so-called unstable physical conditions by a certain external means (e.g., by application of a constant electric field above its critical value); the external source thus feeds energy constantly into the charged particles. A major part of this energy goes to the excitation of oscillations via the wave–particle coupling mechanism which causes the instability; the remainder will be lost to the environment through collisions. The large-amplitude oscillations thus built up in the plasma interact frequently with each other, so that a continuous flow of energy toward the large wave-number region may be established. The oscillation energy is eventually dissipated into heat by collisional damping when the wave number steps out of the domain of instability. A stationary state may thereby be set up in the plasma and a continuous flow of energy is maintained in it.

Existence of frequent interactions among turbulent fluctuations may also make it possible to consider a new mode of wave propagation of second-sound type [15–17]. The propagation of energy-density waves associated with the plasma turbulence may take place when the growing waves collide more frequently with each other than with any other constituents of the plasma.

A quantity of primary interest in the theory of turbulence is, therefore, the energy spectrum $\mathcal{E}_{\mathbf{k}}$ contained in the wave vector $\mathbf{k}$ mode of the fluctuations. For a plasma in thermodynamic equilibrium at temperature T, $\mathcal{E}_{\mathbf{k}}$ in a given collective mode takes on an equipartition value T, as Eq. (9.104a) illustrates. Generally, in a quiescent plasma, $\mathcal{E}_{\mathbf{k}}$ is written in a form proportional to the discreteness parameters of Section 1.3A. We may portray the fluctuations in these circumstances as arising essentially from the random motion of the discrete particles; hence, $\mathcal{E}_{\mathbf{k}}$ vanishes in the fluid limit. A strongly turbulent state of a plasma may be defined as that in which *the energy spectrum of the fluctuations remains finite even in the fluid limit*; the correlations persist in this state with macroscopic intensities. It therefore becomes essential for a theory of plasma turbulence to deal directly with a state with strong correlations, rather than to start from a conventional Vlasov description of the plasma state which would ignore the existence of the correlations.

A number of attempts have been made to formulate a theory of strongly turbulent plasmas. Dupree advanced a perturbation theory which is based on the use of a statistical set of exact particle orbits instead of the unperturbed orbits conventionally used in a solution of the Vlasov equation [18]. It is then argued that the dominant nonlinear effect of low-frequency instabilities is an incoherent scattering of particle orbits by waves, which causes particle diffusion and appears in the theory as an enhanced viscosity. Weinstock and Williams have developed a theory of strong plasma turbulence along a similar line of arguments [19].

In spite of these and many other important developments in the theory of plasma turbulence [20] the present state of our knowledge in these fields is far from satisfactory. In fact, studies of plasma turbulence, theoretical and experimental, are regarded as one of the major frontiers in plasma physics. Both the fundamental theories of plasma turbulence and their applications to specific turbulence problems need to be further developed. Reflecting such a state of art, the presentation in the balance of this chapter will be devoted to those subjects which the author feels to be the most important aspects of plasma turbulence. We stress the qualitative difference existing between the plasmas in quiescent and turbulent states, which cannot be bridged by a simple perturbation-theoretical technique. The theoretical framework of plasma turbulence which we shall describe thereby resembles, to a certain extent, that of a second-order phase transition in statistical mechanics.

11.2 THE SELF-CONSISTENT APPROACH

As we remarked earlier, a turbulent state of a plasma is characterized by the existence of fluctuations and correlations with macroscopic intensities. It thus becomes important to take account of the existence of the finite correlations or fluctuations in the stationary state of the plasma at the very beginning of our theoretical investigation. Yet, at the onset we do not know much about the nature of the fluctuations for a given turbulent plasma; indeed, it is rather a *final* objective of the theory to determine the spectral function of such fluctuations. We are thereby led to adopt a nonperturbative self-consistent approach to investigate the nature of a turbulent state of a plasma. Let us, therefore, consider the contents of this approach [21].

For a theoretical study of the stability of a plasma, a first step that one naturally takes is to specify the unperturbed stationary state of the plasma. In a kinetic-theoretical treatment of a homogeneous plasma, this is usually accomplished by specifying single-particle distribution functions. One then applies a weak external disturbance and studies the characteristic response of the system. If a certain growing disturbance is found possible in the system, the plasma is said to be unstable against that particular kind of disturbance. If the plasma is kept in that unstable state by a suitable external means, this might literally imply that the plasma would collapse by endless development of such disturbances in itself.

In reality, however, we encounter various examples of plasmas in which a stationary state is maintained even in so-called unstable circumstances. Although such a plasma is generally "noisy" and frequently accompanied by anomalous transport phenomena, we must regard such a state as *stable* because it is realized and sustained in a stationary way. We therefore seem to be faced with a gap between the physical reality and what a theory would tell about its stability.

In order to bridge this gap, it may be instructive to recall the physical significance of the stability analysis: It is clear that all that a stability analysis can tell us is whether or not the particular state *originally specified* is stable against external disturbances; if a different state is chosen, a different stability criterion will result. The seeming discrepancy between the physical reality and a theoretical analysis may therefore be traced simply to the inappropriateness of our original selection of the stationary state; a better choice would lead to a stable description of the plasma, in accord with actual physical observations.

How do we then find a new stationary state which should be appropriate to describe the true situation? A clue to this problem is already apparent if we look more closely at what the theory of critical fluctuations in Section 9.3E indicates. In this analysis we started with the given velocity distribution functions to characterize the state of the plasma; fluctuations, or space–time correlations between the physical variables, are implicitly neglected in this description of the plasma state. When the plasma is in thermodynamic equilibrium, we can legitimately disregard the effects of fluctuations on the properties of the system, as we have remarked; in these circumstances a stability analysis which ignores the presence of fluctuations in the stationary state can be well justified. However, when the plasma approaches from the region of stability a critical point corresponding to the onset of an instability, there occurs an enormous increase of fluctuations above the thermal level; the amplitude of the fluctuations would seem to diverge to infinity at the critical point [see Eq. (9.76)]. It is in this divergent behavior of density fluctuations that we sense a danger signal which points to the inadequacy of our original specification of the stationary state by means of the single-particle distribution functions only. The tremendous enhancement of the fluctuations should be regarded as a signal which demands that we take proper account of the existence of fluctuations to describe correctly the properties of the plasma in the vicinity of the critical point and in the turbulent region.

Based upon the considerations described in the foregoing paragraphs, let us now introduce a fundamental assumption: To the extent that a stationary plasma may be realized in reality, it should be possible to determine a state which is *stable* against weak external perturbations even though the plasma be *turbulent*. With the aid of this assumption, we may then establish the following self-consistent scheme for calculating the spectral functions of fluctuations in a turbulent plasma: (1) Assume a stationary state of a uniform turbulent plasma which may be characterized by the existence of finite fluctuation spectra superposed on an ordinary quiescent stationary state; the amplitude of the spectral function $\mathcal{E}_{\mathbf{k}}$ is left undetermined at this stage. (2) Apply a weak perturbation to this turbulent state, and calculate the various linear response functions. (3) Make use of the dielectric superposition principle for a nonequilibrium plasma (Section 9.3C) to write the fluctuation spectrum in terms of these response functions. (4) The response functions in turn contain the spectral functions of fluctuations; the final step is to solve the resulting self-consistent equation for $\mathcal{E}_{\mathbf{k}}$.

We remark that the present approach is nonperturbative in its nature; in this scheme, we can pass continuously from a stable region to a turbulent region. In the stable region, the $\mathcal{E}_{\mathbf{k}}$ thus calculated should turn out to be so small in magnitude (i.e., $\mathcal{E}_{\mathbf{k}} \sim g$) that the resulting corrections are negligible;

the fluctuation spectrum will be essentially equal to that obtained from a quiescent calculation. In the turbulent region, the stationary state will be characterized by macroscopic intensities of fluctuations (i.e., $\mathcal{E}_{\mathbf{k}} \sim g^0$) associated with certain modes of oscillations.

The complexity arising from the plasma turbulence greatly affects the calculation of the dielectric response function involved in Step 2 of the foregoing self-consistent approach. In short, a scheme of perturbation calculations, starting from a quiescent description of the plasma and retaining only a finite number of perturbation terms with respect to the turbulence spectrum, does not provide even qualitatively a correct description of a plasma response in a turbulent state. For a turbulent plasma, an external test charge introduced in the system will induce fluctuations not only from the average quiescent background but also from the turbulent fluctuations, owing to the nonlinear coupling. The induced fluctuations then couple again with the background or the turbulence; such polarization processes thus proceed endlessly. The central problem involved in the calculation of the dielectric response function is therefore to find a way to take account of such infinite series of higher-order interaction processes in as meaningful a way as possible.

A solution to this problem is offered by treating the quiescent background and the turbulent fluctuations as a single entity, namely, a turbulent stationary state. The dielectric response function is then calculated with respect to this actual state rather than to a quiescent state. We are thereby led to the self-consistent scheme described earlier. A result of such a calculation indeed shows that this method enables us to sum an important subset of all the higher-order interaction processes and to obtain a dielectric response function for a turbulent plasma [21].

It is also of interest to note a close connection between this self-consistent approach and a recent development in the theory of electron gas at metallic densities [22]. Here, a standard random-phase approximation (RPA) is known to predict physically unacceptable (i.e., negative) values for the electron correlation function at short distances; a classical counterpart to this effect has been noted in connection with Eq. (9.32). To treat the strong correlations existing between the electrons at short distances in a self-consistent fashion, Hubbard [23] and Singwi and his co-workers [24] independently applied an approach similar to the foregoing. The linear response function relevant to the problem is the dielectric response function expressed as a functional of the correlation function; the fluctuation-dissipation theorem provides the relationship between the correlation function and the dielectric response function. The results of their calculations have shown a remarkable improvement over a conventional calculation based on the RPA for short-range electron correlations.

In spite of these successes, a number of shortcomings have been noted in the self-consistent approach above. First, the calculation of response functions in a turbulent plasma always involves some kind of approximation; in many cases, it is not easy to assess the accuracy of the treatment or the extent of the approximation in a systematic way. Second, there exist no exact theories, such as the fluctuation-dissipation theorem, which would relate the response functions to the fluctuation spectrum of a turbulent plasma; the latter plasma is by definition a system remote from thermodynamic equilibrium. The dielectric superposition principle does offer an approximate relationship to be used for this purpose [25]; the accuracy of this principle as applied to a turbulent plasma, however, remains to be critically examined.

In order to obtain an alternative and more systematic approach to the theory of plasma turbulence, we may return to a kinetic-theoretical treatment and attempt to solve the BBGKY hierarchy of plasma kinetic equations for plasma turbulence [26].

11.3 KINETIC-THEORETICAL APPROACH

With the knowledge of the pair correlation function, or the fluctuation spectrum, the first equation of the BBGKY hierarchy describes the behavior of the single-particle distribution function; the correlation function enters in the collision term of the kinetic equation. Turbulence therefore affects the transport processes of the plasma through this equation.

The second equation of the hierarchy determines the pair correlation function. It is in this sense that we regard the second equation as equivalent in physical content to the self-consistent equation for $\mathcal{E}_k$ which we sought to establish in the previous approach. By working directly on the second equation, therefore, we must be able to arrive at the equation for $\mathcal{E}_k$ without going through the intermediate steps.

In order to determine the pair correlation function from the second equation, the ternary correlation function as well as the single-particle distribution function must be known. Conversely, we note that if the latter two functions are known, we can always obtain a formal solution by following the procedure of Section 2.2. The problem thus reduces to that of finding the expression for the ternary correlation function which is appropriate to the turbulent situation under consideration. The first two equations of the BBGKY hierarchy will then provide a set of coupled equations for the single-particle distribution function and the pair correlation function.

For a quiescent plasma, there exists a well-known hierarchy among the correlation functions in terms of various orders in the plasma parameter g.

We can use this fact to truncate the BBGKY hierarchy systematically, and can thus, in principle, determine the ternary correlation function to an arbitrary degree of accuracy in the plasma parameter [27].

In the case of turbulence, the foregoing procedure is no longer applicable; by our definition of the strongly turbulent state, the pair correlation function must be of the same order in the plasma parameter as the single-particle distribution function. A similar statement may also be true for all the higher-order correlation functions. In fact, for the theory of strong plasma turbulence, we are interested in accurately determining only those parts of the many-particle distribution functions which remain finite in the fluid limit. Therefore, we are led to adopt the following guideline for finding an appropriate ternary correlation function: Many-particle distribution functions as well as the single-particle distribution function remain finite in the limit $g \to 0$ and satisfy the BBGKY hierarchy also in this limit.

It has not been possible to find a general expression for the many-particle distribution functions which exactly satisfies this criterion. However, an inspection of the third equation of the hierarchy in the fluid limit reveals that the ternary correlation function constructed according to Kirkwood's superposition approximation [28] and the higher-order correlation functions similarly constructed indeed satisfy the criterion to a good degree of approximation. We, therefore, adopt this approximation; it is the major approximation involved in this kinetic theory of plasma turbulence. The accuracy of the theory can be assessed and improved in a systematic way from the third equation of the hierarchy.

A. Mathematical Formulation

We begin with a consideration of a single component plasma containing N charged particles in a volume V, as specified in Section 2.1. According to Kirkwood's superposition approximation, the ternary correlation function $H(1,2,3)$ defined through Eqs. (2.11) is written as

$$H(1,2,3) = \frac{G(1,2)G(2,3)}{F(2)} + \frac{G(2,3)G(3,1)}{F(3)} + \frac{G(3,1)G(1,2)}{F(1)} + \frac{G(1,2)G(2,3)G(3,1)}{F(1)F(2)F(3)}. \tag{11.1}$$

Substituting this expression into the second equation of the BBGKY hierarchy, we find

$$\left[\frac{\partial}{\partial t}+L(1)+L(2)\right]G(1,2)=\frac{1}{n}\left[\tilde{V}(1,2)+\tilde{V}(2,1)\right]F(1)F(2)$$

$$+\int \tilde{V}(1,3)\left[F(1)G(2,3)+G(1,2)F(3)+\frac{G(1,2)G(2,3)}{F(2)}\right]d3$$

$$+\int \tilde{V}(2,3)\left[F(2)G(1,3)+G(1,2)F(3)+\frac{G(1,2)G(1,3)}{F(1)}\right]d3. \tag{11.2}$$

Here, $\tilde{V}(1,2)$ represents a modified two-particle operator

$$\tilde{V}(1,2)\equiv V(1,2)\left[1+\frac{G(1,2)}{F(1)F(2)}\right]. \tag{11.3}$$

This operator thereby takes account of the extent to which the finite pair correlations in a turbulent plasma may act to modify the effective interaction between two particles.

The kinetic theory of plasma turbulence resulting from a solution of Eq. (11.2) thus contains an important physical feature by taking account of this modification. Let $\Phi(\mathbf{r})$ denote the interaction potential between two particles at a distance $\mathbf{r}$. When the effects of the particle correlations are taken into account, the effective potential may be written as $\bar{\Phi}(\mathbf{r})=\Phi(\mathbf{r})[1+p(\mathbf{r})]$, where $p(\mathbf{r})$ is the pair correlation function defined by Eq. (9.12). Said another way, the ratio between the Fourier transforms of $\bar{\Phi}(\mathbf{r})$ and $\Phi(\mathbf{r})$ is given by

$$t(\mathbf{k})=1+\frac{1}{N}\sum_{\mathbf{q}}\frac{\mathbf{k}\cdot\mathbf{q}}{q^2}S(\mathbf{k}-\mathbf{q}) \tag{11.4}$$

where $S(\mathbf{k})$ is the static form factor (9.13). In deriving Eq. (11.4), we have assumed Coulomb interaction for $\Phi(\mathbf{r})$; general cases are treated in [22]. It is also clear that $t(\mathbf{k})$ is the ratio between the Fourier transforms of $\tilde{V}(1,2)$ and $V(1,2)$. The appearance of this $\mathbf{k}$-dependent factor $t(\mathbf{k})$ represents a major effect of strong correlations in a turbulent plasma. We emphasize that this factor describes the modification of effective particle interactions arising from the presence of turbulence.

A stationary solution of Eq. (11.2) may be obtained in a way similar to those described in Sections 2.4 and 9.3A. It again takes the form [26]

$$S(\mathbf{k})=\frac{1}{N}\int_{-\infty}^{\infty}d\omega\frac{S^{(0)}(\mathbf{k},\omega)}{|\tilde{\epsilon}(\mathbf{k},\omega)|^2} \tag{11.5}$$

where

$$S^{(0)}(\mathbf{k},\omega)=N\int d\mathbf{v}f(\mathbf{v})\delta(\omega-\mathbf{k}\cdot\mathbf{v})$$
$$\times\left\{1+\frac{1}{N}\sum_{\mathbf{q}}\frac{t(\mathbf{q})}{t(\mathbf{k})}\frac{\mathbf{k}\cdot\mathbf{q}}{q^2}S(\mathbf{k}-\mathbf{q})[S(\mathbf{q})-1][1-|\tilde{\epsilon}(\mathbf{k},\omega)|^2]\right\}. \tag{11.6}$$

$$\tilde{\epsilon}(\mathbf{k},\omega)=1+\frac{\omega_p^2 t(\mathbf{k})}{k^2}\int d\mathbf{v}\frac{1}{\omega-\mathbf{k}\cdot\mathbf{v}}\mathbf{k}\cdot\frac{\partial f(\mathbf{v})}{\partial\mathbf{v}}. \tag{11.7}$$

Equation (11.5) thus represents the self-consistent integral equation which should be solved for $S(\mathbf{k})$.

There are a number of important physical consequences which derive directly from this integral equation; some of them are discussed in the following.

B. Stable Turbulent State

As the form of Eq. (11.5) clearly indicates, the zeros of the function $\tilde{\epsilon}(\mathbf{k},\omega)$ on the complex ω plane give rise to a special kind of density-fluctuation excitation, the collective modes. In this sense, $\tilde{\epsilon}(\mathbf{k},\omega)$ may be regarded as an effective dielectric function of the turbulent plasma, although it fails to satisfy the f-sum rule, (C.9). Examination of the equation

$$\tilde{\epsilon}(\mathbf{k},\omega)=0 \tag{11.8}$$

tells us about stability of the collective modes.

Let us first note that the dielectric function (11.7) depends not only on $f(\mathbf{v})$ but also on $S(\mathbf{k})$ through $t(\mathbf{k})$. The solution of Eq. (11.8), written as

$$\omega=\tilde{\omega}_{\mathbf{k}}+i\tilde{\gamma}_{\mathbf{k}}, \tag{11.9}$$

may now be a functional of both the velocity distribution function and the fluctuation spectrum. Starting from a conventionally unstable state of a plasma, thus, we can conceive of two basically different mechanisms by which the plasma is eventually stabilized.

One is the process usually referred to as a quasilinear effect: In an unstable plasma, the coherent excitation of plasma waves by single fast electrons provides a mechanism of feeding energy from the single particles to the collective modes. It is impossible, however, that this kind of situation persists forever unless an external source is provided to feed energy into the individual particles. Depletion of particles with high energies will take place. The resulting change in the shape of the distribution function may then be described by an equation of the Fokker–Planck type [1–9].

$$\frac{\partial f(\mathbf{v})}{\partial t}=\frac{\partial}{\partial \mathbf{v}}\cdot\left[\mathsf{D}(\mathbf{v})\cdot\frac{\partial}{\partial \mathbf{v}}f(\mathbf{v})\right]+\frac{\partial}{\partial \mathbf{v}}\cdot[\mathsf{F}(\mathbf{v})f(\mathbf{v})]. \tag{11.10}$$

Here,

$$\mathsf{D}(\mathbf{v})=\frac{1}{V}\sum_{\mathbf{k}}\frac{8\pi^2e^2}{m^2\omega_{\mathbf{k}}\epsilon'(\mathbf{k},\omega_{\mathbf{k}})}\frac{\mathbf{kk}}{k^2}\mathsf{E}_{\mathbf{k}}\delta(\omega_{\mathbf{k}}-\mathbf{k}\cdot\mathbf{v}), \tag{11.11}$$

$$\mathsf{F}(\mathbf{v})=\frac{1}{V}\sum_{\mathbf{k}}\frac{8\pi^2e^2}{m\epsilon'(\mathbf{k},\omega_{\mathbf{k}})}\frac{\mathbf{k}}{k^2}\delta(\omega_{\mathbf{k}}-\mathbf{k}\cdot\mathbf{v}) \tag{11.12}$$

and the solution

$$\omega=\omega_{\mathbf{k}}+i\gamma_{\mathbf{k}} \tag{11.13}$$

derives from the conventional dispersion relation

$$\epsilon(\mathbf{k},\omega)\equiv 1+\frac{\omega_{\mathrm{p}}^2}{k^2}\int d\mathbf{v}\frac{1}{\omega-\mathbf{k}\cdot\mathbf{v}}\mathbf{k}\cdot\frac{\partial f(\mathbf{v})}{\partial \mathbf{v}}=0. \tag{11.14}$$

In Eqs. (11.11) and (11.12), $\epsilon'(\mathbf{k},\omega_{\mathbf{k}})$ denotes $[\partial\epsilon/\partial\omega]_{\omega=\omega_{\mathbf{k}}}$; the domain of the **k** summation in these equations is confined so that $\gamma_{\mathbf{k}}>0$. A stabilization may thus be achieved when the shape of $f(\mathbf{v})$ changes according to Eq. (11.10) in such a way that the solution (11.13) takes on negative imaginary values, $\gamma_{\mathbf{k}}<0$, over the entire **k** space.

If the situation is such that the shape of the velocity distribution is kept unstable by some external means (e.g., application of an electric field, injection of a beam, etc.), the second mechanism becomes more important. Here, large-amplitude fluctuations or strong correlations are established in the plasma, so that the effective interactions between particles are drastically modified. This change in turn modifies the properties of the collective modes as given by Eq. (11.9). When the velocity distribution has a form such that $\gamma_{\mathbf{k}}$ of Eq. (11.13) takes on positive values over a certain range of the **k** space, the fluctuation spectrum in a turbulent state may adject itself so that $\tilde{\gamma}_{\mathbf{k}}$ may still remain negative over the entire **k** space. In these circumstances, a conventionally unstable particle distribution and large-amplitude fluctuation spectrum can coexist in a stable manner, forming a turbulent stationary state. We shall later consider explicit examples which demonstrate such a stabilizing action of the fluctuation spectrum.

C. Wave Kinetic Equation

Analogous to the combined result of Eqs. (9.102) and (9.103), the energy $\mathcal{E}_{\mathbf{k}}$ contained in the $\mathbf{k}$ mode of the plasma oscillations is now given by

$$\mathcal{E}_{\mathbf{k}} = \frac{2\pi n e^2}{k^2}\left[\tilde{\omega}_{\mathbf{k}}\tilde{\epsilon}'(\mathbf{k},\tilde{\omega}_{\mathbf{k}})\right]S(\mathbf{k}) \tag{11.15}$$

where

$$\tilde{\epsilon}'(\mathbf{k},\tilde{\omega}_{\mathbf{k}}) = \left[\frac{\partial\tilde{\epsilon}(\mathbf{k},\omega)}{\partial\omega}\right]_{\omega=\tilde{\omega}_{\mathbf{k}}} \tag{11.16}$$

and $S(\mathbf{k})$ is defined as that part of the static form factor coming from the integration only in the vicinity of the collective mode $\omega=\tilde{\omega}_{\mathbf{k}}$. In this vicinity, we also note that $|\tilde{\epsilon}(\mathbf{k},\omega)|^2 \cong 0$, and

$$\frac{1}{|\tilde{\epsilon}(\mathbf{k},\omega)|^2} \cong -\frac{\pi}{\left[\tilde{\epsilon}'(\mathbf{k},\tilde{\omega}_{\mathbf{k}})\right]^2\tilde{\gamma}_{\mathbf{k}}}\delta(\omega-\tilde{\omega}_{\mathbf{k}}). \tag{11.17}$$

With the aid of these formulas as well as Eqs. (11.6) and (11.15), we may rewrite Eq. (11.5) in the following way.

$$\mathcal{E}_{\mathbf{k}} = -\frac{1}{2\tilde{\gamma}_{\mathbf{k}}}\left[\Theta + \sum_{\mathbf{q}}\tilde{B}(\mathbf{k},\mathbf{q})\mathcal{E}_{\mathbf{k}-\mathbf{q}}\mathcal{E}_{\mathbf{q}}\right]$$

$$= \frac{\Theta + \sum_{\mathbf{q}}\tilde{B}(\mathbf{k},\mathbf{q})\mathcal{E}_{\mathbf{k}-\mathbf{q}}\mathcal{E}_{\mathbf{q}}}{-2\gamma_{\mathbf{k}} + \sum_{\mathbf{q}}\tilde{A}(\mathbf{k},\mathbf{q})\mathcal{E}_{\mathbf{q}}}. \tag{11.18}$$

Here,

$$\Theta \equiv \frac{\pi m\omega_p^2\tilde{\omega}_{\mathbf{k}}}{k^2\tilde{\epsilon}'(\mathbf{k},\tilde{\omega}_{\mathbf{k}})}\int d\mathbf{v}\,\delta(\tilde{\omega}_{\mathbf{k}}-\mathbf{k}\cdot\mathbf{v})f(\mathbf{v}) \tag{11.19}$$

is the rate of spontaneous emission of the plasma oscillations [compare Eq. (4.42)]

$$\sum_{\mathbf{q}}\tilde{A}(\mathbf{k},\mathbf{q})\mathcal{E}_{\mathbf{q}} = 2(\gamma_{\mathbf{k}}-\tilde{\gamma}_{\mathbf{k}}), \tag{11.20}$$

which can be seen from Eq. (11.4) and comparison between (11.7) and (11.14); and

$$\tilde{B}(\mathbf{k},\mathbf{q}) \equiv \frac{\tilde{\omega}_{\mathbf{k}}}{Vn^2e^2k^2\tilde{\epsilon}'(\mathbf{k},\tilde{\omega}_{\mathbf{k}})} \int d\mathbf{v}\,\delta(\tilde{\omega}_{\mathbf{k}} - \mathbf{k}\cdot\mathbf{v}) f(\mathbf{v}) \frac{\iota(\mathbf{q})}{\iota(\mathbf{k})} \frac{\mathbf{k}\cdot\mathbf{q}|\mathbf{k}-\mathbf{q}|^2}{\tilde{\omega}_{\mathbf{q}}\tilde{\omega}_{\mathbf{k}-\mathbf{q}}\tilde{\epsilon}'(\mathbf{q},\tilde{\omega}_{\mathbf{q}})\tilde{\epsilon}'(\mathbf{k}-\mathbf{q},\tilde{\omega}_{\mathbf{k}-\mathbf{q}})} \tag{11.21}$$

We now construct a wave kinetic equation for $\mathcal{E}_{\mathbf{k}}$ in such a way that its stationary solution should agree with Eq. (11.18). Since for a quiescent plasma we know that [3]

$$\frac{\partial \mathcal{E}_{\mathbf{k}}}{\partial t} = \Theta + 2\gamma_{\mathbf{k}}\mathcal{E}_{\mathbf{k}},$$

the required wave kinetic equation for a turbulent plasma must be

$$\frac{\partial \mathcal{E}_{\mathbf{k}}}{\partial t} = \Theta + 2\gamma_{\mathbf{k}}\mathcal{E}_{\mathbf{k}} - \sum_{\mathbf{q}} \tilde{A}(\mathbf{k},\mathbf{q})\mathcal{E}_{\mathbf{k}}\mathcal{E}_{\mathbf{q}} + \sum_{\mathbf{q}} \tilde{B}(\mathbf{k},\mathbf{q})\mathcal{E}_{\mathbf{k}-\mathbf{q}}\mathcal{E}_{\mathbf{q}}. \tag{11.22}$$

Thus, $\tilde{A}(\mathbf{k},\mathbf{q})$ and $\tilde{B}(\mathbf{k},\mathbf{q})$ may be interpreted as the mode-coupling constants, which themselves depend on the spectral density of turbulence.

D. Ordering of Various Quantities with Respect to the Plasma Parameter

Equation (11.18) makes it possible to order various physical parameters with respect to the powers of the plasma parameter g. It is easy to note from (11.19) and (11.21) that $\Theta \sim g^1$ and $\tilde{B} \sim g^0$. We may also note from (11.7) and (11.14) that $(\gamma_{\mathbf{k}} - \tilde{\gamma}_{\mathbf{k}}) \sim g^0 S(\mathbf{q})/n$; hence, in the light of (11.15) and (11.20), we find that $\tilde{A} \sim g^0$.

Let us now introduce a formal definition of the three states of a plasma: quiescent, weakly turbulent, and strongly turbulent.

A quiescent state is defined as that in which $\gamma_{\mathbf{k}}$ is negative (i.e., conventionally stable) and $|\gamma_{\mathbf{k}}/\omega_{\mathbf{k}}| \sim g^0$. Equation (11.18) then yields

$$\mathcal{E}_{\mathbf{k}} = -\frac{\Theta}{2\gamma_{\mathbf{k}}} \sim g^1. \tag{11.23}$$

The thermal fluctuations, Eqs. (9.104), are special examples of the general relationship (11.23). The fluctuation spectrum associated with the collective mode in a quiescent plasma is maintained by an equilibrium between the spontaneous emission and the Landau damping. The effective rate of damping $\tilde{\gamma}_k$ is essentially the same as γ_k in this state; hence $|\tilde{\gamma}_k/\tilde{\omega}_k| \sim g^0$, as $\tilde{\omega}_k$ always remains of the order g^0.

A weakly turbulent state is defined as a state in the very vicinity of the critical point ($\gamma_k = 0$), so that $|\gamma_k/\omega_k| \leqslant g^{1/2}$. We must now take account of the entire expression Eq. (11.18); the equation yields $\mathcal{E}_k \sim g^{1/2}$, some enhancement over the thermal-fluctuation level of Eq. (11.23). Each term on the right-hand side of the wave kinetic equation (11.22) is now of the order g^1; the presence of turbulence may be ignored in the evaluation of the mode-coupling constants $\tilde{A}$ and $\tilde{B}$. Hence, as has been postulated in the quasilinear theory, the effects of mode coupling may be treated in a perturbation-theoretical way for a weakly turbulent plasma. The stationary state of a weakly turbulent plasma is maintained by an equilibrium involving the rate of spontaneous emission, the Landau growth rate, and the three-wave mode-coupling processes as depicted in Fig. 11.1. The imaginary part of the frequency $\tilde{\gamma}_k$, which remains negative always, obeys the ordering $|\tilde{\gamma}_k/\tilde{\omega}_k| \sim g^{1/2}$; thus, the coherence, or the inverse of the frequency bandwidth of the fluctuation spectrum, is also enhanced substantially in this state. The ordering described in this paragraph is essentially the same as that discussed originally by Frieman and Rutherford [5].

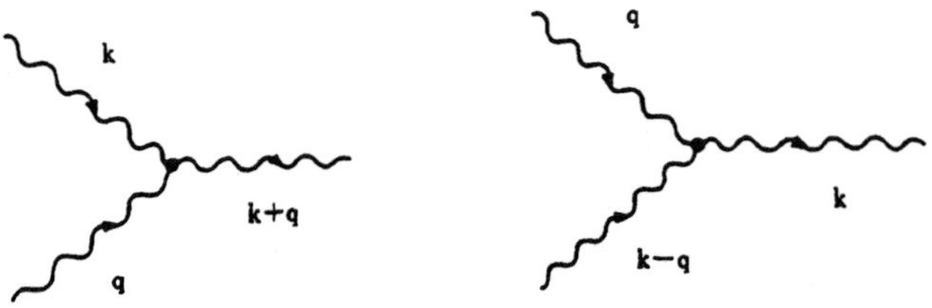

Fig. 11.1 Three-wave processes.

A strongly turbulent state is defined as that in which γ_k is positive (i.e., conventionally unstable) and $|\gamma_k/\omega_k| \sim g^0$. Equation (11.18) then gives $\mathcal{E}_k \sim g^0$; $\sum \tilde{A}(\mathbf{k},\mathbf{q})\mathcal{E}_q$ should be greater than $2\gamma_k$ in this state. We thus find that the spontaneous emission term Θ is negligible in Eq. (11.18). The thermal fluctuations, or discreteness of the particles, have virtually no effect in determining the state of strong turbulence; fluid description of the plasma appears sufficient. The stationary state of a strongly turbulent plasma is maintained by an equilibrium between the Landau growth rate and the three-wave and higher-order mode-coupling processes. $|\tilde{\gamma}_k/\tilde{\omega}_k|$ now goes up to g^0; the bandwidth of the frequency spectrum of turbulence, therefore, greatly increases over the corresponding value in a weakly turbulent plasma.

E. Size-Dependent Effect

We have emphasized, on a number of occasions, the importance of taking into account the modification of the effective interactions between two particles caused by the presence of the strong correlations in a turbulent plasma. This modification is described by the function $t(\mathbf{k})$. It must be noted, however, that this function approaches unity as $\mathbf{k}\to 0$; $\tilde{V}(1,2)$ and $V(1,2)$ become the same for large values of $\mathbf{r}_1-\mathbf{r}_2$. The presence of turbulence can have no effect on the properties of the collective modes in the vicinity of $\mathbf{k}=0$. Turbulence in those plasma systems which may inherently sustain unstable collective modes down to $\mathbf{k}=0$, therefore, requires special consideration in terms of the size effects of a finite plasma.

Physically, the possible existence of such a size-dependent effect may be anticipated from the critical role that the correlation function plays in determining the stability of a turbulent plasma. The finite dimension of the system may impose a boundary condition that $S(\mathbf{k})$ should vanish within a certain (small) domain of $\mathbf{k}$ around $\mathbf{k}=0$. Thus this condition in turn affects the stability of the system, and hence the level and spectral distribution of turbulence.

The size effect enters into the mathematical picture of the theory because of the increasing difficulty in creating the difference between $\tilde{\epsilon}(\mathbf{k},\omega)$ and $\epsilon(\mathbf{k},\omega)$ through $t(\mathbf{k})$ as the wave number decreases. When a typical dimension of the plasma is given by L, it is necessary to secure the stability of the collective modes for all values of the wave numbers down to order $1/L$; the most difficult part to do usually lies in the long-wavelength domain, $k\sim 1/L$. As L increases, the amplitudes and the spectral distribution of the turbulence must adjust themselves so that the resulting value of $|t(1/L)-1|$ is such as to assure a stability of the collective mode at $k=1/L$.

The effectiveness of such a size-dependent effect must generally be judged in relation to the effects of particle collisions. We note that the particle collisions act to impede an organized motion of the plasma to propagate over a distance greater than the mean free path; the collisional effects, therefore, can introduce a lower cutoff for the wave-number domain in which a collective mode may be a well-defined excitation. Such effects can be incorporated into the expression of the dielectric function, and through it, into the integral equation, Eq. (11.5). When the collisional mean free path is much smaller than a characteristic dimension of the system, the size effects will thus become insignificant.

In the theory of hydrodynamic turbulence, a similar size-dependent effect occurs through the Reynolds number; the resemblance may have more than casual physical significance. We may also recall that a size-dependent effect was involved in the stabilization of the Rayleigh–Taylor instability in Section 8.4C and of the drift wave in Section 8.5C.

F. Beam–Plasma Instability

In Section 11.3B, we emphasized the importance of taking account of the modification in the effective interactions caused by the strong correlations existing in a turbulent plasma. We now wish to consider two specific examples of conventionally unstable plasmas which may be brought into a stable turbulent state through the mechanisms mentioned therein. We begin with the case of the beam–plasma system considered in Section 7.2.

When the effects of the fluctuations are neglected, the dielectric response function for such a system is given by Eq. (7.10), i.e.,

$$\epsilon(k,\omega)=1+\chi(k,\omega)=1-\frac{\omega_0^2}{\omega^2}-\frac{\omega_b^2}{(\omega-kv_d)^2}. \tag{11.24}$$

The properties of the collective modes have been investigated through the diagrams in Fig. 7.4; these are reproduced as the dashed lines in Fig. 11.2. The maximum wave number for instability, denoted by K, is given by

$$K=\left[1+(\omega_b/\omega_0)^{2/3}\right]^{3/2}\omega_0/v_d.$$

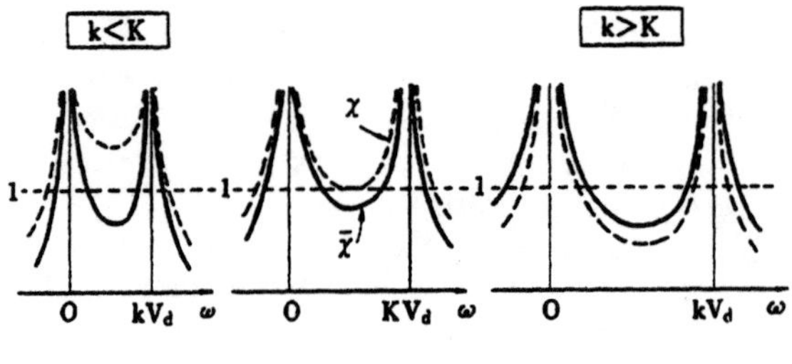

Fig. 11.2 Stabilizing action of the turbulence spectrum of Fig. 11.3 in a beam–plasma system.

Let us now go to a consideration based on the dielectric function, Eq. (11.7); for the beam–plasma system, we have

$$\tilde{\epsilon}(k,\omega)=1+\tilde{\chi}(k,\omega)=1-t(k)\left[\frac{\omega_0^2}{\omega^2}+\frac{\omega_b^2}{(\omega-kv_d)^2}\right] \tag{11.25}$$

where

$$t(k)=1+\frac{L}{2\pi N}\int dq\frac{k}{k-q}S(q). \tag{11.26}$$

In Eq. (11.26), L refers to the length of the plasma along the beam direction. Suppose that $S(k)$ has a peak structure at $k=k_m$, as described in Fig. 11.3; with the aid of Eq. (11.26) we have $t(k)>1$ for $k>k_m$, and $t(k)<1$ for $k<k_m$. If the peak k_m occurs somewhat above K (i.e., $k_m \gtrsim K$), then the resulting curves of $\tilde{\chi}(k,\omega)$ should behave like the solid lines in Fig. 11.2; all four roots may now take on real values in the domain $k \leqslant K$ as well. This clearly demonstrates that a proper inclusion of the correlation effects can lead to a stable description of the beam–plasma system, which is conventionally known to be an unstable one.

An important point to be noted in this connection is that the factor $t(k)$ approaches unity as $k \to 0$; naturally, the turbulence has no effect on the interactions between the particles when they are separated by extremely large distances. Therefore, an inclusion of the correlation effects cannot stabilize a collective mode in the very vicinity of $k=0$. This brings us back to a consideration of the size effects as described in the previous section.

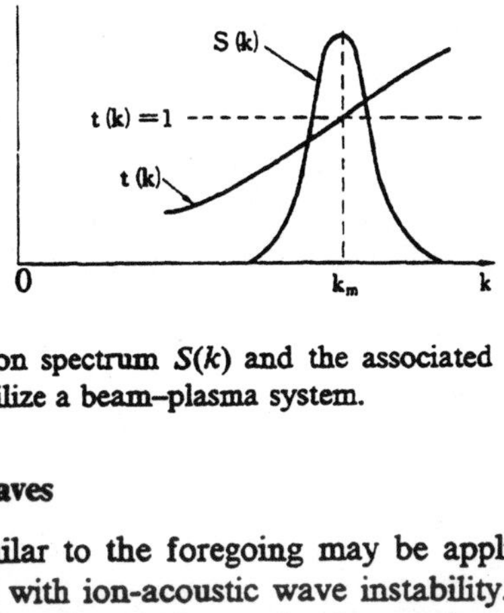

Fig. 11.3 A fluctuation spectrum $S(k)$ and the associated modification factor $t(k)$ which may act to stabilize a beam–plasma system.

G. Ion-Acoustic Waves

A consideration similar to the foregoing may be applied to the fluctuation spectrum associated with ion-acoustic wave instability. The excitation spectrum ω_k in the quiescent description under the usual assumptions that $m_e \ll m_i$ and $T_e \gg T_i$ (the subscripts i and e refer to the ions and the electrons, respectively) is given by

$$\omega_k^2 = \frac{\omega_i^2}{[1+(k_e/k)^2]} \tag{11.27}$$

where

$$\omega_i^2 = \frac{4\pi n e^2}{m_i}, \qquad k_e^2 = \frac{4\pi n e^2}{T_e}. \tag{11.28}$$

When the drift velocity v_d of the electrons relative to the ions exceeds its critical value v_c, the acoustic waves in a certain wave-number domain become unstable.

A situation of this kind may be described graphically in terms of a boundary curve $v_d(k)$ between stable and unstable waves; the curve is generally determined as a sum of three contributions:

$$v_d(k) = (\omega_k/k) + (\text{ionic Landau damping}) + (\text{collisional effects}). \quad (11.29)$$

Equation (7.21) has taken account of the first two contributions only; generally, however, the collisional effects can also influence the boundary curve [29]. Among the three, the first term ω_k/k is usually the predominant one. A schematic example of such a boundary curve is illustrated by the dashed line in Fig. 11.4; when the drift velocity is fixed at a value v_d greater than v_c, the waves with $k_1 < k < k_2$ are unstable.

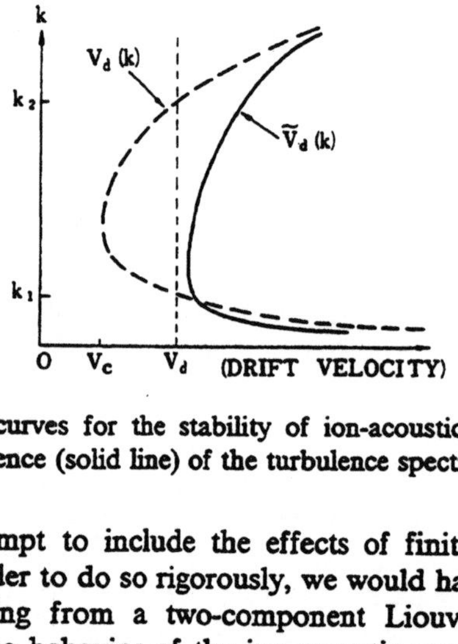

Fig. 11.4 Boundary curves for the stability of ion-acoustic waves in the absence (dashed line) and presence (solid line) of the turbulence spectrum of Fig. 11.5.

Let us now attempt to include the effects of finite correlations in the picture above. In order to do so rigorously, we would have to reconstruct the whole theory, starting from a two-component Liouville equation. When interest is only in the behavior of the ion-acoustic waves, however, we can adopt an adiabatic approximation for the electrons and regard them only as contributing to the static dielectric background for the ions; Eq. (11.27) is derived simply from such an idea.

The turbulence arising from the ion-acoustic instabilities will, therefore, affect the effective ion–ion interaction most strongly within the limit of validity for the foregoing adiabatic approximation; let us describe this effect by $t_+(k)$. Under the turbulent conditions, Eq. (11.29) should thus be modified as

$$\tilde{v}_d(k) = (\tilde{\omega}_k/k) + (\text{ionic Landau damping}) + (\text{collisional effects}), \quad (11.30)$$

where

$$\tilde{\omega}_k^2 = \frac{\omega_i^2 t_+(k)}{\left[1+(k_e/k)^2\right]}. \quad (11.31)$$

If $S(k)$ behaves as shown in Fig. 11.5 with a peak at $k_m(\lesssim k_1)$, then the effective boundary curve $\tilde{v}_d(k)$ of Eq. (11.30) will shift to the solid line of Fig. 11.4; a stabilization of the ion-acoustic waves due to the presence of large-amplitude ion correlations is clearly indicated.

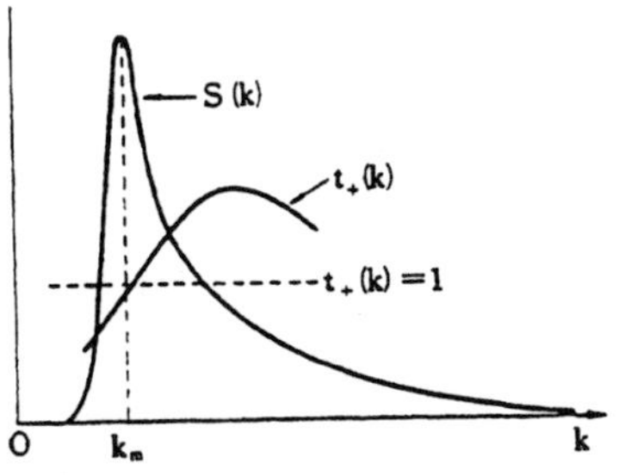

Fig. 11.5 A fluctuation spectrum $S(k)$ associated with the ion-acoustic waves and the modification factor $t_+(k)$ for ion–ion interaction.

11.4 SCALING LAWS AND THE UNIVERSAL SPECTRUM

The discussion in Section 11.3D has revealed the ordering of various physical quantities with respect to the plasma parameter. Such an ordering may be regarded as an indication of a possible existence of a scaling law or a universal spectrum in plasma turbulence. In this section, we wish to consider these problems through an examination of invariance properties of kinetic equations under a set of similarity transformations.

A. Similarity Transformations

We begin with the second equation of the BBGKY hierarchy written in the notation of Chapter 2 as

$$\left[\frac{\partial}{\partial t}+L(1)+L(2)\right]G(1,2)=\frac{1}{n}[V(1,2)+V(2,1)][F(1)F(2)+G(1,2)]$$
$$+\int V(1,3)F(1)G(2,3)d3+\int V(2,3)F(2)G(3,1)d3$$
$$+\int [V(1,3)+V(2,3)][F(3)G(1,2)+H(1,2,3)]d3. \tag{11.32}$$

For the theory of strong plasma turbulence, we are interested in accurately determining only those parts of the many-particle distribution functions which remain finite in the fluid limit; according to the results in Section 11.3D, the fluid description suffices for such a plasma. Hence, Eq. (11.32) may be simplified as

$$\left[\frac{\partial}{\partial t}+L(1)+L(2)\right]G(1,2)=\int V(1,3)F(1)G(2,3)d3+\int V(2,3)F(2)G(3,1)d3$$
$$+\int [V(1,3)+V(2,3)][F(3)G(1,2)+H(1,2,3)]d3. \tag{11.33}$$

Let us now consider a set of similarity transformation for physical parameters,

$$n\to\alpha n, \qquad v\to\beta v,$$
$$m\to\mu m, \qquad e\to\eta e. \tag{11.34}$$

Since the plasma frequency ω_p and the Debye wave number k_D transform as

$$\omega_p\to\alpha^{1/2}\eta\mu^{-1/2}\omega_p,$$
$$k_D\to\alpha^{1/2}\eta\mu^{-1/2}\beta^{-1}k_D, \tag{11.35}$$

the variables ω and k must scale in the same way; these determine the scaling of time and length, respectively.

For a homogeneous system, $F(1)=f(\mathbf{v}_1)$, a velocity distribution function normalized to unity. Hence,

$$F(1)\to\beta^{-3}F(1). \tag{11.36}$$

We may introduce a scaling factor Γ of the pair correlation function in such a way that

$$\frac{G(1,2)}{F(1)} \to \Gamma \beta^{-3} \frac{G(1,2)}{F(1)}. \tag{11.37}$$

Let us then assume that the ternary correlation function scales in a similar manner, so that

$$\frac{H(1,2,3)}{G(1,2)} \to \Gamma \beta^{-3} \frac{H(1,2,3)}{G(1,2)}. \tag{11.38}$$

We now carry out the transformations (11.34)–(11.38) to Eq. (11.33). We then find that Eq. (11.33) remains invariant under these transformations as long as Γ scales so that

$$\Gamma = 1. \tag{11.39}$$

This is, therefore, the scaling law of strong plasma turbulence; this simple scaling law, as we shall see, contains a rich physical content. We might also mention that the ansatz for $H(1,2,3)$ adopted in Eq. (11.1) conforms to the transformation (11.38) when the scaling law is given by (11.39).

Let us carry out a similarity transformation for $S(\mathbf{k})$ so that

$$S \to \zeta S. \tag{11.40}$$

For a strongly turbulent plasma, $S(\mathbf{k}) \gg 1$, so that the first term, unity, on the right-hand side of (9.11) may be neglected. In the light of (11.39), we then find that Eq. (9.11), with the omission of 1, is kept invariant when a scaling law

$$\alpha \eta^6 \mu^{-3} \beta^{-6} \zeta^2 = 1 \tag{11.41}$$

is satisfied.

The energy $\mathcal{E}_{\mathbf{k}}$ contained in the $\mathbf{k}$ mode of the plasma oscillations is given by Eq. (11.15). Since the dielectric function (11.7) remains invariant under the similarity transformation, we find that the energy spectrum transforms as

$$\mathcal{E}_{\mathbf{k}} \to \mu \beta^2 \zeta \mathcal{E}_{\mathbf{k}}. \tag{11.42}$$

We pick up the scaling factors of the energy spectrum, the wave number, and the kinetic energy $(\frac{1}{2}) mv^2$ per particle; and define them, respectively, as

$$E \equiv \mu \beta^2 \zeta, \qquad \xi \equiv \alpha^{1/2} \eta \mu^{-1/2} \beta^{-1}, \qquad \theta \equiv \mu \beta^2. \tag{11.43}$$

Thus, Eq. (11.41) reduces to

$$E=\alpha\theta\xi^{-3}. \tag{11.44}$$

The scaling laws, (11.39), (11.41), and (11.44), are mutually equivalent.

For a quiescent plasma, the ordering (2.12) and (2.13) applies; hence, we have $\Gamma=\alpha^{1/2}\eta^{3}\mu^{-3/2}\beta^{-3}$ or $\zeta=1$. We thus find from (11.43)

$$E=\theta. \tag{11.45}$$

Similarly, for a weakly turbulent plasma, the ordering of Section 11.3D indicates

$$E=\alpha^{1/2}\theta\xi^{-3/2}. \tag{11.46}$$

The results of Eqs. (11.44)–(11.46) may then be compactly expressed as

$$E=\alpha^{a}\theta\xi^{-3a} \tag{11.47}$$

where the index a takes on values

$$a=\begin{cases} 0 & \text{quiescent,} \\ \frac{1}{2} & \text{weakly turbulent,} \\ 1 & \text{strongly turbulent.} \end{cases} \tag{11.48}$$

B. Universal Spectrum

We may conclude from the scaling law (11.47) that the energy spectrum $\mathcal{E}_k$ of turbulence must be expressed in the form

$$\mathcal{E}_k=n^{a}\langle\tfrac{1}{2}mv^2\rangle k^{-3a}y(k/k_D, \cos\chi; g, \{z\}). \tag{11.49}$$

Here, $\langle\frac{1}{2}mv^2\rangle$ is an average kinetic energy per particle; y is a nondimensional function of dimensionless variables, k/k_D and $\cos\chi$, direction cosine of $\mathbf{k}$ with respect to an axis of anisotropy of the system (if such exists). Generally, the function y may contain as parameters any dimensionless parameters which can be constructed from the physical parameters describing the plasma. The plasma parameter g is certainly one of them. If the drift motion of the particles is involved, the ratio between the drift velocity and the average random velocity, for example, may also be such a parameter; $\{z\}$ in (11.49) collectively denotes the dimensionless parameters involved.

The ordering of $\mathcal{E}_k$ with respect to $g\equiv(n\lambda_D^3)^{-1}$ has been determined in Section 11.3D for each of the three plasma states; the factor in front of the function y in (11.49) is in accord with that ordering. Hence, we may take the limit of $g \to 0$ in Eq. (11.49); defining $y_0(k/k_D, \cos\chi; \{z\}) \equiv \lim_{g\to 0} y(k/k_D, \cos\chi; g, \{z\})$, we have

$$\mathcal{E}_k = n^a \langle \tfrac{1}{2}mv^2 \rangle k^{-3a} y_0(k/k_D, \cos\chi; \{z\}). \qquad (11.50)$$

Generally, a collective mode represents a well-defined elementary excitation in the wave-vector domain such that

$$1/l_m < k < k_D \qquad (11.51)$$

where l_m is the Coulomb mean-free path (see Problem 1.5). The domain of applicability for the spectrum (11.50) should thus be restricted to (11.51).

Although Eq. (11.50) is suggestive of k^{-3a} spectrum, it is premature to conclude so, because any power of (k/k_D) can be factored out of the function y_0. Thus, writing

$$\mathcal{E}_k \sim n^\rho \langle \tfrac{1}{2}mv^2 \rangle^\tau k^{-\kappa}, \qquad (11.52)$$

we find the scaling relationships

$$3\rho + \tau - \kappa = 1, \qquad \rho + \tau = 1 + a. \qquad (11.53)$$

The parameter a takes on values (11.48), depending on the state of the plasma.

If, however, we can define in the k space of the turbulence spectrum something similar to the inertial region in hydrodynamic turbulence, then in such a domain, it is possible to establish from Eq. (11.50) a universal spectrum of plasma turbulence, analogous to the Kolmogorov–Obukhov law of hydrodynamics [30–32]. Let us first note that the Debye wave number k_D gives a measure of wave number above which the collective modes are substantially damped; there the fluctuations are dissipated into heat through interaction with particles. The mathematical limiting procedure of $k_D \to \infty$ in plasma turbulence should therefore correspond to the case of (kinematic viscosity)$\to 0$ in hydrodynamic turbulence. So we may look for an inertial region of strong plasma turbulence in a long-wavelength domain away from k_D so that the turbulence spectrum may not depend on k_D. If such a region exists, we can then write (11.50) as

$$\mathcal{E}_k = n^a \langle \tfrac{1}{2}mv^2 \rangle k^{-3a} y_1(\cos\chi; \{z\}) \qquad (11.54)$$

where y_1 is another nondimensional function. A universal spectrum proportional to k^{-3a} is, therefore, indicated in the inertial region of strong plasma turbulence. For a quiescent plasma, $a=0$; Eqs. (9.104) are thus seen to be specific examples of the spectrum (11.54). It, in fact, represents a Rayleigh–

Jeans spectrum. With additional assumptions, such as isotropy of the spectrum, we could further simplify Eq. (11.54).

Thus far, we have considered the scaling laws of plasma turbulence with a three-dimensional spectrum. When a magnetic field is applied to the plasma, the particles move with helical orbits along the magnetic lines of force. If we pass to the limit of an extremely strong magnetic field, then the plasma behaves as if it were a one-dimensional gas. We can repeat the entire procedure of similarity transformation as applied to such a one-dimensional plasma. In place of Eq. (11.44), we now have

$$E=\alpha\theta\xi^{-1}.$$

Consequently, Eq. (11.50) is replaced by

$$\mathcal{E}_k=n^a\langle\tfrac{1}{2}mv^2\rangle k^{-3a+2}y_2(k/k_D;\{z\}). \tag{11.55}$$

11.5 ANOMALOUS TRANSPORT

Various transport processes in the plasma are greatly affected by the presence of plasma turbulence. These are referred to as anomalous transport phenomena, which may also be analyzed in terms of the scaling laws of the previous section. We consider here a number of examples of such anomalous transport processes by assuming that the plasma is in a strongly turbulent state; extension of the results to a weakly turbulent situation should be straightforward.

A. Turbulent Heating

The energy spectrum of strong plasma turbulence has been given by Eq. (11.50) with $a=1$; this quantity exceeds by far the values determined from equipartition with the particle kinetic energy in the plasma. A relatively small number of charged particles which couple resonantly with the turbulence will thereby be heated selectively and acquire suprathermal energies.

To analyze such a phenomenon, we begin with the first equation of the BBGKY hierarchy; its collision term is generally written as Eq. (2.40). For a strongly turbulent plasma, we may investigate this equation with the aid of the similarity transformations described in Section 11.4A. Introducing the collision frequency $\nu(\mathbf{v})$ via $[\partial f(\mathbf{v})/\partial t]_c\equiv-\nu(\mathbf{v})f(\mathbf{v})$, we find

$$\nu(\mathbf{v})=\frac{e^2\omega_p^2}{m^2v^5}\langle\mathcal{E}_k\rangle y_3\left(\frac{vk_D}{\omega_p},\cos\varphi;\{z\}\right), \tag{11.56}$$

where $\cos\varphi$ represents the direction cosine of $\mathbf{v}$ with respect to the axis of anisotropy.

It follows from Eq. (11.56) that the average rate of increase in particle energy $w=\frac{1}{2}mv^2$ through resonant interaction with turbulent fluctuations, may be written as

$$\frac{dw}{dt}=\frac{m^{1/2}e^2\omega_p^2}{w^{3/2}}\langle \mathcal{E}_k\rangle y_4. \tag{11.57}$$

An expression equivalent to this has been obtained by Tsytovich [33]; the dimensionless factor y_4 is shown to be equal to $(2)^{-3/2}$ in his calculation.

A charged particle accelerated to a suprathermal energy now suffers a retarding force of the plasma as considered in Section 4.3B. The rate of acceleration should thus be given by the balance between the heating rate (11.57) and the stopping power (4.36). Assuming $y_4=2^{-3/2}$ in (11.57), we have†

$$\frac{dw}{dt}=\frac{m^{1/2}e^2\omega_p^2}{2^{3/2}w^{3/2}}\langle \mathcal{E}_k\rangle-\frac{m^{1/2}e^2\omega_p^2}{2^{1/2}w^{1/2}}\ln\Lambda. \tag{11.58}$$

This equation approximately describes the rate of turbulent heating in the plasma.

The particles accelerated through these stochastic heating processes acquire a certain energy spectrum in the suprathermal domain away from the bulk of the particles; the spectral shape depends on the nature of the collective modes and the spectral distribution of turbulence. The average energy w^* of such suprathermal particles can be determined from Eq. (11.58) by setting $dw/dt=0$; we thus find

$$w^*=\frac{\langle \mathcal{E}_k\rangle}{2\ln\Lambda}. \tag{11.59}$$

For a strongly turbulent plasma, we calculate from (11.50)

$$\langle \mathcal{E}_k\rangle_s=3Ag^{-1}\langle \tfrac{1}{2}mv^2\rangle\ln\Lambda. \tag{11.60}$$

where the Coulomb logarithm $\ln\Lambda$ stems from the restriction in the $\mathbf{k}$ space as specified by (11.51), and A is a dimensionless factor of order unity arising from the function y_0 in Eq. (11.50). The quantity $\langle\frac{1}{2}mv^2\rangle$ refers to the average

†In obtaining the last term of Eq. (11.58) from (4.36), we have approximated $\ln(k_{max}v/\omega_p)=\ln\Lambda$, the Coulomb logarithm. Actually for a suprathermal particle, $\ln(k_{max}v/\omega_p)\gtrsim\ln\Lambda$. Such an error may be permissible within the approximation inherent in the derivation of Eq. (11.58).

kinetic energy per particle for the bulk of the plasma. Substitution of (11.60) into (11.59) yields

$$w_s^* = A(3/2g)\langle \tfrac{1}{2}mv^2 \rangle. \tag{11.61}$$

Thus the average energy of a suprathermal particle in a strongly turbulent plasma is approximately g^{-1} times greater than the average kinetic energy per particle; the number of suprathermal particles should be correspondingly small.

For a weakly turbulent plasma, we similarly obtain

$$\langle \varepsilon_k \rangle_w = 2Ag^{-1/2}\langle \tfrac{1}{2}mv^2 \rangle, \tag{11.62}$$

whence

$$w_w^* = \frac{A}{g^{1/2}\ln\Lambda}\langle \tfrac{1}{2}mv^2 \rangle. \tag{11.63}$$

B. Anomalous Resistivity

In addition to accelerating a small number of selected particles to high energies, the plasma turbulence acts to scatter the particles in the plasma. The effective collision frequency thus depends on the level of turbulence. For a strongly turbulent plasma, we may apply the scaling law (11.41) to Eq. (11.56); the average collision frequency $\langle \nu \rangle$ is thus expressed as

$$\langle \nu \rangle = \omega_p y_5(\{z\}). \tag{11.64}$$

In strong turbulence, the average collision frequency is enhanced by a factor of order g^{-1} over the value (10.102) for a quiescent plasma. The electrical conductivity is then calculated to be

$$\begin{aligned}\sigma &= \omega_p^2/4\pi\langle \nu \rangle \\ &= \omega_p y_6(\{z\}).\end{aligned} \tag{11.65}$$

A similar expression has been obtained by O'Neil [34].

C. Diffusion across the Magnetic Field

It has been widely speculated that enhanced fluctuations or turbulence may be the cause of the anomalous diffusion phenomena of Bohm type [35]. According to the calculations in Section 10.5 and especially Eq. (10.106), the diffusion coefficient for a Maxwellian plasma contains an anomalous term

$D_\perp^*$ in addition to the usual classical term. It has the expression

$$D_\perp^* = \frac{(2\pi)^{1/2} n e^2 c^2 m_e^{1/2}}{B^2 T_e^{1/2}} \left(1 + \frac{T_i}{T_e}\right) \ln\left(\frac{m_i}{m_e}\right)$$
$$\times \left\{ \exp(\alpha^*) K_0(\alpha^*) - \frac{1}{2}\left(1 - \frac{T_i}{T_e}\right)\left[\exp(\alpha^*) K_1(\alpha^*) - \frac{1}{\alpha^*}\right]\right\} \tag{11.66}$$

where K_0 and K_1 are modified Bessel functions of the second kind, and α^* has been defined by Eq. (10.107). The electron fluctuations in the low-frequency domain such that $\omega < |\Omega_e|$ give rise to the anomalous term.

When such fluctuations are enhanced above the thermal level owing to the onset of an instability, the anomalous term will grow accordingly, resulting in an enhanced diffusion. As we have discussed in Sections 11.3D and 11.4B and have in particular demonstrated in Eqs. (11.50) and (11.60), the enhancement of fluctuations in strong plasma turbulence is measured by g^{-1}. We also note that, because of the factor $(1 + T_e/T_i)$, which is usually quite large, $\alpha^* \gg 1$ over a wide and physically important range of the magnetic field. In these circumstances, we can expand $K_0(\alpha^*)$ and $K_1(\alpha^*)$ in Eq. (11.66) in asymptotic series and keep only the leading terms. Taking account of the enhancement of fluctuations by a factor g^{-1} for strong turbulence, we thus have

$$\tilde{D}_\perp^* \cong \frac{1}{32\pi} \frac{c}{eB} (T_e T_i)^{1/2} \left(1 + \frac{T_i}{T_e}\right)^{3/2} \ln\left(\frac{m_i}{m_e}\right) \qquad (\alpha^* \gg 1). \tag{11.67}$$

This is essentially equivalent to the value predicted by Bohm.

When the magnetic field is so strong that $\alpha^* \ll 1$, Eq. (11.67) should be replaced by

$$\tilde{D}_\perp^* \cong \frac{1}{8\pi(2\pi)^{1/2}} \frac{c}{eB} (T_e + T_i) \frac{\omega_e}{|\Omega_e|} \ln\left(\frac{m_i}{m_e}\right) \ln\left[\frac{\Omega_e^2}{\omega_e^2} \frac{T_i}{T_e + T_i}\right] \qquad (\alpha^* \ll 1). \tag{11.68}$$

For a weakly turbulent plasma, we similarly estimate the diffusion coefficient by multiplying Eq. (11.67) or (11.68) by a factor $g^{1/2}$.

D. Conductivity across the Magnetic Field

In the previous section, the enhanced rate of plasma diffusion across the magnetic field was interpreted essentially in terms of the enhanced rate of particle scattering in a turbulent plasma. Such an enhanced rate of particle scattering should then affect the electrical conductivity of the plasma across the magnetic field.

When the effective rate of collisions is given by ν_e, the conductivity is calculated as

$$\sigma_\perp = \frac{\omega_p^2}{4\pi}\frac{\nu_e}{\Omega_e^2+\nu_e^2}. \tag{11.69}$$

For a strongly turbulent plasma, the effective collision frequency may be obtained from Eq. (11.67) or (11.68) with the aid of Eq. (10.99); hence

$$\nu_e \cong \frac{|\Omega_e|}{32\pi}\left(\frac{T_i}{T_e}\right)^{1/2}\left(1+\frac{T_i}{T_e}\right)^{3/2}\ln\left(\frac{m_i}{m_e}\right) \qquad (\alpha^* \gg 1) \tag{11.70}$$

$$\nu_e \cong \frac{\omega_e}{8\pi(2\pi)^{1/2}}\left(1+\frac{T_i}{T_e}\right)\ln\left(\frac{m_i}{m_e}\right)\ln\left[\frac{\Omega_e^2}{\omega_e^2}\frac{T_i}{T_e+T_i}\right] \qquad (\alpha^* \ll 1). \tag{11.71}$$

For a weakly turbulent plasma, we multiply Eq. (11.70) or Eq. (11.71) by a factor $g^{1/2}$.

E. Radiation from Turbulence

It has been supposed that plasma turbulence, being produced by the instability of longitudinal collective modes, would not emit electromagnetic radiation unless nonlinear coupling mechanisms with transverse modes were taken into consideration. Such coupling processes can be studied through a nonlinear analysis of the Klimontovich equation. [See, e.g., Eqs. (9.105) and (9.106).] Theoretical studies have been made in these directions [36].

It is important, however, to recognize the existence of a radiation mechanism of plasma turbulence which stems solely from the anisotropy of the plasma [37, 38]; such an anisotropy may be provided by a drift motion of particles or by an external magnetic field. Consider the frequency-wave-vector dispersion relationship of the electromagnetic waves, Eq. (5.23). When the system is isotropic, this equation gives two transverse and one longitudinal modes. For an anisotropic system, such a decoupling is not generally possible; instead, we find two quasitransverse and one quasilongitudinal modes. These modes approach the corresponding pure modes when the parameters characterizing the strength of the anisotropy vanish or when **k** falls in the direction of the anisotropy. The quasilongitudinal mode contains, although small in magnitude, a component of the electromagnetic field perpendicular to **k**; this component thereby contributes to the radiation spectrum.

These radiation mechanisms may thus offer an additional diagnostic technique for plasma turbulence. Consideration and understanding of the radiation emanating from plasma turbulence are also regarded as central problems in attempting to interpret various astrophysical radio phenomena, including pulsars.

References

1. W. E. Drummond and D. Pines, *Nucl. Fusion Suppl.* 3, 1049 (1962); *Ann. Phys. (N.Y.)* **28**, 478 (1964).
2. A. A. Vedenov, E. P. Velikhov, and R. Z. Sagdeev, *Nucl. Fusion Suppl.* 2, 465 (1962); A. A. Vedenov, *J. Nucl. Energy* **C5**, 169 (1963).
3. D. Pines and J. R. Schrieffer, *Phys. Rev.* **125**, 804 (1962).
4. R. Balescu, *J. Math. Phys.* **4**, 1009 (1963).
5. P. H. Rutherford and E. A. Frieman, *Phys. Fluids* **6**, 1139 (1963); E. A. Frieman and P. H. Rutherford, *Ann. Phys. (N.Y.)* **28**, 134 (1964).
6. K. Nishikawa and Y. Osaka, *Prog. Theoret. Phys. (Kyoto)* **33**, 402 (1965).
7. I. B. Bernstein and F. Engelmann, *Phys. Fluids* **9**, 937 (1966).
8. A. Rogister and C. Oberman, *J. Plasma Phys.* **2**, 33 (1968); *J. Plasma Phys.* 3, 119 (1969).
9. S. Ichimaru, *Phys. Rev.* **174**, 289, 300 (1968).
10. V. N. Tsytovich, *Usp. Fiz. Nauk* **90**, 435 (1966) [*Soviet Phys.-Uspekhi* **9**, 805 (1967)]; *Nonlinear Effects in Plasma* (J. S. Wood, transl.; S. M. Hamberger, ed.) (Plenum, New York, 1970).
11. R. Z. Sagdeev and A. A. Galeev, *Nonlinear Plasma Theory* (T. M. O'Neil and D. L. Book, eds.) (W. A. Benjamin, New York, 1969).
12. R. C. Davidson, *Methods in Nonlinear Plasma Theory* (Academic, New York, 1972).
13. M. N. Rosenbluth, B. Coppi, and R. N. Sudan, in *Proc. Third Intern. Conf. Plasma Physics and Controlled Nuclear Fusion Research, Novosibirsk, 1968*, p. 771 (IAEA, Vienna, 1969).
14. R. E. Aamodt and M. L. Sloan, *Phys. Fluids* **11**, 2218 (1968).
15. V. A. Liperovskii and V. N. Tsytovich, *Zh. Tekh. Fiz.* **36**, 575 (1966) [*Soviet Phys.-Tech. Phys.* **11**, 432 (1966)].
16. S. Ichimaru, *Phys. Rev.* **165**, 251 (1968).
17. E. A. Kaner and V. M. Yakovenko, *Zh. Eksp. Teor. Fiz.* **58**, 582 (1970) [*Soviet Phys. JETP* **31**, 316 (1970)].
18. T. H. Dupree, *Phys. Fluids* **9**, 1773 (1966); *Phys. Fluids* **10**, 1049 (1967); *Phys. Fluids* **11**, 2680 (1968).
19. J. Weinstock, *Phys. Fluids* **12**, 1045 (1969); J. Weinstock and R. H. Williams, *Phys. Fluids* **14**, 1472 (1971).
20. B. B. Kadomtsev, *Plasma Turbulence* (L. C. Ronson and M. G. Rusbridge, transls.) (Academic, New York, 1965).
21. S. Ichimaru and T. Nakano, *Phys. Rev.* **165**, 231 (1968).
22. S. Ichimaru, *Phys. Rev.* **A2**, 494 (1970).
23. J. Hubbard, *Phys. Letters* **25A**, 709 (1967).
24. K. S. Singwi, M. P. Tosi, R. H. Land, and A. S. Sjölander, *Phys. Rev.* **176**, 589 (1968); *Solid State Commun.* **7**, 1503 (1969); *Phys. Rev.* **B1**, 1044 (1971).

25. S. Ichimaru, in *1968 Tokyo Summer Lectures in Theoretical Physics: Statistical Physics of Charged Particle Systems* (R. Kubo and T. Kihara, eds.), p. 69 (Syokabo, Tokyo and W. A. Benjamin, New York, 1969).
26. S. Ichimaru, *Phys. Fluids* **13**, 1560 (1970).
27. T. O'Neil and N. Rostoker, *Phys. Fluids* **8**, 1109 (1965).
28. S. A. Rice and P. Gray, *The Statistical Mechanics of Simple Liquids*, p. 74 (Interscience, New York, 1965).
29. S. Ichimaru, *Ann. Phys. (N.Y.)* **20**, 78 (1962).
30. A. N. Kolmogorov, *Compt. Rend Acad. Sci. USSR* **30**, 301 (1941) [*Soviet Phys.-Uspekhi* **10**, 734 (1968)].
31. A. M. Obukhov, *Compt. Rend. Acad. Sci. USSR* **32**, 19 (1941).
32. L. D. Landau and E. M. Lifshitz, *Fluid Mechanics* (J. B. Sykes and W. H. Reid, transls.) (Addison-Wesley, Reading, Mass., 1959).
33. V. N. Tsytovich, *Usp. Fiz. Nauk* **89**, 89 (1966) [*Soviet Phys.-Uspekhi* **9**, 370 (1966)].
34. T. M. O'Neil, *Phys. Rev. Letters* **25**, 995 (1970).
35. D. Bohm, in *The Characteristics of Electrical Discharges in Magnetic Fields* (A. Guthrie and R. K. Wakerling, eds.), Chapter 2, Section 5 (McGraw-Hill, New York, 1949).
36. S. A. Kaplan and V. N. Tsytovich, *Usp. Fiz. Nauk* **97**, 77 (1969) [*Soviet Phys.-Uspekhi* **12**, 42 (1969)].
37. H. W. Wyld, *Phys. Fluids* **3**, 408 (1960).
38. S. Ichimaru and S. H. Starr, *Phys. Rev.* **A2**, 821 (1970).

Problems

11.1. Taking explicit account of the expansion of $\bar{\epsilon}(\mathbf{k},\omega)$ in the vicinity of $\omega=\bar{\omega}_\mathbf{k}$ as expressed in Eq. (4.16), obtain Eq. (11.19) through the approach described in Section 4.3C.

11.2. Suppose a situation in which the three-wave processes as illustrated in Fig. 11.1 are prohibited among the fluctuations in the turbulence spectrum. How then should we modify the ordering arguments of the weakly turbulent plasma in Section 11.3D, which assume availability of the three-wave processes?

11.3. Consider the effects and physical implication of the similarity transformations (11.34) and (11.40) on $t(\mathbf{k})$ in Eq. (11.4) under condition (11.41).

11.4. Derive Eqs. (11.60) and (11.62).

APPENDIXES

APPENDIX A

STRUCTURE OF KINETIC EQUATIONS

The development and exposition of the kinetic theories unavoidably will turn out to be quite formal and perhaps unappetizing to most readers. The basic aims of this appendix are an introduction of Klimontovich's microscopic description and a derivation of the Bogoliubov–Born–Green–Kirkwood–Yvon (BBGKY) hierarchy equations.

1. KLIMONTOVICH EQUATION

We consider a classical system containing N identical particles in a box of volume $V; n \equiv N/V$ denotes the average number density. Each particle is characterized by the electric charge q and the mass m; we assume a smeared-out background of opposite charges so that the average space-charge field of the system may be canceled.

In the six-dimensional phase space consisting of the position $\mathbf{r}$ and the velocity $\mathbf{v}$, each particle has its own trajectory; for the ith particle, we write

$$X_i(t) \equiv [\mathbf{r}_i(t), \mathbf{v}_i(t)]. \tag{A.1}$$

Since we are dealing with point particles, the microscopic density of the particles in the phase space may be expressed by the summation of the six-dimensional delta functions as

$$N(X;t) \equiv (1/n) \sum_{i=1}^{n} \delta[X - X_i(t)] \tag{A.2}$$

where $X \equiv (\mathbf{r}, \mathbf{v})$. $N(X;t)$ may be called the *Klimontovich distribution function.*

The distribution function satisfies the continuity equation in the phase space:

$$\frac{dN}{dt} = \frac{\partial N}{\partial t} + \dot{X} \cdot \frac{\partial N}{\partial X} = 0.$$

Writing the phase-space coordinates explicitly, we have

$$\frac{\partial N}{\partial t} + \mathbf{v} \cdot \frac{\partial N}{\partial \mathbf{r}} + \dot{\mathbf{v}} \cdot \frac{\partial N}{\partial \mathbf{v}} = 0 \tag{A.3}$$

where $\dot{\mathbf{v}}$ is the acceleration at the point $(\mathbf{r},\mathbf{v})$.

In plasma physics, the electromagnetic acceleration is the most important one to be considered. Thus,

$$\dot{\mathbf{v}}=\frac{q}{m}\left[\mathbf{E}(\mathbf{r},t)+\frac{\mathbf{v}}{c}\times\mathbf{B}(\mathbf{r},t)\right]. \tag{A.4}$$

The local electric and magnetic fields $\mathbf{E}(\mathbf{r},t)$ and $\mathbf{B}(\mathbf{r},t)$ consist of two separate contributions: those applied from external sources, and those produced by the microscopic fine-grained distribution of the charged particles, (A.2).

$$\mathbf{E}(\mathbf{r},t)=\mathbf{E}_{\text{ext}}(\mathbf{r},t)+\mathbf{e}(\mathbf{r},t),\qquad \mathbf{B}(\mathbf{r},t)=\mathbf{B}_{\text{ext}}(\mathbf{r},t)+\mathbf{b}(\mathbf{r},t). \tag{A.5}$$

The microscopic fine-grained fields $\mathbf{e}(\mathbf{r},t)$ and $\mathbf{b}(\mathbf{r},t)$ are to be determined from a solution of the Maxwell equations:

$$\nabla\times\mathbf{e}+\frac{1}{c}\frac{\partial\mathbf{b}}{\partial t}=0,$$

$$\nabla\times\mathbf{b}-\frac{1}{c}\frac{\partial\mathbf{e}}{\partial t}=\frac{4\pi}{c}qn\int\mathbf{v}N(X;t)\,d\mathbf{v},$$

$$\nabla\cdot\mathbf{e}=4\pi qn\left[\int N(X;t)\,d\mathbf{v}-1\right],$$

$$\nabla\cdot\mathbf{b}=0. \tag{A.6}$$

For a given $N(X;t)$ we can generally write down a solution to this set of equations; the solution, when substituted in (A.4), would then amount to taking account of both electrostatic and electromagnetic interactions between the particles. For a nonrelativistic plasma, however, the electromagnetic interactions are usually negligible as compared with the electrostatic interactions; we shall henceforth adopt such an electrostatic approximation. The microscopic fields are then written as

$$\mathbf{e}(\mathbf{r},t)=-qn\frac{\partial}{\partial\mathbf{r}}\int\frac{N(X';t)}{|\mathbf{r}-\mathbf{r}'|}dX',\qquad \mathbf{b}(\mathbf{r},t)=0. \tag{A.7}$$

Substituting (A.7) into (A.5), we find an expression for the acceleration (A.4) in terms of $N(X;t)$. With the aid of such an expression, Eq. (A.3) may be rewritten in the form

$$\left[\frac{\partial}{\partial t}+L(X)-\int V(X,X')N(X';t)\,dX'\right]N(X;t)=0. \tag{A.8}$$

Here, $\mathsf{L}(X)$ is a single-particle operator defined by

$$\mathsf{L}(X) \equiv \mathbf{v}\cdot\frac{\partial}{\partial \mathbf{r}} + \frac{q}{m}\left[\mathbf{E}_{\text{ext}}(\mathbf{r},t) + \frac{\mathbf{v}}{c}\times\mathbf{B}_{\text{ext}}(\mathbf{r},t)\right]\cdot\frac{\partial}{\partial \mathbf{v}} \tag{A.9}$$

and $\mathsf{V}(X,X')$ is a two-particle operator arising from the Coulomb interaction, which is defined by

$$\mathsf{V}(X,X') \equiv \frac{q^2 n}{m}\left[\frac{\partial}{\partial \mathbf{r}}\frac{1}{|\mathbf{r}-\mathbf{r}'|}\right]\cdot\frac{\partial}{\partial \mathbf{v}}. \tag{A.10}$$

Equation (A.8) is called the *Klimontovich equation;* this equation describes space–time evolution of the microscopic distribution function.

2. LIOUVILLE DISTRIBUTION

The fine-grained distribution function, although precise in describing microscopic states of the many-particle system, would not by itself correspond to the coarse-grained quantities which we observe in the macroscopic world. To establish a connection between them, we need to introduce an averaging process based on the Liouville distribution over the $6N$-dimensional phase space (the Γ space).

The microscopic state of the system is expressed in the Γ space by a point

$$\{X_i\} \equiv (X_1, X_2, \ldots, X_N),$$

which we shall call a system point. Following a formal procedure of the ensemble theory in statistical mechanics, we may imagine N replicas which are macroscopically identical to the system under consideration; the number N may be chosen as large as we like, so that we may let it approach infinity whenever convenient.

Although identical from the macroscopic point of view, these N replicas are generally characterized by different microscopic configurations; the system points are scattered over the Γ space. We may then define the Liouville distribution function $D(\{X_i\};t)$ in the Γ space according to

$$D(\{X_i\};t)\,d\{X_i\} \equiv \lim_{\mathsf{N}\to\infty} \frac{\left[\begin{array}{c}\text{number of system points in the}\\ \text{infinitesimal volume } d\{X_i\} \text{ in}\\ \text{the } \Gamma \text{ space around } \{X_i\}\end{array}\right]}{\mathsf{N}}.$$

By definition, it satisfies the normalization condition

$$\int D(\{X_i\};t)\,d\{X_i\}=1.$$

The N system points distributed in the Γ space apparently do not interact each other; they behave like an ideal gas. The distribution function $D(\{X_i\};t)$ thus satisfies a grand continuity equation of Liouville type:

$$\frac{\partial D}{\partial t}+\{\dot{X}_i\}\cdot\frac{\partial D}{\partial\{X_i\}}=0. \tag{A.11}$$

Along a trajectory in the phase space, the distribution is conserved.

With the aid of the Liouville distribution, we may now carry out a statistical averaging of a fine-grained quantity $A(X,X',\ldots;\{X_i(t)\})$, defined at a set of points $(X,X',\ldots)$ in the six-dimensional phase space in the following way.

$$\langle A(X,X',\ldots;t)\rangle=\int d\{X_i\}D(\{X_i\};t)A(X,X',\ldots;\{X_i\}). \tag{A.12a}$$

In view of the conservation property, this average may equivalently be transformed into an average over the initial distribution, so that

$$\langle A(X,X',\ldots;t)\rangle=\int d\{X_i(0)\}D(\{X_i(0)\};0)A(X,X',\ldots;\{X_i(\{X_i(0)\};t)\}) \tag{A.12b}$$

where $\{X_i(\{X_i(0)\};t)\}$ represents the coordinates of the system points in the Γ space at t under the condition that it was located at $\{X_i(0)\}$ when $t=0$. Let us consider a number of examples.

Example 1.

$$N(X;\{X_i(t)\})=\frac{1}{n}\sum_{i=1}^{N}\delta[X-X_i(t)].$$

This is the Klimontovich function (A.2). Upon averaging this fine-grained quantity with respect to the Liouville distribution, we obtain the single-particle distribution function $f_1(X;t)$:

$$\langle N(X;t)\rangle=f_1(X;t). \tag{A.13}$$

Example 2.

$$N(X;\{X_i(t)\})N(X';\{X_j(t)\})=\frac{1}{n^2}\left\{\sum_{i=1}^{N}\delta[X-X_i(t)]\right\}\left\{\sum_{j=1}^{N}\delta[X'-X_j(t)]\right\}$$

$$=\frac{1}{n^2}\sum_{i=1}^{N}\delta(X-X')\delta[X-X_i(t)]$$

$$+\frac{1}{n^2}\sum_{i\neq j}^{N}\delta[X-X_i(t)]\delta[X'-X_j(t)]. \qquad (A.14)$$

The average of the first term on the right-hand side reduces to the single-particle distribution function (A.13); the average of the second term defines the two-particle distribution function $f_2(X,X';t)$:

$$\langle N(X;t)N(X';t)\rangle=\frac{1}{n}\delta(X-X')f_1(X;t)+f_2(X,X';t). \qquad (A.15)$$

Example 3.

$$N(X;\{X_i(t)\})N(X';\{X_j(t)\})N(X'';\{X_k(t)\})$$

$$=(1/n^3)\sum_{i,j,k}\delta[X-X_i(t)]\delta[X'-X_j(t)]\delta[X''-X_k(t)].$$

We may classify the various terms into three groups: those with $i=j=k$; with only a pair of the subscripts equal; and with all the subscripts different. The average of the last group produces the three-particle distribution function $f_3(X,X',X'';t)$; hence

$$\langle N(X;t)N(X';t)N(X'';t)\rangle=\frac{1}{n^2}\delta(X-X')\delta(X-X'')f_1(X;t)$$

$$+\frac{1}{n}[\delta(X-X')f_2(X',X'';t)+\delta(X'-X'')f_2(X'',X;t)$$

$$+\delta(X''-X)f_2(X,X';t)]+f_3(X,X',X'';t). \qquad (A.16)$$

A joint distribution function $f_s(X,X',\ldots,X^{(s-1)};t)$ involving an arbitrary number (s) of particles may be defined similarly. As may be clear from the construction, those distribution functions are symmetric with respect to interchange of the coordinates for any pair of particles.

3. BBGKY HIERARCHY

As we saw in the examples described in the last section, many-particle distribution functions can be obtained through a statistical average of products of Klimontovich functions. We wish now to study the equations governing the evolution of those distribution functions.

As we may also have discovered in the examples, our notation describing the phase-space coordinates quickly becomes clumsy, with many primes on the X when three or more particles are involved. To simplify the presentation, we therefore adopt a shorthand notation and use the numerals 1, 2, 3, etc. in place of X, X', X'', etc.

The Klimontovich equation (A.8) may thus be written as

$$\left[\frac{\partial}{\partial t}+\mathrm{L}(1)\right]N(1;t)=\int \mathrm{V}(1,2)N(1;t)N(2;t)\,d2. \qquad \text{(A.17)}$$

We now carry out the Liouville average of this equation. Since we may use the scheme of (A.12b), the averaging process commutes with the differential operators involved in (A.17). Hence with the aid of (A.13) and (A.15), we obtain

$$\left[\frac{\partial}{\partial t}+\mathrm{L}(1)\right]f_1(1;t)=\int \mathrm{V}(1,2)\left\{\frac{1}{n}\delta(1-2)f_1(1;t)+f_2(1,2;t)\right\}d2.$$

For an arbitrary function $y(1,2,\ldots;t)$, we can prove from symmetry considerations that

$$\int \mathrm{V}(1,2)\delta(1-2)y(1,2,\ldots;t)\,d2=0. \qquad \text{(A.18)}$$

Consequently, we find

$$\left[\frac{\partial}{\partial t}+\mathrm{L}(1)\right]f_1(1;t)=\int \mathrm{V}(1,2)f_2(1,2;t)\,d2. \qquad \text{(A.19)}$$

We may likewise start from an equation

$$\left[\frac{\partial}{\partial t}+\mathrm{L}(1)+\mathrm{L}(2)\right]N(1;t)N(2;t)$$

$$=\int [\mathrm{V}(1,3)+\mathrm{V}(2,3)]N(1;t)N(2;t)N(3;t)\,d3, \qquad \text{(A.20)}$$

which may be derived from a combination of Klimontovich equations. Upon averaging this equation with respect to the Liouville distribution and making use of (A.15) and (A.16), we obtain an equation involving f_1, f_2, and f_3. This equation may then be simplified with the aid of (A.18) and (A.19); the result of such calculations yields

$$\left\{\frac{\partial}{\partial t}+L(1)+L(2)-\frac{1}{n}[V(1,2)+V(2,1)]\right\}f_2(1,2;t)$$
$$=\int[V(1,3)+V(2,3)]f_3(1,2,3;t)\,d3. \qquad \text{(A.21)}$$

We can similarly consider a Klimontovich equation for a product of an arbitrary number of the Klimontovich functions and carry out a statistical average of that equation. We thus obtain the *Bogoliubov–Born–Green–Kirkwood–Yvon* (BBGKY) hierarchy equations, which may be expressed in the following way.

$$\left[\frac{\partial}{\partial t}+\sum_{i=1}^{s}L(i)-\frac{1}{n}\sum_{i\neq j}^{s}V(i,j)\right]f_s(1,\ldots,s;t)$$

$$=\sum_{i=1}^{s}\int V(i,s+1)f_{s+1}(1,\ldots,s+1;t)\,d(s+1). \qquad \text{(A.22)}$$

This coupled set of equations provides a basis for the kinetic theory of plasmas.

APPENDIX B

DERIVATION OF THE BALESCU–LENARD COLLISION TERM

Substituting (2.42) into (2.40), we have

$$\left.\frac{\partial f(\mathbf{v}_1)}{\partial t}\right]_c = \frac{\omega_p^4}{2\pi n}\sum_{\mathbf{k}}\frac{\mathbf{k}}{k^2}\cdot\frac{\partial}{\partial \mathbf{v}_1}\int d\mathbf{v}_2\int d\mathbf{v}_1'\int d\mathbf{v}_2'\int_{-\infty}^{\infty} d\omega\frac{\mathbf{k}}{k^2}\cdot\left[\left(\frac{\partial}{\partial \mathbf{v}_1'}-\frac{\partial}{\partial \mathbf{v}_2'}\right)f(\mathbf{v}_1')f(\mathbf{v}_2')\right]$$

$$\times\, U_{\mathbf{k}}(\mathbf{v}_1,\mathbf{v}_1';\omega)U_{-\mathbf{k}}(\mathbf{v}_2,\mathbf{v}_2';-\omega). \qquad \text{(B.1)}$$

As we described in the paragraph following Eq. (2.42), it is essential to specify the boundary conditions of the functions involved in the ω integration. For clarity, therefore, we write $\omega+i\eta$ in place of ω for those functions which are well defined and analytic in the upper half-plane; we use $\omega-i\eta$ for those functions well defined in the lower half-plane. Having taken these precautions, we substitute (2.36) and (2.8) into (B.1):

$$\left.\frac{\partial f(\mathbf{v}_1)}{\partial t}\right]_c = \frac{\omega_p^4}{2\pi n}\sum_{\mathbf{k}}\frac{\mathbf{k}}{k^2}\cdot\frac{\partial}{\partial \mathbf{v}_1}\int d\mathbf{v}_1'\int d\mathbf{v}_2'\int_{-\infty}^{\infty} d\omega\frac{\mathbf{k}}{k^2}\cdot\left[\left(\frac{\partial}{\partial \mathbf{v}_1'}-\frac{\partial}{\partial \mathbf{v}_2'}\right)f(\mathbf{v}_1')f(\mathbf{v}_2')\right]$$

$$\times\left[\frac{\delta(\mathbf{v}_1-\mathbf{v}_1')}{\omega-\mathbf{k}\cdot\mathbf{v}_1+i\eta}-\frac{\omega_p^2}{k^2}\frac{1}{(\omega-\mathbf{k}\cdot\mathbf{v}_1+i\eta)(\omega-\mathbf{k}\cdot\mathbf{v}_1'+i\eta)\epsilon(\mathbf{k},\omega)}\mathbf{k}\frac{\partial}{\partial \mathbf{v}_1}f(\mathbf{v}_1)\right]$$

$$\times\frac{1}{(\omega-\mathbf{k}\cdot\mathbf{v}_2'-i\eta)\epsilon(-\mathbf{k},-\omega)}. \qquad \text{(B.2)}$$

With specification of the boundary conditions, the dielectric response functions are

$$\epsilon(\pm\mathbf{k},\pm\omega)=1+\frac{\omega_p^2}{k^2}\int d\mathbf{v}\frac{1}{\omega-\mathbf{k}\cdot\mathbf{v}\pm i\eta}\mathbf{k}\cdot\frac{\partial}{\partial \mathbf{v}}f(\mathbf{v}). \qquad \text{(B.3)}$$

Clearly, $\epsilon(-\mathbf{k},-\omega)$ is complex conjugate to $\epsilon(\mathbf{k},\omega)$.

The rest of the calculation is straightforward: with the aid of (B.3), we immediately find from (B.2)

$$\left.\frac{\partial f(\mathbf{v}_1)}{\partial t}\right]_c = \frac{\omega_p^4}{2\pi n}\sum_{\mathbf{k}}\frac{\mathbf{k}}{k^2}\cdot\frac{\partial}{\partial \mathbf{v}_1}\int d\mathbf{v}_2'\int_{-\infty}^{\infty} d\omega\frac{\mathbf{k}}{k^2}\cdot\left[\left(\frac{\partial}{\partial \mathbf{v}_1}-\frac{\partial}{\partial \mathbf{v}_2'}\right)f(\mathbf{v}_1)f(\mathbf{v}_2')\right]$$

$$\times\frac{1}{(\omega-\mathbf{k}\cdot\mathbf{v}_1+i\eta)(\omega-\mathbf{k}\cdot\mathbf{v}_2'-i\eta)\epsilon(-\mathbf{k},-\omega)} \qquad \text{(i)}$$

$$-\frac{\omega_p^4}{2\pi n}\sum_{\mathbf{k}}\frac{\mathbf{k}}{k^2}\cdot\frac{\partial}{\partial \mathbf{v}_1}\int d\mathbf{v}_2'\int_{-\infty}^{\infty} d\omega\frac{\mathbf{k}}{k^2}\cdot\left[\frac{\partial}{\partial \mathbf{v}_1}f(\mathbf{v}_1)f(\mathbf{v}_2')\right]$$

$$\times\frac{[\epsilon(\mathbf{k},\omega)-1]}{(\omega-\mathbf{k}\cdot\mathbf{v}_1+i\eta)(\omega-\mathbf{k}\cdot\mathbf{v}_2'-i\eta)|\epsilon(\mathbf{k},\omega)|^2} \qquad \text{(ii)}$$

$$+\frac{\omega_p^4}{2\pi n}\sum_{\mathbf{k}}\frac{\mathbf{k}}{k^2}\cdot\frac{\partial}{\partial \mathbf{v}_1}\int d\mathbf{v}_1'\int_{-\infty}^{\infty} d\omega\frac{\mathbf{k}}{k^2}\cdot\left[\frac{\partial}{\partial \mathbf{v}_1}f(\mathbf{v}_1)f(\mathbf{v}_1')\right]$$

$$\times\frac{[\epsilon(-\mathbf{k},-\omega)-1]}{(\omega-\mathbf{k}\cdot\mathbf{v}_1+i\eta)(\omega-\mathbf{k}\cdot\mathbf{v}_1'+i\eta)|\epsilon(\mathbf{k},\omega)|^2}. \qquad \text{(iii)}$$

Parts of the terms (i) and (ii) cancel each other; the rest of (i) and (ii) go over to the terms (iv) and (v), respectively, in the following equation. The term (iii) is split into two parts to produce (vi) and (vii). Hence, we have

$$\left.\frac{\partial f(\mathbf{v}_1)}{\partial t}\right]_c = -\frac{\omega_p^2}{2\pi n}\sum_{\mathbf{k}}\frac{\mathbf{k}}{k^2}\cdot\frac{\partial}{\partial \mathbf{v}_1}\int_{-\infty}^{\infty} d\omega f(\mathbf{v}_1)\frac{\epsilon(-\mathbf{k},-\omega)-1}{(\omega-\mathbf{k}\cdot\mathbf{v}_1+i\eta)\epsilon(-\mathbf{k},-\omega)} \qquad \text{(iv)}$$

$$+\frac{\omega_p^4}{2\pi n}\sum_{\mathbf{k}}\frac{\mathbf{k}}{k^2}\cdot\frac{\partial}{\partial \mathbf{v}_1}\int d\mathbf{v}_2'\int_{-\infty}^{\infty} d\omega\left[\frac{\mathbf{k}}{k^2}\cdot\frac{\partial}{\partial \mathbf{v}_1}f(\mathbf{v}_1)f(\mathbf{v}_2')\right]$$

$$\times\frac{1}{(\omega-\mathbf{k}\cdot\mathbf{v}_1+i\eta)(\omega-\mathbf{k}\cdot\mathbf{v}_2'-i\eta)|\epsilon(\mathbf{k},\omega)|^2} \qquad \text{(v)}$$

$$+\frac{\omega_p^4}{2\pi n}\sum_{\mathbf{k}}\frac{\mathbf{k}}{k^2}\cdot\frac{\partial}{\partial \mathbf{v}_1}\int d\mathbf{v}_1'\int_{-\infty}^{\infty} d\omega\left[\frac{\mathbf{k}}{k^2}\cdot\frac{\partial}{\partial \mathbf{v}_1}f(\mathbf{v}_1)f(\mathbf{v}_1')\right]$$

$$\times\frac{1}{(\omega-\mathbf{k}\cdot\mathbf{v}_1+i\eta)(\omega-\mathbf{k}\cdot\mathbf{v}_1'+i\eta)\epsilon(\mathbf{k},\omega)} \qquad \text{(vi)}$$

$$-\frac{\omega_p^4}{2\pi n}\sum_{\mathbf{k}}\frac{\mathbf{k}}{k^2}\cdot\frac{\partial}{\partial \mathbf{v}_1}\int d\mathbf{v}_1'\int_{-\infty}^{\infty} d\omega\left[\frac{\mathbf{k}}{k^2}\cdot\frac{\partial}{\partial \mathbf{v}_1}f(\mathbf{v}_1)f(\mathbf{v}_1')\right]$$

$$\times\frac{1}{(\omega-\mathbf{k}\cdot\mathbf{v}_1+i\eta)(\omega-\mathbf{k}\cdot\mathbf{v}_1'+i\eta)|\epsilon(\mathbf{k},\omega)|^2}. \qquad \text{(vii)}$$

The integration in (iv) may be carried out by closing the contour with an infinite semicircle in the lower half-plane; only the pole at $\omega = \mathbf{k}\cdot\mathbf{v}_1 - i\eta$ contributes. Similarly, by closing the contour in the upper half-plane, we find that (vi) vanishes. The terms (v) and (vii) may be combined with the aid of the Dirac identities

$$\lim_{\eta\to+0} \frac{1}{x \pm i\eta} = \frac{P}{x} \mp i\pi\delta(x) \tag{B.4}$$

where P stands for the principal part. We thus obtain

$$\left.\frac{\partial f(\mathbf{v}_1)}{\partial t}\right]_c = i\frac{\omega_p^2}{n}\sum_{\mathbf{k}} \frac{\mathbf{k}}{k^2}\cdot\frac{\partial}{\partial \mathbf{v}_1} f(\mathbf{v}_1) \frac{\epsilon(-\mathbf{k},-\mathbf{k}\cdot\mathbf{v}_1)-1}{\epsilon(-\mathbf{k},-\mathbf{k}\cdot\mathbf{v}_1)}$$

$$+ i\frac{\omega_p^4}{n}\sum_{\mathbf{k}} \frac{\mathbf{k}}{k^2}\cdot\frac{\partial}{\partial \mathbf{v}_1}\int d\mathbf{v}_2 \left[\frac{\mathbf{k}}{k^2}\cdot\frac{\partial}{\partial \mathbf{v}_1} f(\mathbf{v}_1) f(\mathbf{v}_2)\right] \frac{1}{(\mathbf{k}\cdot\mathbf{v}_2 - \mathbf{k}\cdot\mathbf{v}_1 + i\eta)|\epsilon(\mathbf{k},\mathbf{k}\cdot\mathbf{v}_2)|^2}.$$

Checking the symmetry of the terms with respect to inversion of $\mathbf{k}$ and again with the aid of (B.4), we find

$$\left.\frac{\partial f(\mathbf{v}_1)}{\partial t}\right]_c = -\frac{\omega_p^2}{n}\sum_{\mathbf{k}} \frac{\mathbf{k}}{k^2}\cdot\frac{\partial}{\partial \mathbf{v}_1} f(\mathbf{v}_1) \frac{\operatorname{Im}\epsilon(\mathbf{k},\mathbf{k}\cdot\mathbf{v}_1)}{|\epsilon(\mathbf{k},\mathbf{k}\cdot\mathbf{v}_1)|^2}$$

$$+ \frac{\pi\omega_p^4}{n}\sum_{\mathbf{k}} \frac{\mathbf{k}}{k^2}\cdot\frac{\partial}{\partial \mathbf{v}_1}\int d\mathbf{v}_2 \left[\frac{\mathbf{k}}{k^2}\cdot\frac{\partial}{\partial \mathbf{v}_1} f(\mathbf{v}_1) f(\mathbf{v}_2)\right] \frac{\delta(\mathbf{k}\cdot\mathbf{v}_1 - \mathbf{k}\cdot\mathbf{v}_2)}{|\epsilon(\mathbf{k},\mathbf{k}\cdot\mathbf{v}_1)|^2} \tag{B.5}$$

where

$$\operatorname{Im}\epsilon(\mathbf{k},\omega) = -\frac{\pi\omega_p^2}{k^2}\int d\mathbf{v}\,\delta(\omega - \mathbf{k}\cdot\mathbf{v})\mathbf{k}\cdot\frac{\partial}{\partial \mathbf{v}} f(\mathbf{v}). \tag{B.6}$$

Rearrangement of the terms in (B.5) leads us to (2.43).

APPENDIX C

ANALYTIC PROPERTIES OF THE DIELECTRIC RESPONSE FUNCTION

The dielectric response function satisfies a number of general relationships stemming from its analytic properties; we summarize some of them here. First, we consider the Kramers–Kronig relations; these arise as a consequence of the physical principle of causality discussed in Section 3.1A. We then take up a sum rule, which is closely related to the high-frequency asymptotic behavior (3.51). Although we limit ourselves to the case of the dielectric response function alone, similar relationships can be derived for other response functions as well.

1. THE KRAMERS–KRONIG RELATIONS

Since $\sigma(\mathbf{k},\omega)$ has been defined in terms of a one-sided Fourier transformation (3.5), it is analytic in the upper half of the complex ω plane. Following definitions (3.11) and (3.36), we then conclude that $\epsilon(\mathbf{k},\omega)$ should have the same analytic property. Since

$$\lim_{\omega\to\infty} \epsilon(\mathbf{k},\omega) = 1, \tag{C.1}$$

we note that

$$\int_{-\infty}^{\infty} d\omega' \frac{\epsilon(\mathbf{k},\omega')-1}{\omega'-\omega+i\eta} = 0. \tag{C.2}$$

We prove this equality by closing the contour with an infinite semicircle in the upper half-plane and then by using Cauchy's theorem for integration; contribution from the semicircle vanishes because of (C.1) and the factor $(\omega'-\omega+i\eta)^{-1}$ in the integrand.

With the aid of the Dirac formula (B.4), we may now split (C.2) into real and imaginary parts; we thus obtain

$$\mathrm{Re}\,\epsilon(\mathbf{k},\omega) - 1 = \frac{1}{\pi}\int_{-\infty}^{\infty} d\omega' \frac{P}{\omega'-\omega}\mathrm{Im}\,\epsilon(\mathbf{k},\omega'), \tag{C.3}$$

$$\mathrm{Im}\,\epsilon(\mathbf{k},\omega) = \frac{1}{\pi}\int_{-\infty}^{\infty} d\omega' \frac{P}{\omega'-\omega}[1-\mathrm{Re}\,\epsilon(\mathbf{k},\omega')]. \tag{C.4}$$

These are the *Kramers–Kronig relations* for $\epsilon(\mathbf{k},\omega)$. As may be clear from the derivation above, the Kramers–Kronig relations are a direct consequence of causality.

We can prove that when $\epsilon(\mathbf{k},\omega)$ has no poles in the upper half–plane, $1/\epsilon(\mathbf{k},\omega)$ likewise has no poles in that domain [L. D. Landau and E. M. Lifshitz, *Statistical Physics*, 2nd ed., Sec. 125 (Addison-Wesley, Reading, Mass., 1969)]. Hence, an application of the foregoing procedure to the function $1/\epsilon(\mathbf{k},\omega)$ yields another set of Kramers–Kronig relations:

$$\operatorname{Re}\frac{1}{\epsilon(\mathbf{k},\omega)}-1=\frac{1}{\pi}\int_{-\infty}^{\infty}d\omega'\frac{P}{\omega'-\omega}\operatorname{Im}\frac{1}{\epsilon(\mathbf{k},\omega')},\tag{C.5}$$

$$\operatorname{Im}\frac{1}{\epsilon(\mathbf{k},\omega)}=\frac{1}{\pi}\int_{-\infty}^{\infty}d\omega'\frac{P}{\omega'-\omega}\left[1-\operatorname{Re}\frac{1}{\epsilon(\mathbf{k},\omega)}\right].\tag{C.6}$$

2. *f*-SUM RULE

We have remarked in connection with (3.51) that a correct dielectric function must behave as

$$\epsilon(\mathbf{k},\omega)\rightarrow 1-\left(\omega_{\mathrm{p}}^2/\omega^2\right)\tag{C.7}$$

at high frequencies. Coupled with the Kramers–Kronig relations, this asymptotic behavior leads to an important sum rule for the imaginary part of $\epsilon(\mathbf{k},\omega)$, which is a longitudinal analogue of a standard *f*-sum rule for the optical oscillator strengths.

To derive the sum rule, we first rewrite (C.3) so that

$$\tfrac{1}{2}\operatorname{Re}[\epsilon(\mathbf{k},\omega)+\epsilon(\mathbf{k},-\omega)]-1=\frac{1}{\pi}\int_{-\infty}^{\infty}d\omega'\frac{P}{(\omega')^2-\omega^2}\omega'\operatorname{Im}\epsilon(\mathbf{k},\omega').$$

In the asymptotic limit of large ω, we thus find

$$\tfrac{1}{2}\operatorname{Re}[\epsilon(\mathbf{k},\omega)+\epsilon(\mathbf{k},-\omega)]\rightarrow 1-(\pi\omega^2)^{-1}\int_{-\infty}^{\infty}\omega\operatorname{Im}\epsilon(\mathbf{k},\omega)\,d\omega.\tag{C.8}$$

A comparison between (C.7) and (C.8) yields a sum rule

$$\int_{-\infty}^{\infty}\omega\operatorname{Im}\epsilon(\mathbf{k},\omega)\,d\omega=\pi\omega_{\mathrm{p}}^2.\tag{C.9}$$

This is, therefore, the *f-sum rule* for the dielectric response function. Similarly, starting from (C.5) and combining it with (C.7), we find

$$\int_{-\infty}^{\infty}\omega\operatorname{Im}[\epsilon(\mathbf{k},\omega)]^{-1}d\omega=-\pi\omega_{\mathrm{p}}^2.\tag{C.10}$$

APPENDIX D

FLUCTUATION-DISSIPATION THEOREM

In the theory of the many-particle system in thermodynamic equilibrium, the fluctuation-dissipation theorem provides a rigorous connection between the power spectrum of fluctuations and the imaginary part of the relevant linear response function. The theorem relates the canonically (or grand canonically) averaged commutator and anticommutator of any pair of Hermitian operators, such as the number densities evaluated at two different points in space and time. The average of such a commutator is related to the response function, while the average of an anticommutator gives rise to a correlation function which turns into a fluctuation spectrum after Fourier transformations. It may, therefore, be said that the theorem possesses a form unique in physics in that it correlates the properties of the system in equilibrium (i.e., fluctuations) with the parameters which characterize the irreversible processes, that is, the imaginary parts of the response functions. In this appendix, we present a derivation of this fluctuation-dissipation theorem.

1. LINEAR RESPONSE FUNCTIONS

Consider the application of an external disturbance field,

$$a(\mathbf{r},t)=a\exp[i(\mathbf{k}\cdot\mathbf{r}-\omega t)+\eta t]+\mathrm{cc} \tag{D.1}$$

to the many-particle system which is assumed to be translationally invariant in space and time. The disturbance will create perturbations in the system; the increment $\delta B(\mathbf{r},t)$ of a physical quantity B may be calculated as the difference between its expectation values with and without the disturbance,

$$\delta B(\mathbf{r},t)=\langle B(\mathbf{r},t;a)\rangle-\langle B(\mathbf{r},t;0)\rangle. \tag{D.2}$$

Within the framework of the linear response theory elucidated in Section 3.1A, Eq. (D.2) should be expressed as

$$\delta B(\mathbf{r},t)=b\exp[i(\mathbf{k}\cdot\mathbf{r}-\omega t)+\eta t]+\mathrm{cc}. \tag{D.3}$$

The linear response function $K_{BA}(\mathbf{k},\omega)$ describing this particular response is then given by

$$K_{BA}(\mathbf{k},\omega)=b/a. \tag{D.4}$$

We evaluate such a response function with the retarded boundary conditions.

The response of the system can be calculated with the aid of a standard perturbation theory of quantum mechanics. The coupling between the external disturbance and the physical variables of the system produces an extra Hamiltonian which acts to drive the system out of its unperturbed state. Such a situation can be analyzed through a quantum-mechanical equation of motion.

We thus begin with the expression for the total Hamiltonian of the system as a summation of the exact Hamiltonian H of the system without the external disturbance and the extra Hamiltonian $H'(t)$,

$$H_t=H+H'(t). \tag{D.5}$$

The extra Hamiltonian may be explicitly written as†

$$H'(t)=-\int d\mathbf{r}A(\mathbf{r})a(\mathbf{r},t). \tag{D.6}$$

This equation in fact serves as the definition for the operator $A(\mathbf{r})$ which represents the physical variable of the system coupling with the external disturbance.

The wave function $\Psi_S(t)$ of the system obeys the Schrödinger equation

$$i\hbar\frac{\partial\Psi_S(t)}{\partial t}=[H_S+H_S'(t)]\Psi_S(t). \tag{D.7}$$

The subscript S is used here to imply that we are in the Schrödinger representation. One can formally solve Eq. (D.7) by going over to the interaction representation. In this representation, the wave function and the operator O are transformed from those in the Schrödinger representation according to

$$\Psi(t)=\exp\left(\frac{iHt}{\hbar}\right)\Psi_S(t), \tag{D.8}$$

$$O(t)=\exp\left(\frac{iHt}{\hbar}\right)O_S\exp\left(-\frac{iHt}{\hbar}\right). \tag{D.9}$$

†In this appendix, we are working with the periodic boundary conditions appropriate to the cube of unit volume.

The equation for the wave function is then given by

$$i\hbar\frac{\partial\Psi(t)}{\partial t}=\mathrm{H}'(t)\Psi(t) \tag{D.10}$$

where

$$\mathrm{H}'(t)=-\int d\mathbf{r}\mathrm{A}(\mathbf{r},t)a(\mathbf{r},t) \tag{D.11}$$

is the extra Hamiltonian in the interaction representation. The differential equation (D.10) with an initial condition

$$\Psi(-\infty)=\Psi_0 \tag{D.12}$$

may be written in the form of the integral equation as

$$\Psi(t)=\Psi_0-\frac{i}{\hbar}\int_{-\infty}^{t}dt'\mathrm{H}'(t')\Psi(t'). \tag{D.13}$$

In principle, we can solve this equation by iteration.

Since we are here interested in the linear response calculation, it is sufficient to retain only up to the first-order effects of the perturbation in (D.13); hence, we have

$$\Psi(t)=\Psi_0-\frac{i}{\hbar}\int_{-\infty}^{t}dt'\mathrm{H}'(t')\Psi_0. \tag{D.14}$$

Assuming an appropriate normalization for the wave functions and adopting a standard bracket notation to express the quantum-mechanical expectation values, we may now calculate the induced perturbation, Eq. (D.2), according to

$$\delta B(\mathbf{r},t)=\langle\Psi(t)|\mathrm{B}(\mathbf{r},t)|\Psi(t)\rangle-\langle\Psi_0|\mathrm{B}(\mathbf{r},t)|\Psi_0\rangle.$$

With the aid of (D.11) and (D.14), we thus obtain

$$\delta B(\mathbf{r},t)=\frac{i}{\hbar}\int d\mathbf{r}'\int_{-\infty}^{t}dt'\langle\Psi_0|[\mathrm{B}(\mathbf{r},t),\mathrm{A}(\mathbf{r}',t')]|\Psi_0\rangle a(\mathbf{r}',t') \tag{D.15}$$

where the square bracket denotes the commutator $[\mathrm{A},\mathrm{B}]=\mathrm{AB}-\mathrm{BA}$. Recalling a similar calculation involved in the derivation of Eq. (3.5) from Eq. (3.1), we find the linear response function (D.4) from (D.15) as

$$K_{\mathrm{BA}}(\mathbf{k},\omega)=\frac{i}{\hbar}\int d(\mathbf{r}-\mathbf{r}')\int_0^{\infty}d(t-t')\langle\Psi_0|[\mathrm{B}(\mathbf{r},t),\mathrm{A}(\mathbf{r}',t')]|\Psi_0\rangle$$
$$\times\exp\{-i[\mathbf{k}\cdot(\mathbf{r}-\mathbf{r}')-\omega(t-t')]\}. \tag{D.16}$$

The response function is thus expressed in terms of an expectation value of the commutator between two Hermitian operators, $\mathsf{B}(\mathbf{r},t)$ and $\mathsf{A}(\mathbf{r}',t')$.

2. AVERAGE OVER THE CANONICAL ENSEMBLE

Equation (D.16) is now averaged over the unperturbed states of the system with the aid of the statistical operator of the canonical ensemble

$$\mathsf{D}=\exp[(F-\mathsf{H})/T]. \tag{D.17}$$

The free energy F is to be determined from the normalization

$$\exp(-F/T)=\mathrm{Tr}\{\exp(-\mathsf{H}/T)\}$$

where Tr means the summation over the diagonal elements of the matrix. We thus have

$$K_{\mathrm{BA}}(\mathbf{k},\omega)=\frac{i}{\hbar}\int d(\mathbf{r}-\mathbf{r}')\int_0^{\infty}d(t-t')\,\mathrm{Tr}\{\mathsf{D}[\mathsf{B}(\mathbf{r},t),\mathsf{A}(\mathbf{r}',t')]\}$$

$$\times\exp\{-i[\mathbf{k}\cdot(\mathbf{r}-\mathbf{r}')-\omega(t-t')]\} \tag{D.18}$$

To calculate (D.18) further, we introduce a set of eigenstates $|n\rangle$ of the unperturbed system with energy eigenvalues E_n; we have

$$\mathsf{H}|n\rangle=E_n|n\rangle. \tag{D.19}$$

The explicit time dependence of the operators such as $\mathsf{A}(\mathbf{r},t)$ and $\mathsf{B}(\mathbf{r},t)$ may be recovered through Eq. (D.9). We then decompose their time-independent parts into Fourier series so that

$$\mathsf{A}(\mathbf{r})=\sum_{\mathbf{k}}\mathsf{A}_{\mathbf{k}}\exp(i\mathbf{k}\cdot\mathbf{r}). \tag{D.20}$$

After a straightforward calculation, Eq. (D.18) reduces to

$$K_{\mathrm{BA}}(\mathbf{k},\omega)=-\frac{1}{\hbar}\sum_{m,n}D(E_n)\left[1-\exp\left(-\frac{\hbar\omega_{mn}}{T}\right)\right]\frac{(B_{\mathbf{k}})_{nm}(A_{-\mathbf{k}})_{mn}}{\omega-\omega_{mn}+i\eta} \tag{D.21}$$

where

$$D(E_n)\equiv\exp\left(\frac{F-E_n}{T}\right), \tag{D.22}$$

$$\omega_{mn}\equiv\frac{E_m-E_n}{\hbar}, \tag{D.23}$$

and the matrix elements are

$$(A_{\mathbf{k}})_{mn}=\langle m|\mathsf{A}_{\mathbf{k}}|n\rangle. \tag{D.24}$$

3. FLUCTUATIONS

Let us consider a generalized dynamic form factor $S_{BA}(\mathbf{k},\omega)$, which is defined in terms of a statistical average of the anticommutator between the two operators $A(\mathbf{r}',t')$ and $B(\mathbf{r},t)$ as

$$S_{BA}(\mathbf{k},\omega)=\frac{1}{4\pi}\int d(\mathbf{r}-\mathbf{r}')\int_{-\infty}^{\infty}d(t-t')\,\mathrm{Tr}\{D[B(\mathbf{r},t)A(\mathbf{r}',t')+A(\mathbf{r}',t')B(\mathbf{r},t)]\}$$
$$\times\exp\{-i[\mathbf{k}\cdot(\mathbf{r}-\mathbf{r}')-\omega(t-t')]\}. \tag{D.25}$$

In a classical system, the anticommutator reduces to a simple product. The dynamic form factor (9.17) is a special case of Eq. (D.25) in that both A and B represent the density fluctuations $\delta\rho$.

The calculation of Eq. (D.25) proceeds in much the same way as that of Eq. (D.21); we obtain

$$S_{BA}(\mathbf{k},\omega)=\frac{1}{2}\left[1+\exp\left(-\frac{\hbar\omega}{T}\right)\right]\sum_{m,n}D(E_n)[(B_{\mathbf{k}})_{nm}(A_{-\mathbf{k}})_{mn}]\delta(\omega-\omega_{mn}). \tag{D.26}$$

On the other hand, from Eq. (D.21), we find

$$-(i/2)[K_{BA}(\mathbf{k},\omega)-K_{AB}(-\mathbf{k},-\omega)]$$
$$=\frac{\pi}{\hbar}\left[1-\exp\left(-\frac{\hbar\omega}{T}\right)\right]\sum_{m,n}D(E_n)[(B_{\mathbf{k}})_{nm}(A_{-\mathbf{k}})_{mn}]\delta(\omega-\omega_{mn}). \tag{D.27}$$

Comparison between Eqs. (D.26) and (D.27) yields

$$S_{BA}(\mathbf{k},\omega)=-\frac{i\hbar}{4\pi}\coth\left(\frac{\hbar\omega}{2T}\right)[K_{BA}(\mathbf{k},\omega)-K_{AB}(-\mathbf{k},-\omega)]. \tag{D.28}$$

This relationship is the *fluctuation-dissipation theorem*; the fluctuation spectrum described by the dynamic form factor $S_{BA}(\mathbf{k},\omega)$ is related directly to the imaginary parts of the relevant response functions $K_{BA}(\mathbf{k},\omega)$ and $K_{AB}(-\mathbf{k},-\omega)$.

$$S_{BA}(\mathbf{k},\omega)=-(iT/2\pi\omega)[K_{BA}(\mathbf{k},\omega)-K_{AB}(-\mathbf{k},-\omega)]. \tag{D.29}$$

This is, therefore, the classical expression of the fluctuation-dissipation theorem; sometimes this relationship is also called the generalized Nyquist theorem.

APPENDIX E

SCATTERING CROSS SECTION OF A MANY-PARTICLE SYSTEM IN THE BORN APPROXIMATION

The dynamic form factor is a quantity which can be measured directly by a proper scattering experiment. The probe usually takes the form of a beam of particles or radiation. The incident beam interacts with the particles that make up the many-particle system; it thereby exchanges energy and momentum with the system and is scattered. The differential cross section of such scattering acts then reveals the dynamic structure of the correlations in the system. In this appendix we calculate the cross section of the many-particle system against the scattering of a beam of particles incident on it. Such a calculation is relevant, for example, to the interpretation of the characteristic energy loss experiments of high-energy electron beams scattered by solid state plasmas [e.g., D. Pines, *Elementary Excitations in Solids* (W. A. Benjamin, New York, 1963)].

1. THE HAMILTONIAN

Consider a monochromatic beam of particles with mass M and momentum $\mathbf{P}_1$. The Hamiltonian of the system consisting of the many-particle system under observation (we shall call it "plasma") and the incident particle of the beam may be written as†

$$\mathcal{H}_t = \mathcal{H} + (P^2/2M) + \sum_{i=1}^{n} \Phi(|\mathbf{R}-\mathbf{r}_i|). \tag{E.1}$$

Here, $\mathcal{H}$ is the Hamiltonian of the plasma; $\mathbf{R}$ and $\mathbf{P}$ are the position and the momentum of the incident particle. The last term of (E.1) describes the interaction between the beam and the plasma; a binary interaction with a potential $\Phi(r)$ is assumed. We may Fourier-analyze this interaction Hamiltonian so that

$$\mathcal{H}' = \sum_{\mathbf{k}}{}' \Phi_k \rho_{-\mathbf{k}} \exp(-i\mathbf{k}\cdot\mathbf{R}) \tag{E.2}$$

where (9.3) and (9.15) have been used.

†In this appendix, we are working with the periodic boundary conditions appropriate to a unit volume. The particle density in the incident beam is normalized to one particle per unit volume.

2. TRANSITION PROBABILITIES

We now evoke the "golden rule" of quantum-mechanical perturbation theory to calculate the transition probability from an initial state I to a final state F; we thus write

$$W(I\to F)=(2\pi/\hbar)|\langle F|H'|I\rangle|^2\delta(E_F-E_I). \tag{E.3}$$

The initial state consists of the combination of a plane-wave state $\exp(i\mathbf{P}_1\cdot\mathbf{R}/\hbar)$ for the incident particle and an energy eigenstate $|n\rangle$ for the plasma [see Eq. (D.19)]. Similarly, the final state is given by the product between $\exp(i\mathbf{P}_2\cdot\mathbf{R}/\hbar)$ for the scattered particle and $|m\rangle$ for the plasma. The matrix element is then calculated as

$$\begin{aligned}\langle F|H'|I\rangle &= \int d\mathbf{R}\exp\left(-\frac{i\mathbf{P}_2\cdot\mathbf{R}}{\hbar}\right)\langle m|\sum_{\mathbf{k}}{}'\Phi_k\rho_{-\mathbf{k}}\exp(-\mathbf{k}\cdot\mathbf{R})|n\rangle\exp\left(\frac{i\mathbf{P}_1\cdot\mathbf{R}}{\hbar}\right)\\ &=\Phi_k(\rho_{-\mathbf{k}})_{mn}\end{aligned} \tag{E.4}$$

with the constraint

$$\hbar\mathbf{k}=\mathbf{P}_1-\mathbf{P}_2. \tag{E.5}$$

We also calculate

$$E_F-E_I=E_m-E_n+\frac{P_2^2}{2M}-\frac{P_1^2}{2M}=\hbar(\omega_{mn}-\omega) \tag{E.6}$$

where

$$\hbar\omega=(P_1^2-P_2^2)/2M \tag{E.7}$$

and ω_{mn} are the excitation frequencies defined by (D.23). Substitution of (E.4) and (E.6) into Eq. (E.3) yields

$$W(I\to F)=(2\pi/\hbar^2)\Phi_k^2|(\rho_{-\mathbf{k}})_{mn}|^2\delta(\omega_{mn}-\omega). \tag{E.8}$$

The total probability for the beam particle to be scattered from $\mathbf{P}_1$ to $\mathbf{P}_2$ under the condition that the initial state of the plasma is known to be $|n\rangle$ is then given by

$$W_n(\mathbf{P}_1\to\mathbf{P}_2)=(2\pi/\hbar^2)\Phi_k^2\sum_m|(\rho_{-\mathbf{k}})_{mn}|^2\delta(\omega_{mn}-\omega). \tag{E.9}$$

We now average this quantity over the initial states of the plasma with the aid of the canonical distribution (D.22); we thus have

$$W(\mathbf{P}_1\to\mathbf{P}_2)=\sum_n D(E_n)W_n(\mathbf{P}_1\to\mathbf{P}_2)$$

$$=\frac{2\pi}{\hbar^2}\Phi_k^2\sum_{m,n}D(E_n)|(\rho_{-\mathbf{k}})_{mn}|^2\delta(\omega_{mn}-\omega).$$

The dynamic form factor of the plasma is expressed as†

$$S(\mathbf{k},\omega)=\sum_{m,n}D(E_n)|(\rho_{-\mathbf{k}})_{mn}|^2\delta(\omega_{mn}-\omega). \tag{E.10}$$

Hence, we find

$$W(\mathbf{P}_1\to\mathbf{P}_2)=(2\pi/\hbar^2)\Phi_k^2 S(\mathbf{k},\omega). \tag{E.11}$$

The scattering experiments, therefore, amount to a direct measurement of the dynamic form factor of the plasma.

3. SCATTERING CROSS SECTION

In many cases it is more convenient to describe the scattering processes in terms of the differential cross section rather than the transition probability. The differential cross section dQ for the scattering into a solid angle do and an energy interval $\hbar\,d\omega$ may be calculated according to

$$dQ=\frac{(\text{time rate of transition into } do\ d\omega)}{(\text{incident flux})}. \tag{E.12}$$

The incident flux is simply the product between the density and the flow velocity of the beam particles, that is,

$$(\text{incident flux})=\frac{1}{(\text{unit volume})}\frac{P_1}{M}=\frac{P_1}{M}.$$

†This formula derives not from the anticommutator expression (D.25) of the dynamic form factor, but from the expression involving a simple product

$$S(\mathbf{k},\omega)=(2\pi)^{-1}\int d(\mathbf{r}-\mathbf{r}')\int_{-\infty}^{\infty}d(t-t')\,\mathrm{Tr}\{\mathrm{D}[\delta\rho(\mathbf{r},t)\delta\rho(\mathbf{r}',t')]\}$$

$$\times\exp\{-i[\mathbf{k}\cdot(\mathbf{r}-\mathbf{r}')-\omega(t-t')]\}.$$

The dynamic form factor may be defined in either way; for a classical system, no differences result.

The time rate of transition into $do\,d\omega$ is the transition probability $W(\mathbf{P}_1 \to \mathbf{P}_2)$ multiplied by the number of states in $do\,d\omega$; the latter is calculated with the aid of (E.7) as

$$\frac{P_2^2 dP_2 do \cdot (\text{unit volume})}{(2\pi)^3 \hbar^3} = \frac{MP_2}{(2\pi)^3 \hbar^2} do\,d\omega.$$

Substituting these calculations into (E.12), we finally obtain

$$\frac{d^2Q}{do\,d\omega} = \frac{M^2 P_2}{4\pi^2 \hbar^4 P_1} \Phi_k^2 S(\mathbf{k}, \omega). \tag{E.13}$$

This formula together with the momentum and energy relationships (E.5) and (E.7) provides a complete description of the scattering events as long as the validity of the Born approximation is assured. It is clear that $S(\mathbf{k}, \omega)$ and $M^2 P_2 / P_1$ are related to the properties of the plasma and of the beam, respectively; the coupling between them is described by Φ_k^2.

When the energy of the incident beam is high, $P_1^2/2M$ may be much greater than those excitation energies $\hbar\omega$ of the plasma for which $S(\mathbf{k}, \omega)$ takes on appreciable strengths. In these circumstances, P_2 ($\cong P_1$) may be regarded as independent of ω; the cross section (E.13) depends on ω only through $S(\mathbf{k}, \omega)$. We may thus use the sum rule (9.20) to integrate the cross section over the frequencies:

$$\frac{dQ}{do} = \frac{NM^2}{4\pi^2 \hbar^4} \Phi_k^2 S(\mathbf{k}). \tag{E.14}$$

If, therefore, we measure the total intensity of the beam scattered into a given solid angle without resolving the particle energies, then the static form factor of the system may be obtained according to this formula.

INDEXES

AUTHOR INDEX

Numbers in parentheses indicate the numbers of the references when these are cited in the text without the names of the authors.

Numbers set in *italics* designate the page numbers on which the complete literature citation is given.

SUBJECT INDEX

CPSIA information can be obtained
at www.ICGtesting.com
Printed in the USA
FFOW03n0043010318
45325086-45988FF

9 780805 387537